STUDY OF LINEAR
AND
NONLINEAR MODELS
WITH
"MATHEMATICA"

STUDY OF LINEAR AND NONLINEAR MODELS WITH "MATHEMATICA"

Czesław Mączka
AGH University of Science and Technology, Poland

Sergii Skurativskyi
National Academy of Science of Ukraine, Ukraine

Vsevolod Vladimirov
AGH University of Science and Technology, Poland

NEW JERSEY • LONDON • SINGAPORE • BEIJING • SHANGHAI • HONG KONG • TAIPEI • CHENNAI • TOKYO

Published by

World Scientific Publishing Co. Pte. Ltd.
5 Toh Tuck Link, Singapore 596224
USA office: 27 Warren Street, Suite 401-402, Hackensack, NJ 07601
UK office: 57 Shelton Street, Covent Garden, London WC2H 9HE

Library of Congress Control Number: 2022047085

British Library Cataloguing-in-Publication Data
A catalogue record for this book is available from the British Library.

STUDY OF LINEAR AND NONLINEAR MODELS WITH "MATHEMATICA"

ISBN 978-981-126-622-5 (hardcover)
ISBN 978-981-126-623-2 (ebook for institutions)
ISBN 978-981-126-624-9 (ebook for individuals)

For any available supplementary material, please visit
https://www.worldscientific.com/worldscibooks/10.1142/13139#t=suppl

To our students

Preface

The material of this book was originally conceived as a textbook on modeling nonlinear processes, addressed to senior students of physical and mathematical specialties. Such a course has been taught by the authors for many years at technical universities. However, in the process of writing, the material of the book grew significantly and was enriched with new chapters. The current version of the book includes, in particular, an extensive section devoted to linear models of mathematical physics and methods for their solutions. We decided not to limit ourselves to only nonlinear models, since it is very useful to compare the solutions of nonlinear models with the solutions of the corresponding linear models. In addition, from the point of view of the advantages provided by the use of software packages such as *Mathematica* or *Maple*, linear models represent the most fertile field for application, because they solve, as a rule, in a matter of seconds those problem that, being calculated manually, take a lot of time and require significant dexterity. In the process of writing, there was also a temptation to include in the book, in addition to the classical and well-known nonlinear models that traditionally form the backbone of any book conceived as a textbook on nonlinear science, to devote some part of the material to topics more suitable for a monograph. The point is that good ideas, as a rule, have a long life not limited by the solution of a specific problem. These include the Hirota method, which initially found its application in the theory of solitons. Subsequently, the Hirota method was applied for obtaining exact solutions to a large number of nonlinear models in no way related to soliton topics. Progress in the wide and successful application of the Hirota method and its modifications to various problems of mathematical physics was due precisely to the possibility of using special software packages, so the inclusion of the relevant sections in this book seemed quite natural to us.

From the very beginning, the authors of the book faced the difficult task of finding their place among the abundance of books devoted to modeling based on the use of computer software packages. When choosing the material, we were guided to a greater extent by our scientific preferences. And since the interests of the authors include on one hand models of nonlinear oscillations, stability and bifurcations of their solutions, and, on the other hand, nonlinear wave structures, their interaction, as well as the influence of the properties of the carrier medium on the formation of such structures, it is precisely these aspects that occupy the dominant position in the book. Let us note that of the numerous books known to the authors devoted to the presentation of the basics of modeling supported by the *Mathematica* package, the most consonant are the books of [Dubin (2003)] and [Enns and Mc Gurie (2001)]. However, despite some intersection of the topics covered, which, apparently, cannot be avoided in the presentation of the classical problems of nonlinear mechanics and mathematical physics, we tried to disassociate ourselves from the other authors as much as possible, offering our range of problems, models, examples and methods for solving them.

This book is addressed to graduate and post-graduate students of applied mathematics and physics, as well as to engineers and scientists. Referring to academic background, it is assumed that the reader is familiar with the calculus and is at least superficially familiar with the differential equations. Due to the presence of a large amount of literature devoted to the basics of working with the *Mathematica* package, we do not pay due attention to the presentation of these fundamentals. At the beginning of the presentation, we do not use any complex commands or multi-line procedures, and therefore we hope that even a reader with no experience in working with this package can easily acquire the necessary experience on his own.

When covering the topic of modeling and its role in cognition of the material world, it is worth noting that modeling is practiced in all theoretical sections of the natural sciences. The hitherto development of science, and in particular of physics, confirms our belief that every empirical law is an approximate law, having a limited scope of applicability, and the theorist's work consists in the construction of a mathematical model, which is then tested using appropriate tools. Thus, in fact, the subject of theoretical research is never the phenomenon as such, but a more or less accurate approximation of it presented as a mathematical model. The process of creating a model is not unequivocal, and the art of modeling is to find a compromise between accuracy and simplicity of mathematical description. A useful model takes into account the most important features of the phenomenon under study, disregarding the less important features, making it simple enough that it can be carefully analyzed. It is worth noting that the correctness and usefulness of the model (theory) is conditioned by the following factors:

- compliance with empirical data;
- compliance with fundamental laws in the given field of knowledge;
- internal consistency;
- possibility to draw new conclusions.

The theoretical conclusions, in turn, should be consistent with the experimental data. The general relationship between the phenomenon and the creation of an appropriate model can be presented in the form of a diagram:

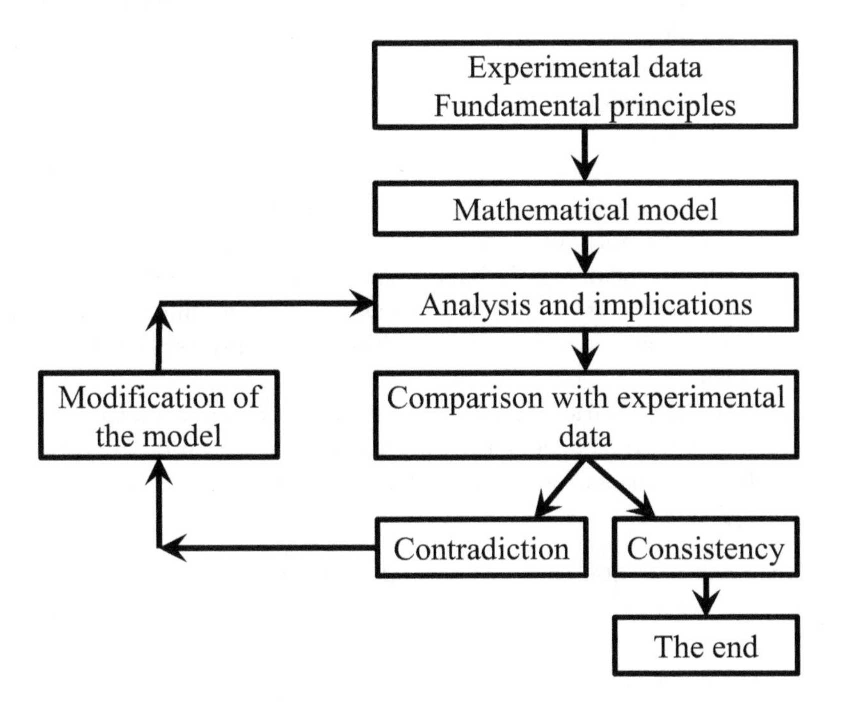

In connection with what has been said above, the question immediately arises: since representatives of all natural sciences (and increasingly also of social sciences) deal with the construction and study of models, does it make sense to single out modeling as a separate discipline? Thus, there are many arguments that justify such a separation, and some of them are presented below.

- Mathematical models are universal and do not depend on the content put into them by representatives of various fields of science. For example, the equation

$$\ddot{x}(t) + \omega^2 x(t) = 0$$

describes small oscillations of a mathematical pendulum in the gravitational field ($x(t)$ is the angle of pendulum deviation from the vertical), as well as movements of a material point attached to the spring ($x(t)$ is the spring deviation from equilibrium), electric charge oscillations in LC circuit ($x(t)$ is the charge on the capacitor), changes in the population size and many other processes.
- Nonlinear evolutionary models have universal characteristics, for example, a large number of models described by the multidimensional nonlinear dynamical systems demonstrate chaotic features and the scenarios of transition to chaos in such systems are also universal.
- Models described in terms of nonlinear evolutionary partial differential equations very often describe stable coherent patterns, taking the form of solitary waves, moving fronts, oscillating or periodic structures. Solutions of this type, which are very important from the point of view of applications, have a number of common features and are studied by means of the universal mathematical tools.
- The mathematical modeling does not aim at solving a specific theoretical problem and thus allows to look at various phenomena from a broader perspective, thus noticing their common repeating features.

This book presents models of natural phenomena described in terms of differential and difference equations, as well as the methods allowing to analyze these models. In case of numerous nonlinear models considered in the book, one cannot expect to obtain an exact solution to a specific problem. Therefore, the researches of nonlinear models in addition to searching for exact solutions where possible, include the methods of qualitative analysis and numerical simulations. The *Mathematica* package was created from the very beginning with a view of being used in widely understood modeling. It contains a large number of ready-made procedures, and within this package it is easy to move from analytical formulas to numerical schemes. The use of the instruments and graphical tools of the *Mathematica* package allows one to present the basic facts of the modeling of nonlinear phenomena in an elegant form. In this book, we've stuck to *Mathematica 12.1*, but the vast majority of the routines can be implemented in earlier releases of this package, starting at least with *Mathematica 7.0*.

Modeling distinguishes between the systems whose elements are placed at separate points and the distributed systems. If the sizes of the elements of a physical system are negligible in comparison with the distances between them, then we can approximately consider these elements as material points (sometimes they are called the point masses). Examples of such systems are planets orbiting around the Sun; a pendulum hanged in the gravitational field; charged particles moving in the electromagnetic fields; electrical circuits, etc. Examples of systems with distributed parameters are liquid, gaseous and solid media when the description concerns mass and heat transport or wave propagation.

The most common models of systems with concentrated parameters are ordinary differential equations. Systems with distributed parameters are modeled as a rule by partial differential equations. Systems with concentrated parameters will be considered in the first part, while the second part will be entirely devoted to systems with distributed parameters. Below we present the contents of the book chapter by chapter.

The first part consists of four chapters. Chapter 1 contains examples of models described in terms of ordinary differential equations and their discrete analogs. In particular, the model of the evolution of the bacterial population due to Malthus is considered, and it is shown that its predictions do not correspond to the experimental data. Further, a more realistic model called the logistic one is analyzed, and the analytical solutions of this model are compared with the corresponding numerical solutions. In addition, we analyze numerically discrete analogue of the logistic equation and demonstrate the differences between the discrete and continuous models. Later in this chapter, Hamilton's principle of least action is presented and the Lagrange formalism is developed that allows one to obtain equations of motion for complex mechanical systems. At the end of the chapter, various nonlinear models are constructed using this scheme and the properties of their solutions are numerically studied.

The first part of Chapter 2 is devoted to the technique of studying systems of ordinary differential equations of the first order, called autonomous dynamical systems. The concepts of phase space, trajectory, stationary point, and linearization of a system in its small neighborhood are introduced. The list of issues that form the basis of a qualitative analysis of a dynamical system, also called qualitative integration, is touched upon. The so-called conservative systems with one degree of freedom are considered in more detail. For such systems all problems of qualitative analysis can be completely solved obtaining this way comprehensive information about the nature of the entire set of solutions of any particular system belonging to this class. In the further part of this chapter, the concepts of invariant subspaces of linear dynamical systems and local invariant manifolds are introduced. The described conceptual apparatus is used at the end of the chapter to solve problems on the loss of stability by the lower equilibrium position of the pendulum under the influence of small perturbations of a certain type, as well as on the acquisition of stability by the upper equilibrium position of the pendulum under the influence of small high-frequency perturbations applied to the suspension point.

Chapter 3 is devoted to the study of nonlinear periodic solutions of dynamical systems, called limit cycles. In particular, local bifurcations leading to the emergence of limit cycles are considered, and a mathematical apparatus has been developed that allows one to study their stability. At the end of the chapter, numerous examples of the study of the occurrence of periodic solutions in specific systems are given.

In Chapter 4, examples of the evolution of limit cycles in the case of multidimensional and non-autonomous dynamical systems are considered. In particular, the scenario of transition to chaotic oscillations through a cascade of period doubling bifurcations is analyzed. We also analyze the issue of extreme sensitivity of numerical solutions describing chaotic oscillations to rounding errors and ways to distinguish chaotic solutions from outwardly similar multi-periodic solutions. At the end of the chapter, a bridge connecting continuous dynamical systems with the corresponding discrete mappings is discussed, and on the example of the latter, the mechanism of the period doubling cascade is analyzed.

In Chapter 5, the fundamental models are considered that serve to describe various processes occurring in a continuous media, in particular, derivations of the hydro- and gas-dynamics equations based on the laws of conservation of mass, momentum and energy are presented, as well as the heat transport equation. In addition, generalized models are derived that take into account the structure of the medium, energy sources, inhomogeneities and other features of the systems under study. There are also introduced simplified equations that serve as submodels for the fundamental equations of continuous media and put forward universal methods for obtaining exact solutions based on the theory of similarity and dimensions. At the end of the chapter the classification of scalar quasilinear second order partial differential equations is given.

Chapter 6 is devoted to methods for solving linear partial differential equations. It includes methods based on Fourier and Laplace transforms, as well as the method of separation of variables. In terms of software, this chapter widely uses standard programs and commands that carry out computer implementation of integral transformations, as well as commands that interpret special functions arising in a natural way when solving the Sturm-Liouville problem.

Chapter 7 is devoted to the construction of various finite-difference schemes and their application to solving problems of mathematical physics. Both explicit and implicit finite difference schemes, the method of lines, the Galerkin method, and the finite element method are described. Methods for studying the stability of numerical schemes are also touched upon.

Chapter 8 deals with nonlinear PDEs called completely integrable. In particular, we consider the Hopf equation, being the submodel for the hydrodynamic-type equations, the Burgers equation being the submodel for the Navier-Stokes equations, and the celebrated Korteweg-de Vries equation. The valuability of the submodels is connected with the fact that, being completely integrable, they possess a number of solutions characteristic to more realistic models, in particular, the shock wave solutions, diffuse shock waves, and other patterns.

We distinguish in a special way the Korteweg-de Vries equation, which describes numerous physical systems and has a number of unique properties. In particular, it is shown that the Korteweg-de Vries equation has an infinite set of conservation laws and is in a certain sense equivalent to the linear Schrodinger equation. It possesses a one-parameter family of exponentially localized solutions which evolve without changing of their shape and interact like elastic balls.

In Chapter 9, the Hirota's method is analyzed in detail, which makes it possible to obtain analytical solutions to the KdV equations describing an arbitrary number of soliton solutions interacting with each other. Recently, due to the widespread use of computer algebra methods, the Hirota's method and its modifications have been actively used in the search for solutions to non-linear evolutionary equations which are not completely integrable. Hirota's method is used in this chapter to obtain complex wave solutions supported by the convection-reaction-diffusion equations.

In Chapter 10 three non-integrable models are considered, possessing generalized solutions called compactons. Properties of compactons resemble those of the soliton solutions. In particular, they interact with each other almost elastically and restore their primary shape after the interaction. The first of the considered models, proposed in the paper [Rosenau and Hyman (1993)] is obtained by the formal generalization of the Korteweg-de Vries equation onto the case of nonlinear dispersion. The other two models are more realistic because they are derived from physical considerations. The second model describes the pre-stressed elastic medium in the continuum approximation. The third model is a hydrodynamic-type system that takes into account the effects of nonlocality. The presence of compacton solutions among the solutions of the models under consideration and the fact that they differ little in properties from "true" solitons is an important observation that allows one to challenge the repeatedly expressed opinion that the properties of solitons are due to the fact that they are supported by the KdV equation, which has an infinite set of conservation laws.

The authors also considered it appropriate to place three appendices at the end of the book, intended to contribute to the completeness of the presentation.

The material presented in the book can be divided into two independent one-semester courses, which, apart from classroom lectures, include practical classes in the computer room (or individual work with a computer) as an indispensable element.

Contents

PART 1

Models described in terms of ordinary differential equations and their discrete analogs

Chapter 1

Examples of models described in terms of ordinary differential equations. Lagrange's formalism and its applications

1.1 Malthusian model and logistic model

In a highly idealized Malthusian model, describing, for example, the development of the bacterial population, the function sought is the size of the population $P(t)$. The food base is assumed to be unlimited and there are no growth inhibitors. In such a situation, it can be assumed that the rate of population change is proportional to the size of the population, i.e. the following equation holds:

$$\frac{d\,P(t)}{d\,t} = rP(t), \tag{1.1}$$

where r is the reproducibility coefficient which, in the simplest case, is inversely proportional to the average period of reproduction. Adding to this the initial condition $P(0) = P_0 > 0$, we get the solution

$$P(t) = P_0\, e^{rt}. \tag{1.2}$$

Due to the simplicity of the solution, we can easily draw some conclusions from it. Let us assume that $P(t)$ describes a population of unicellular bacteria and that under favorable circumstances (moderate constant temperature, abundance of food base, etc.), each bacterium divides on average every 20 minutes. Let's also assume that the process started with the reproduction of one bacterium, that is, $P_0 = 1$. Then the solution will be given by the formula $P(t) = e^{t/1200}$, where t is time expressed in seconds. It can be assumed that the bacterium has the shape of a cube with linear dimensions of $10^{-6}\ m$ and, respectively, volumes of $10^{-18}\ m^3$. Let us answer the question: after what time T the bacterial population developing according to the Malthusian model will cover the Earth's surface with a layer of one meter? The radius of the Earth is approximately $6400 \cdot 10^3\ m$, so the time can be obtained by solving the equation

$$4\pi \cdot (6400 \cdot 10^3)^2 = 10^{-18}\, e^{T/1200}.$$

Using the standard tools of the *Mathematica* package, we will get the following answer (in hours):

Cell 1.1.

In[1]: N[Log[4 Pi (6400 $10^3)^2$ 10^{18}] 1200/3600]
Out[1]: 25.1071

Calculations performed in the *Mathematica* package show that this will happen approximately 25 hours after a single cell begins to divide. This conclusion is, of course, absurd, even though some bacteria actually take 20 to 40 minutes to reproduce. The reason for the complete incompatibility of the Malthusian model with actual observations lies in the fact that the Malthusian model in no way takes into account the finiteness of food resources. It also fails to take into account the fact that constant and rapid reproduction is possible in a narrow temperature range and in the complete absence of factors that slow down population growth.

A more realistic model taking into account the finite dimensions of the area that a population may occupy takes the form

$$\frac{d\,P(t)}{d\,t} = r P(t) \frac{K - P(t)}{K},$$

where K is proportional to the maximum size of the population (this size is related to the volume of the area where the population is reproducing). By introducing a new normalized variable $N(t) = P(t)/K$ and using the symbol $\lambda = r$, we get the standard *logistic equation*

$$\frac{dN}{dt} = \lambda N(1 - N). \tag{1.3}$$

Assuming additionally that $N(0) = N_0$, we will get (in other variables, because the symbol N is reserved in the *Mathematica* package for a numeric value) a solution using the standard DSolve procedure:

Cell 1.2.

In[1]: DSolve [{Z'[x] − λ Z[x] (1 − Z[x]) == 0, Z[0] == Z_0 }, Z[x], x]
Out[1]: {{Z[x] → $\frac{e^{x\lambda} Z_0}{1 - Z_0 + e^{x\lambda} Z_0}$ }}

So the *Mathematica* package gives us the solution

$$N(t) = \frac{e^{x\lambda}\, N_0}{1 - N_0 + e^{x\lambda}\, N_0}, \tag{1.4}$$

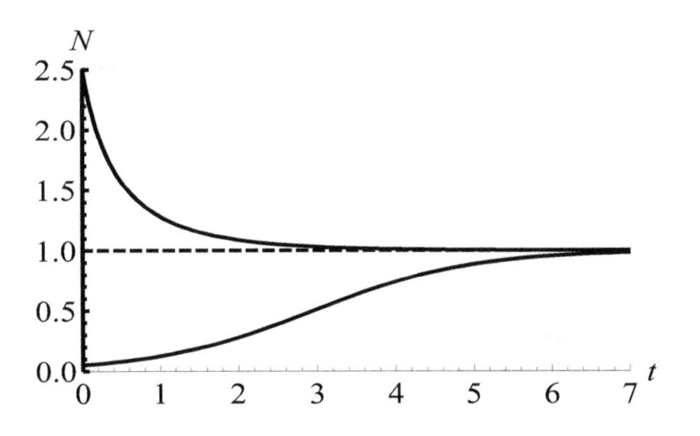

Fig. 1.1 Solutions of the logistic equation.

which, regardless of the initial conditions assumed and the value of the parameter $\lambda > 0$, tends asymptotically towards the normalized value $N = 1$, see Fig. 1.1.

1.2 Discrete analogs of the logistic equation

The discrete analog of the equation (1.3) can be obtained in different ways. The first is to go from the differential equation to the corresponding difference equation, which we get by dividing $[0, T]$ into intervals of Δt. Using the substitution $N(t) = N(m\Delta t) = N_m$, $N(t \pm \Delta t) = N_{m \pm 1}$ we get:

$$\frac{N_{m+1} - N_m}{\Delta t} = \lambda N_m(1 - N_m). \tag{1.5}$$

This equation can be presented as

$$N_{m+1} = (1 + \lambda \Delta t)N_m - \lambda \Delta t N_m^2. \tag{1.6}$$

Introducing scaling $N_m = \alpha Y_m$ and putting $\alpha = (1 + \lambda \Delta t)/(\lambda \Delta t)$, we get from (1.6) the standard form of the discrete logistic equation

$$Y_{m+1} = r Y_m (1 - Y_m), \tag{1.7}$$

where $r = 1 + \lambda \Delta t$. We assume that $0 < \lambda \Delta t \ll 1$, so the parameter r is slightly larger than one. Under this assumption, the numerical solution of the difference equation (1.7) is very close to the analytical solution of the equation (1.3), see Fig. 1.2. Numerical solution of discrete logistic equation and its comparison with analytical results is implemented in the file PM1_1.nb.

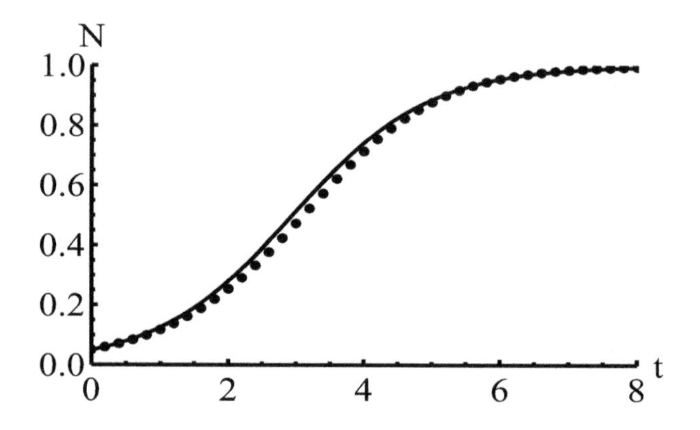

Fig. 1.2 Analytical solution of the equation (1.7) obtained with $\lambda = 1$, $N_0 = 0.05$ (solid line) and the numerical solution of the equation (1.3) after appropriate scaling (dots).

Discrete logistic mapping occurs not only as a result of the discretization of the differential model. Let us consider the following bank savings issue ([Peitgen and Richter (1984)], [Schuster (2005)], Chapter 3). Assume that the cash contribution Z_0 increases as follows:

$$z_{n+1} = (1+\epsilon)z_n = (1+\epsilon)^2 z_{n-1} = \cdots = (1+\epsilon)^{n+1} z_0,$$

where ϵ is the percentage of capital growth over a specified period of time (the amount may increase annually, once a month, etc., depending on the terms of the deposit). To prevent hyperinflation, politicians could submit a bill that the percentage should decline as capital grows, which will be ensured if, for example, the rate of capital growth changes as follows: $\epsilon \to \epsilon_0 \left(1 - z_n/z_{max}\right)$. Then the bank account would change in accordance with the equation:

$$z_{n+1} = \left[1 + \epsilon_0 \left(1 - z_n/z_{max}\right)\right] z_n.$$

By introducing the scaling $z_n = \alpha\, x_n$, where $\alpha = (1+\epsilon_0)z_{max}/\epsilon_0$, we get the standard form of the logistic equation:

$$x_{n+1} = r\, x_n \left(1 - x_n\right), \quad r = 1 + \epsilon_0. \tag{1.8}$$

Based on the analysis of the equation (1.3) and its numerical analogue, it could seem that the proposed mechanism will lead to the stabilization of capital growth. Note, however, that in this case there is no reason to assume that $\epsilon_0 \ll 1$, as interest rates may reach any value during hyperinflation. On the other hand, numerical simulations, carried out in the PM1_1.nb file using the code quoted below, show that in a situation where $r \in (3,4)$ the solution is cardinally different from the solutions corresponding to the r parameter values close to one (see Fig. 1.3).

Cell 1.3.

```
In[1]: Clear["Global'*"];
In[2]: t[0] = 0; x[0] = 0.05; r = 2.12; total = 50;
In[3]: x[n_] := x[n] = r x[n - 1] (1 - x[n - 1]);
In[4]: t[n_] := t[n] = t[n - 1] + 1;
In[5]: Mon2 = Table[{t[n], x[n]}, {n, 0, total}];
In[6]: LogMon2 = ListLinePlot[Mon2, PlotRange → Full, AxesLabel → {n, Xn}]
```

Remark. Note that employment of the sign "=" after the function definition is not mandatory, but without doing so we'll obtain the code which works very slowly when $n \gg 1$, because every time we ask, e.g. for $v[n]$, it evaluates the recursion realization back to $v[0]$. An extra sign "=" makes *Mathematica* remember values of the function $v[n]$ that has been evaluated previously.

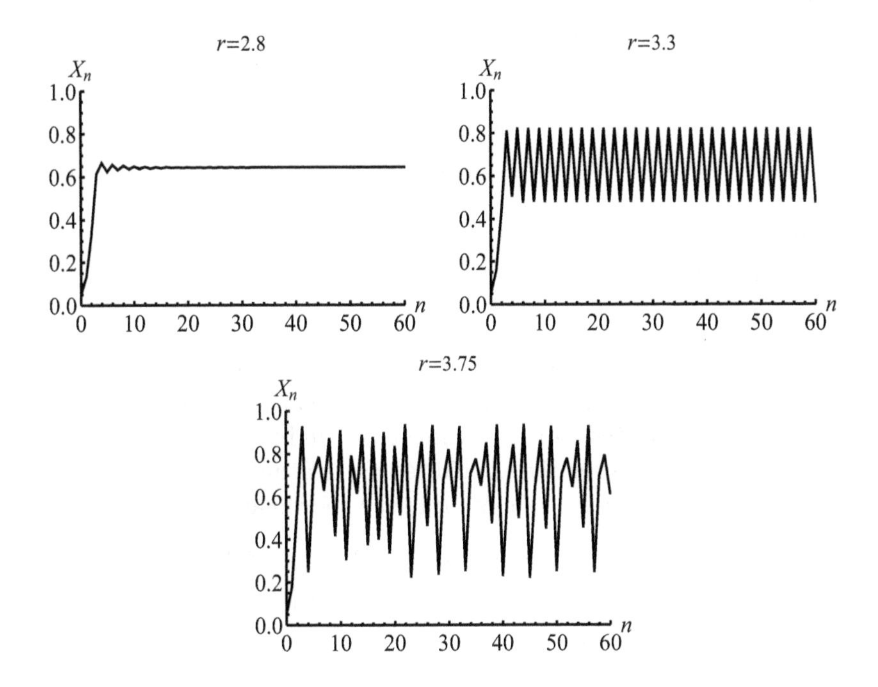

Fig. 1.3 Logistic mapping plots obtained with $x_0 = 0.05$ and different values of the parameter r.

We present in Fig. 1.3 a fragment of the numerical analysis of the equation (1.8). This equation will be discussed in more detail in later chapters. In the context of this review, it is worth paying attention to the qualitative differences in the solutions to the equation (1.8) corresponding to the different ranges of the variability of the r parameter. Note that the solutions to the equation (1.3) behave qualitatively identically for any values of the parameter λ, so in this respect the continuous logistic equation differs significantly from its discrete analog.

Problem. The code contained in the cell 1.3 after minor modifications can be used to examine the generalized logistic mapping $x_{n+1} = r\, x_n \left(1 - x_n^3\right)$. Modify it accordingly, plot the x_n function graphs for $r \in (0,\, 2)$. Obtain an analytical solution of its continual analog

$$\frac{d\,x}{d\,t} = \lambda\, x \left(1 - x^3\right), \qquad 0 < \lambda$$

using the procedure DSolve. Compare the behavior of solutions of the continual analog with that of the discrete equation for different values of the parameter r.

1.3 Model of a pendulum moving in a gravitational field

Let the material point with the mass m, suspended on a weightless unstretchable thread of length l, perform plane movements in the gravitational field (see Fig. 1.4).

The coordinates of the point with mass m shown in Fig. 1.4 are expressed by the formulas

$$X = l\, \cos[\alpha], \qquad Y = l\, \sin[\alpha]. \tag{1.9}$$

Two forces act on the material point: the gravitational force $m\, \vec{g}$, where $\vec{g}$ is the acceleration directed along the OX axis, and the tension force T (which is always directed along the thread axis and takes the value equal to the projection of gravity in the same direction). Using the Newton's second law, we write the equations of motion in the coordinates X, Y:

$$m\ddot{X} = m\, g - T \cos[\alpha], \qquad m\ddot{Y} = -T\, \sin[\alpha]. \tag{1.10}$$

By differentiating the formulas (1.9) with respect to time twice and substituting the result into (1.10), we get

$$-ml\left(\ddot{\alpha}\sin[\alpha] + \dot{\alpha}^2\cos[\alpha]\right) = mg - T\cos[\alpha],$$

$$ml\left(\ddot{\alpha}\cos[\alpha] - \dot{\alpha}^2\sin[\alpha]\right) = -T\sin[\alpha].$$

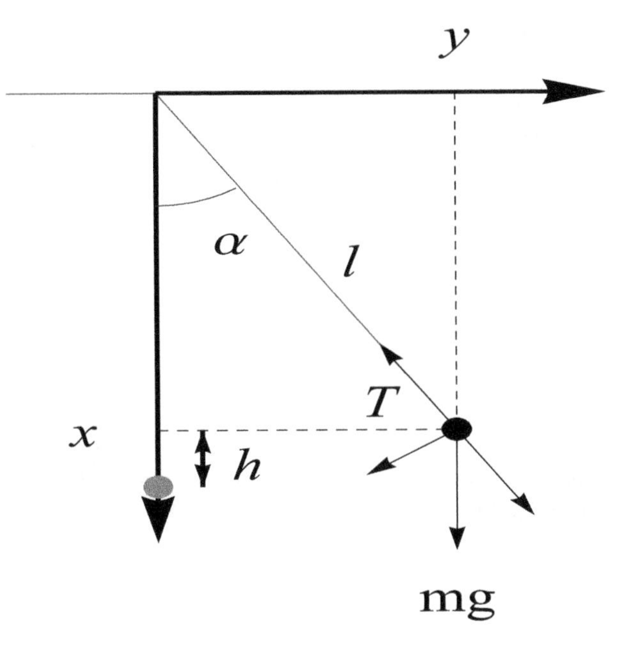

Fig. 1.4 A pendulum in the gravitational field.

Multiplying the first equation by $-\sin[\alpha]$ and the second equation by $\cos[\alpha]$ and adding the sides, we get, after elementary transformations, the equation

$$\ddot{\alpha} + \Omega^2 \sin[\alpha] = 0, \tag{1.11}$$

where $\Omega = \sqrt{g/l}$.

1.4 Lagrange's formalism

The flat pendulum model presented in the previous section is an example of one of the simplest mechanical systems. Therefore, deriving the equation (1.11) is not too difficult. In the case of more complex mechanical systems, the very derivation of the equations of motion often becomes a big challenge. The procedure for deriving the equations of dynamics of an ideal mechanical system can be significantly simplified using the following reasoning. The position of a material point in space is uniquely defined if the time dependence of the three Cartesian coordinates $\{x[t], y[t], z[t]\}$ is known. If there is N points, then the full information about the layout is given by knowing $3\,N$ coordinates $\{x_i[t], y_i[t], z_i[t]\}$, $i = 1, 2, ..., N$. In general, the number of s independent parameters, the knowledge of which is necessary to determine the position of the system at any time, is called *the number of degrees of freedom.*

It is obvious that from s Cartesian variables it is possible to pass by diffeomorphism to another variables $(q_1, ..., q_s)$. The new variables are called *the generalized coordinates*, and their derivatives $(\dot{q}_1, ..., \dot{q}_s)$ are called the generalized velocities. In order to derive the equations of motion of mechanical systems, the *principle of least action* (or variational principle) is used. According to this principle, any mechanical system with s degrees of freedom is characterized by a certain function

$$L\left(q_1, ..., q_s; \, \dot{q}_1, ..., \dot{q}_s; \, t\right),$$

called the Lagrange function (or Lagrangian). Let us denote it briefly as $L(q, \, \dot{q}, \, t)$. Suppose that at the initial moment of time t_0 the system is in the position characterized by the parameters q_0, $\dot{q}_0$, while at the ending moment of time t_1 — by the parameters q_1, $\dot{q}_1$. The initial and final configurations can, of course, be connected with the parametrized curves (trajectories) in an infinite number of ways. The principle of least action (also called *the Hamilton's principle*) says that the solutions that correctly describe the dynamics of a mechanical system are the functions $q(t)$, $\dot{q}(t)$, on which the action integral

$$S = \int_{t_0}^{t_1} L(q, \, \dot{q}, \, t) \, dt \tag{1.12}$$

reaches an extreme value. Using the symmetry principles, it is possible to show [Landau and Lifszyc (1965)] that

$$L(q, \, \dot{q}, \, t) = T - U, \tag{1.13}$$

where T is the total kinetic energy of the system, U is the total potential energy of the system. Both functions can be expressed in any generalized coordinates. The principle of least action states that if $q(t)$ is the "true" trajectory of the system, while $y(t) = q(t) + \Delta q(t)$ is any disturbed trajectory, see Fig. 1.5, lying in close vicinity of $q(t)$ ("in close vicinity" means that $||\Delta q|| \ll 1$ in some functional norm), then the inequality

$$\int_{t_0}^{t_1} L(q, \, \dot{q}, \, t) \, dt \leq \int_{t_0}^{t_1} L(y, \, \dot{y}, \, t) \, dt$$

holds. The above inequality, together with the knowledge of the Lagrangian, allows to obtain the equations governing the movement of the mechanical system on virtue of the following statement.

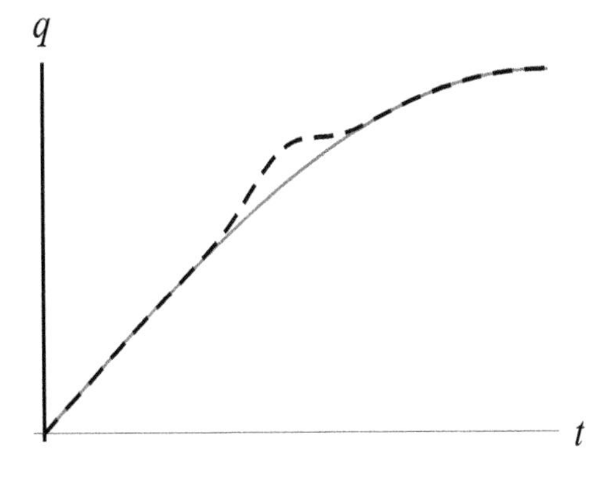

Fig. 1.5 The solid line represents the "true" trajectory $q(t)$, while dashed line — the perturbed trajectory $y(t) = q(t) + \Delta q(t)$.

Theorem 1.1. *The functional (1.12) takes the critical value on the trajectory $q(t)$ if and only if the system of equations is satisfied*

$$\frac{\partial L}{\partial q_i} - \frac{d}{dt}\frac{\partial L}{\partial \dot{q}_i} = 0, \quad i = 1, 2, \ldots, s. \tag{1.14}$$

Proof. Suppose that the i-th generalized variable is disturbed:

$$y(t) = (q_1(t), \ldots, q_{i-1}(t), q_i(t) + \Delta q_i(t), q_{i+1}(t), \ldots, q_s(t)).$$

We restrict our consideration to the disturbances that nullify at the end of the segment $[t_0, t_1]$:

$$\delta q_k(t_0) = \delta q_k(t_1) = \delta \dot{q}_k(t_0) = \delta \dot{q}_k(t_1) = 0, \quad k = 1, \ldots, s.$$

Let us note that $\delta\dot{q}_i(t) = d\,\delta q_i(t)/dt$. As in the calculus of functions of one variable, the necessary condition for the extremum of the functional is the disappearance of the term proportional to $\delta q_i(t)$ in the variational formula. Up to $(O(||\delta q_i||^2)$, the variation of the functional is expressed as:

$$\delta S = \int_{t_0}^{t_1} L\big[q_1(t), \ldots, q_{i-1}(t), q_i(t) + \Delta q_i(t), q_{i+1}(t), \ldots$$

$$q_s(t), \ldots, \dot{q}_i(t) + \Delta\dot{q}_i(t), \ldots\big] dt$$

$$- \int_{t_0}^{t_1} L\left[q_1(t), \ldots, q_{i-1}(t), q_i(t), q_{i+1}(t), \ldots, q_s(t), \ldots, \dot{q}_i(t), \ldots\right] dt$$

$$= \int_{t_0}^{t_1} \frac{\partial}{\partial q_i} L(q, \dot{q}; t)\, \delta q_i + \frac{\partial}{\partial \dot{q}_i} L(q, \dot{q}; t)\, \delta\dot{q}_i\, dt$$

$$= \frac{\partial}{\partial \dot{q}_i} L(q, \dot{q}; t) \delta_{q_i}|_{t_0}^{t_1} + \int_{t_0}^{t_1} \left(\frac{\partial}{\partial q_i} - \frac{d}{dt}\frac{\partial}{\partial \dot{q}_i}\right) L(q, \dot{q}; t)\, \delta q_i\, dt.$$

Due to the nullifying of variations at the ends of the segment $[t_0,\, t_1]$, the first term in the last equality is identically equal to zero. Since the variation $\delta\, q_i$ is arbitrary, the first order term is zero if and only if the integral expression in the last integral is reset to zero at each point on the segment $[t_0,\, t_1]$, and this ends the proof.

Let us apply the principle of least action to obtain the equation of motion of a mathematical pendulum (see Fig. 1.4). The kinetic energy of the pendulum is expressed by the formula

$$T = \frac{m}{2}\left(\dot{X}^2 + \dot{Y}^2\right) = \frac{m\,l^2}{2}\dot{\alpha}^2, \tag{1.15}$$

while its potential energy is expressed as follows:

$$U = m\,g\,h = m\,g\,l\,(1 - \cos[\alpha])\,. \tag{1.16}$$

Applying the formula (1.14) to the expression

$$L = \frac{m\,l^2}{2}\dot{\alpha}^2 - m\,g\,l\,(1 - \cos[\alpha])\,, \tag{1.17}$$

we obtain:

$$0 = \frac{\partial\, L}{\partial\, \alpha} - \frac{d}{d\,t}\frac{\partial\, L}{\partial\, \dot{\alpha}} = -ml^2\left(\ddot{\alpha} + \frac{g}{l}\sin[\alpha]\right).$$

The result derived coincides with that obtained earlier on the basis of the second Newton's law (cf. with (1.11)).

1.5 Examples of obtaining the governing equations using the Lagrange formalism

1.5.1 *Double pendulum*

The double pendulum is a system of two material points, the first of which, with the mass m_1, is suspended on a weightless, non-stretchy rod of length L_1, while the second material point with the mass m_2 is attached to the first with a weightless, non-stretchable rod of the length L_2 (Fig. 1.6). Both masses are subjected to the action of the gravitation force directed downwards. We also assume that the movement of both pendulums takes place in one plane.

We calculate the coordinates of the masses presented in Fig. 1.6:

$$x_1 = L_1 \cos[\alpha], \qquad y_1 = L_1 \sin[\alpha],$$
$$x_2 = x_1 + L_2 \cos[\beta], \qquad y_2 = y_1 + L_2 \sin[\beta].$$

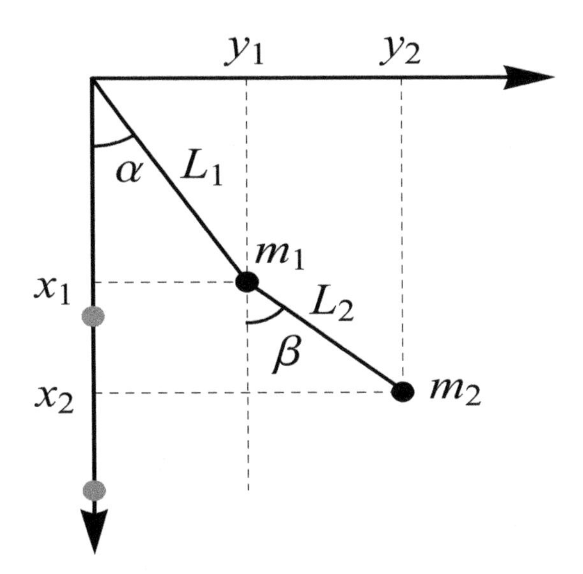

Fig. 1.6 Double pendulum. The gray circles on the vertical axis correspond to the positions of the masses in the state of equilibrium.

Next we calculate the squares of the velocity of both material points:

$$v_1^2 = \dot{\alpha}^2 L_1^2, \quad v_2^2 = \dot{\alpha}^2 L_1^2 + \dot{\beta}^2 L_2^2 + 2\dot{\alpha}\dot{\beta}\cos[\alpha - \beta].$$

The Lagrange function of the system is expressed by the formula $L = T - U$, where

$$T = \frac{1}{2}\dot{\alpha}^2 L_1^2(m_1 + m_2) + \frac{1}{2}\dot{\beta}^2 L_2^2 m_2 + m_2 L_1 L_2 \dot{\alpha}\dot{\beta}\cos[\alpha - \beta]$$

is the kinetic energy, while

$$U = (m_1 + m_2)gL_1(1 - \cos[\alpha]) + m_2 g L_2(1 - \cos[\beta])$$

is the potential energy. Employing the equation (1.14), we get the equation of the form:

$$\begin{aligned} &\ddot{\alpha} + \frac{g}{L_1}\sin[\alpha] + \frac{m_2 L_2}{(m_1 + m_2)L_1}\left\{\ddot{\beta}\cos[\beta - \alpha] - \dot{\beta}^2\sin[\beta - \alpha]\right\} = 0, \\ &\ddot{\beta} + \frac{g}{L_2}\sin[\beta] + \frac{L_1}{L_2}\left\{\ddot{\alpha}\cos[\beta - \alpha] + \dot{\alpha}^2\sin[\beta - \alpha]\right\} = 0. \end{aligned} \tag{1.18}$$

Remark. Obtaining the equations of motion of a mechanical system can only be reduced to looking for formulas expressing kinetic and potential energy, because in the *Mathematica* package there is a standard tool that allows to calculate variational derivatives [Bellomo *et al.* (2002)]. The diagram which allows to obtain the equations of motion of the double pendulum is given below.

Cell 1.4.

In[1]: Lagr = $\frac{1}{2}\alpha'[t]^2 L_1{}^2 (m_1+m_2)+\frac{1}{2}\beta'[t]^2 L_2{}^2 m_2+m_2\alpha'[t]\beta'[t]L_1L_2\mathsf{Cos}[\beta[t]-\alpha[t]]-$
$-((m_1+m_2)\,gL_1(1-\mathsf{Cos}[\alpha[t]])+m_2gL_2(1-\mathsf{Cos}[\beta[t]]))$;
In[2]: eqalf = Simplify[Expand[VariationalD[Lagr,α[t],t]]]
Out[2]: $-L_1(m_1\,(g\mathsf{Sin}[\alpha[t]]+L_1\alpha''[t])+m_2(g\mathsf{Sin}[\alpha[t]]+$
$+L_1\alpha''[t]+L_2\,(\mathsf{Sin}[\alpha[t]-\beta[t]]\beta'[t]^2+\mathsf{Cos}[\alpha[t]-\beta[t]]\beta''[t])))$
In[3]: eqbet= Simplify[Expand[VariationalD[Lagr, β[t],t]]]
Out[3]: $-L_2m_2(g\mathsf{Sin}[\beta[t]]+L_1(-\mathsf{Sin}[\alpha[t]-$
$-\beta[t]]\alpha'[t]^2+\mathsf{Cos}[\alpha[t]-\beta[t]]\alpha''[t])+L_2\beta''[t])$

Dividing the expressions "eqalf", "eqbet" by $-L_1^2(m_1+m_2)$ and $-L_2^2m_2$, respectively, and equating the results to zero, after elementary algebraic transformations we get the system of equations (1.18). The above layout is solved numerically in the PM_2.nb file using the standard NDSolve command. This file contains the animation that allows to observe the movements of the mechanical system at different values of parameters, and to map the phase trajectory into the configuration space of the problem, which is a two-dimensional torus.

1.5.2 *Flat pendulum with a movable suspension point*

A flat pendulum having the mass M is suspended on an inextensible rod L. The pendulum suspension point is the mass m which moves without friction along a horizontal line.

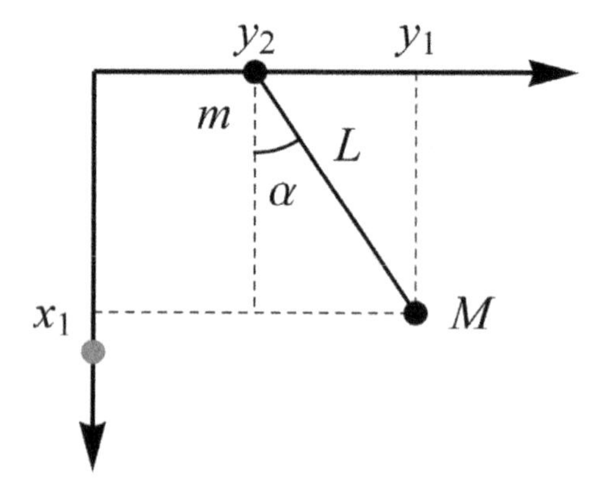

Fig. 1.7 Flat pendulum with a movable suspension point.

The OX axis is identified with the downward pointing vertical axis. For OY we choose a horizontal line pointing to the right (see Fig. 1.7); α is the angle measuring the deviation of the mass M from the vertical. Let y_2 denotes the coordinate of mass m. Then the position of the mass M is given by the formulas

$$x_1 = L\cos[\alpha(t)];$$
$$y_1 = y_2(t) + L\sin[\alpha(t)].$$

Calculating the kinetic and potential energy of the system, we get:

$$E_{kin} = \frac{M}{2}\left[(-L\sin[\alpha(t)]\dot{\alpha}(t))^2 + (y_2'[t] + L\cos[\alpha(t)]\dot{\alpha}(t))^2\right] + \frac{m}{2}(y_2'[t])^2$$
$$= \frac{M}{2}L^2(\alpha'(t))^2 + \frac{M+m}{2}(y_2'[t])^2 + ML\cos[\alpha(t)]\alpha'(t)y_2'(t);$$

$$E_{pot} = Mg(L - x_1) = -MgL\cos[\alpha(t)] + \text{const}.$$

Then with the command VariationalD[·, · t] we calculate the variational derivative of the function $Lagr = E_{kin} - E_{pot}$:

Cell 1.5.

```
In[1]:
  Lagr:=M/2((- L Sin[α[t]]D[α[t],t])^2+(D[y2[t],t]+L Cos[α[t]]D[α[t],t])^2+
  m/2(D[y2[t],t])^2+MgLCos[α[t]]
In[2]: eqalf=Simplify[VariationalD[Lagr, α[t],t]]
Out[2]:
  - LM (gSin[α[t]]+Cos[α[t]] y2''[t]+Lα''[t])
In[3]: eqy2=Simplify[VariationalD[Lagr,y2[t],t]]
Out[3]:
  LM Sin[α[t]] α'[t]^2 - (m+M) y2''[t] - LM Cos[α[t]] α''[t]
```

After simplifying and equating the variational derivatives to zero, we get a system

$$\frac{g}{L}\sin[\alpha(t)] + \frac{1}{L}\cos[\alpha(t)]y_2''(t) + \alpha''(t) = 0,$$

$$\frac{d}{dt}\left\{y_2'(t) + \frac{ML}{m+M}\alpha'(t)\cos[\alpha(t)]\right\} = 0.$$

Eliminating $y_2''[t]$ from the first equation, we finally get the equation

$$\alpha''(t)\left[1 - \frac{M}{m+M}\cos^2[\alpha(t)]\right] + \frac{g}{L}\sin[\alpha(t)] + \frac{M}{2(m+M)}\alpha'(t)^2\sin[2\alpha(t)] = 0.$$

1.5.3 *A system of two masses: one mass is suspended in a gravitational field, and the other can move along a horizontal surface without friction*

The hanging ball with the mass m_2 is tied to a ball with the mass m_1 ($m_1 > m_2$) and the radius ρ, which slides without friction on the horizontal surface (see Fig. 1.8). The coordinate axes are selected in such a way that the horizontal axis (OX axis) points to the right and the OY axis points vertically upwards. We assume that the thread connecting the masses is inextensible and has a length c. The blocks are mounted at the distance h from the axis OX and at the distance L from each other. The parameter defining the geometry of the system is the angle θ, which is formed by the line going from the block to the ball of mass m_1 with the axis parallel to the axis OX, passing through the center of mass of this ball. We solve the problem of finding the equations of motion of this system using the above definitions and doing all the calculations in the *Mathematica* package.

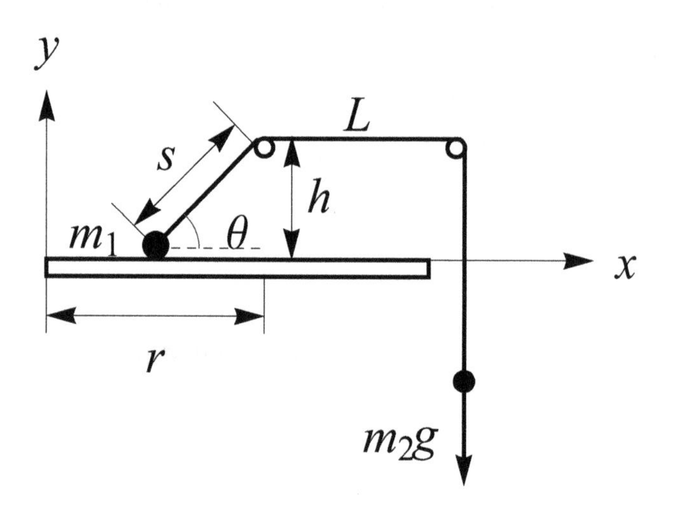

Fig. 1.8 A system of two masses connected by an inextensible weightless thread.

We find the equation of motion by writing down the dependence of the kinetic energy T and the potential energy U in terms of $\theta(t)$ variable and its derivative, and then calculating the variational derivative of the $L = T - U$ function.

Cell 1.6.

In[1]:
```
x1 = r - (h-ρ) Cot[θ[t]];
y2 = Sqrt[(h-ρ)^2/Sin^2[θ[t]]] + L + h - c;
T = m/2 Simplify[D[x1, t]^2] + M/2 Simplify[D[y2, t]^2];
U = Simplify[M g h/Sin[θ[t]]];
Lagr=T-U;
```
In[2]: `eqteta = Simplify[VariationalD[Lagr, θ[t], t]];`

After simplification, we get the equation with respect to the variable $\theta[t]$:

$$\left[m_2 - (m_1 + m_2)\csc^2\theta[t]\right]\ddot{\theta}[t] + \cot\theta[t]\left[2\,(m_1 + m_2)\csc^2\theta[t] - m_2\right]\dot{\theta}[t]^2 + \\ + \frac{g\, m_2 \cot\theta[t]}{\sqrt{(h-\rho)^2\, \csc^2\theta[t]}} = 0.$$

The numerical solution to the above equation and the animation showing the motion of the dynamical system can be found in the file PM1_3.nb.

Problem.

(1) A point of mass m, suspended on a weightless, inextensible rod of length L and capable of making spatial movements is called *a spherical pendulum.*

- Using the Lagrange formalism, derive the equation of motion of such a pendulum in spherical coordinates.
- Solve the problem numerically and present the animation of the movement of the pendulum in space.

(2) Two point masses m_1 and m_2 are suspended on weightless inextensible rods of the same length L at the same height at a distance of H from each other. The masses are connected by a spring which is unstretched when the pendulums hang freely.

- Derive the equations of motion of the system, assuming that:
  - angles of the deviations from the vertical $\alpha_1(t)$ i $\alpha_2(t)$ are not very large and therefore the deformation of the spring can be expressed as

$$L\,\{\sin[\alpha_2(t)] - \sin[\alpha_1(t)]\}\,;$$

  - assume that the potential energy of the spring is expressed by the formula

$$E_{pot} = \frac{k}{2}L^2(\sin[\alpha_2(t)] - \sin[\alpha_1(t)])^2.$$

- Solve the problem numerically and present the animation of the movement of the system.

(3) A bead of mass m slides without friction along the parabola $y = a\,x^2$ under the influence of gravitational force (see Fig. 1.9).

- Write the equation of motion.
- Present the animation choosing the value of m in an arbitrary way and treating the gravitational acceleration $g \in [1, 20]$ and the parameter $a \in [0.5,\ 2]$ as variable control parameters.

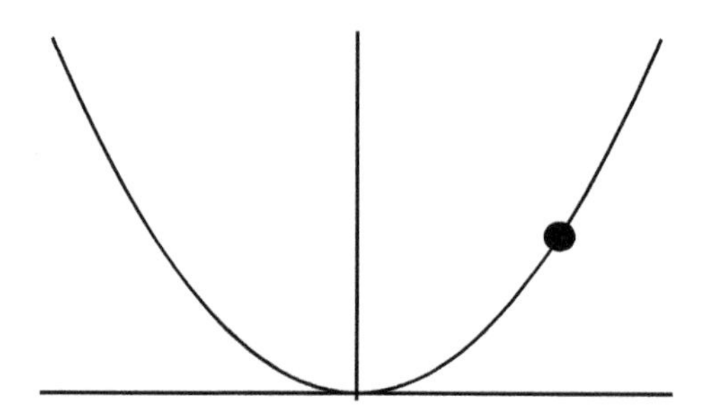

Fig. 1.9 A bead sliding along the parabola.

Chapter 2

Qualitative methods in the study of dynamical systems

2.1 Basic concepts. Linearization. Classification of stationary points on a plane

As is known, any nth-order scalar ordinary differential equation can be presented by change of variables in the form of a system of first-order ordinary differential equations:

$$\frac{dx_i}{dt} = F_i\left(x_1, x_2, \ldots, x_n; t\right), \qquad i = 1, 2, \ldots, n. \tag{2.1}$$

It is obvious that we can present in the similar form any system of ordinary differential equations containing derivatives of higher order. In the case when the functions F_i do not explicitly depend on the variable t, we use the term *the autonomous dynamical system* or simply *dynamical system*. Below we present it in vector form:

$$\frac{dX}{dt} = F(X), \quad X \in \mathbb{R}^n, \quad F: \mathbb{R}^n \supset U \to \mathbb{R}^n. \tag{2.2}$$

In the previous sections, we've presented several models describing the dynamics of systems of material points. The models we are interested in are mainly non-linear and it cannot be expected that we will be able to obtain an analytical solution in every particular case. In general, nonlinear differential equations can only be solved by numerical methods. An alternative to these methods is *the qualitative analysis*, which can be successfully applied to certain classes of dynamical systems. In the qualitative theory, an attempt is made to look at the set of all possible solutions of a dynamical system from a uniform point of view, and analyzing their features on the basis of knowledge of the functions F_i, $i = 1, 2, \ldots, n$. The elements of qualitative analysis presented here are largely based on geometric representation of solutions. The qualitative analysis uses a specific language, so we introduce some necessary concepts and symbols at the very beginning [Enns and Mc Gurie (2001)].

Definition 2.1. A map $\mathbb{R}^n \supset U \stackrel{F}{\Rightarrow} \mathbb{R}^n$, where U is an open set belonging to the domain of functions F_i, $i = 1, 2, \ldots, n$, is called the vector field defined by the right-hand side (r.h.s.) of (2.2).

Theorem 2.1. *Suppose that the functions F_i, $i = 1, 2, \ldots, n$ appearing in the r.h.s. of the system (2.2) are continuously differentiable on an open set $U \subset \mathbb{R}^n$. Then for each point $X_0 \in U$ such that $F(X_0)$ is nonzero vector there exists $c > 0$ such that the system (2.2) has a unique solution $(-c, c) \ni t \to \phi_t(X_0) \in \mathbb{R}^n$, satisfying the initial condition $\phi_0(X_0) = X_0$.*

The proof of this theorem can be found in any standard textbook on ordinary differential equations.

Definition 2.2. A point $X_0 = (x_{1\,0}, x_{2\,0}, \ldots, x_{n\,0}) \in \mathbb{R}^n$ is called *the stationary point* of the dynamical system (2.2), if $F_i(X_0) = 0$, $i = 1, \ldots, n$.

It is possible to replace the full system (2.2) in a small neighborhood of the stationary point with its *linearization.* Let us consider the right side of this system. Every function $F_i(X)$, differentiable in some open set $U \in \mathbb{R}^n$ containing X_0 can be presented as

$$F_i\,(x_1,\, x_2, \ldots,\, x_n) = \sum_{k=1}^{n} \frac{\partial F_i}{\partial x_k}(X_0)\,(x_k - x_{k\,0}) + O(|X - X_0|^2).$$

So, up to $O(|X - X_0|^2)$, one can present the dynamical system in vicinity of the stationary point in the form

$$\frac{d\xi}{dt} = \hat{M}\xi, \tag{2.3}$$

where $\xi = (\xi_1, ..., \xi_n) \in \mathbb{R}^n$ is the set of new variables, $\xi_i = x_i - x_{i\,0}$, $\hat{M}$ is a square matrix with n rows and columns:

$$\hat{M}_{i\,k} = \frac{\partial F_i}{\partial x_k}(X_0), \qquad i,\, k \,=\, 1,\, 2, \ldots, n.$$

Definition 2.3. The system (2.3) is called the linearization of the dynamical system (2.2) in vicinity of the stationary point X_0.

It turns out that under certain conditions the solutions of the system (2.2) and the solutions of the linear system (2.3) in small vicinity of the stationary point are "qualitatively identical". Since solving (analyzing) a linear system is usually easier than a full system, it is worth defining the conditions that allow such a substitution. Whether the behavior of the full system in the vicinity of a stationary point is really represented by its linearization depends on the eigenvalues of the matrix $\hat{M}$. We will analyze it on the example of the system in $\mathbb{R}^2$, giving the classification of simple stationary points.

Let us consider a system of equations

$$\frac{d}{dt}\begin{pmatrix} \xi \\ \eta \end{pmatrix} = \begin{pmatrix} a_{11} & a_{12} \\ a_{21} & a_{22} \end{pmatrix}\begin{pmatrix} \xi \\ \eta \end{pmatrix} = \hat{M}\begin{pmatrix} \xi \\ \eta \end{pmatrix}. \tag{2.4}$$

Following the tradition established in the theory of dynamical systems, we say the set of points $(\xi, \eta) \in \mathbb{R}^2$ forms the *phase plane* (in the case of multidimensional dynamical systems, the corresponding set of points is usually called the phase space). The set of all solutions $(\xi(t), \eta(t))$, $t \in I \subset \mathbb{R}$, presented in the form of parametrized curves, is called the *phase portrait* of the system (2.4). In qualitative theory, a synonym for the parameterized curve $(\xi(t), \eta(t))$ is the term *phase trajectory* or *orbit*. They differ from the commonly used designations in that the trajectories, depicted as continuous curves on the phase plane, are also equipped with arrows indicating the direction of movement along the trajectories when t increases.

We proceed to the classification of the stationary points.

(1) If the eigenvalues λ_1, λ_2 of the matrix $\hat{M}$ are real, positive and different (let, for definiteness $0 < \lambda_1 < \lambda_2$), then the stationary point $(0, 0)$ of the phase plane (ξ, η) is called the *unstable node*, or *source*. In this case, it is possible to define such a non-singular change of variables

$$\begin{pmatrix} y_1 \\ y_2 \end{pmatrix} = \hat{B}\begin{pmatrix} \xi \\ \eta \end{pmatrix} = \begin{pmatrix} b_{11} & b_{12} \\ b_{21} & b_{22} \end{pmatrix}\begin{pmatrix} \xi \\ \eta \end{pmatrix}$$

that the system takes the form

$$\frac{d}{dt}\begin{pmatrix} y_1 \\ y_2 \end{pmatrix} = \begin{pmatrix} \lambda_1 & 0 \\ 0 & \lambda_2 \end{pmatrix}\begin{pmatrix} y_1 \\ y_2 \end{pmatrix}. \tag{2.5}$$

Remark. $\hat{B}$ is the inverse to the matrix $\hat{V} = \left(\vec{V}_1, \vec{V}_2\right)$, composed of the eigenvectors of matrix $\hat{M}$ corresponding to the eigenvalues λ_1 and λ_1.

It is easy to see that solutions of the system (2.5) take the form

$$y_1 = C_1 e^{\lambda_1 t}, \qquad y_2 = C_2 e^{\lambda_2 t}.$$

In case when $C_1 \neq 0$, these solutions can be presented in the form

$$y_2 = C_3 |y_1|^{\frac{\lambda_2}{\lambda_1}}, \tag{2.6}$$

where

$$C_3 = C_2 \begin{cases} \frac{1}{C_1^{\lambda_2/\lambda_1}} & \text{when } C_1 > 0, \\ -\frac{1}{|C_1|^{\lambda_2/\lambda_1}} & \text{when } C_1 < 0. \end{cases}$$

In this case the phase portrait of the system (2.5) looks as shown in Fig. 2.1.

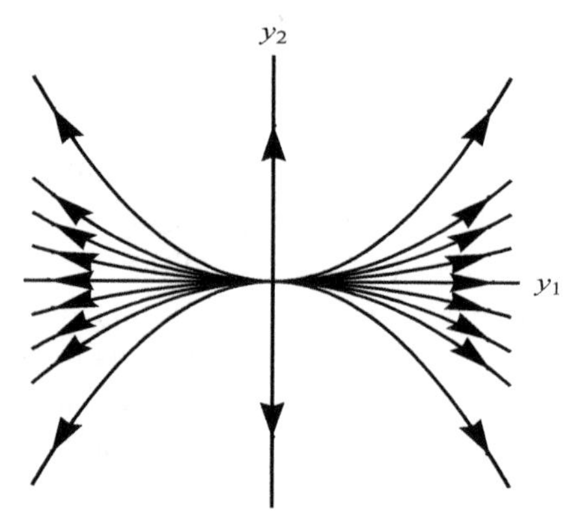

Fig. 2.1 Unstable node (source).

Note that the coordinate axes coincide with the selected phase trajectories of the dynamical system. The horizontal axis corresponds to the case of $C_2 = 0$; the vertical axis corresponds to the case of $C_1 = 0$. These phase trajectories cannot be expressed by the formula (2.6).

(2) If the eigenvalues λ_1, λ_2 of the matrix $\hat{M}$ are real, different and negative (let us assume that $0 > \lambda_1 > \lambda_2$), then the stationary point $(0, 0)$ of the phase plane (ξ, η) is called the *stable node*, or *sink*. Like in the previous case, it is possible to define a non-singular linear change of variables $(\xi, \eta) \to (y_1, y_2)$ such that in new variables the system (2.4) will be formally identical with (2.5). The phase portrait in the vicinity of such a point differs from the previous one only in the direction of movement along the phase trajectories, see Fig. 2.2.

(3) If the eigenvalues λ_1, λ_2 of the matrix $\hat{M}$ are real, non-zero and have different signs (let $\lambda_1 > 0 > \lambda_2$) then the stationary point $(0, 0)$ is called *the saddle point*. There is a linear change of variables $(\xi, \eta) \to$

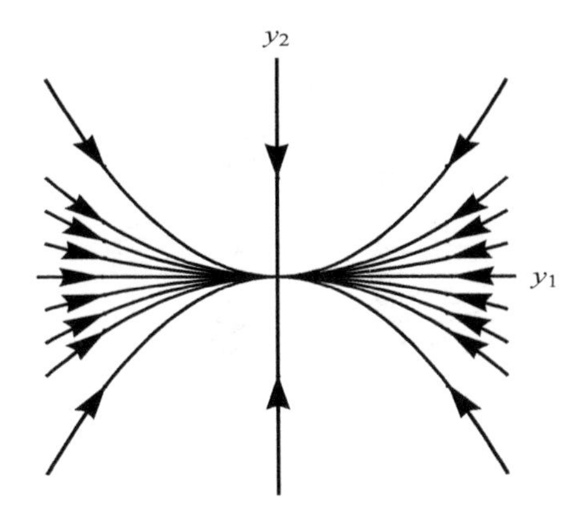

Fig. 2.2 Stable node (sink).

(y_1, y_2) such that in the new variables the system (2.4) takes the form (2.5). Solutions of this system can be presented in the form

$$y_1 = C_1 e^{|\lambda_1|t}, \qquad y_2 = C_2 e^{-|\lambda_2|t}.$$

In case when $C_1 \neq 0$ solutions can also be presented as

$$y_2 = C_3 |y_1|^{-|\lambda_2/\lambda_1|}, \tag{2.7}$$

where

$$C_3 = C_2 \begin{cases} \frac{1}{C_1^{\lambda_2/\lambda_1}} & \text{when } C_1 > 0, \\ -\frac{1}{|C_1|^{\lambda_2/\lambda_1}} & \text{when } C_1 < 0. \end{cases}$$

The phase portrait of the system in this case looks as shown in Fig. 2.3. Note that the coordinate axes are the phase trajectories of the system as well. The horizontal axis corresponds to the case of $C_2 = 0$; the vertical axis corresponds to the case of $C_1 = 0$. The four phase trajectories that start or end at $(0, 0)$, called *the saddle separatrices*, cannot be expressed by the formula (2.7).

(4) If eigenvalues λ_1, λ_2 of the matrix $\hat{M}$ are purely imaginary, i.e. $\lambda_{1,2} = \pm i\omega$, then the stationary point $(0, 0)$ is called *the center.* There is a linear change of variables $(\xi, \eta) \to (y_1, y_2)$ such that in the new variables the system (2.4) takes the form

$$\frac{d}{dt}\begin{pmatrix} y_1 \\ y_2 \end{pmatrix} = \begin{pmatrix} 0 & -\omega \\ \omega & 0 \end{pmatrix}\begin{pmatrix} y_1 \\ y_2 \end{pmatrix}. \tag{2.8}$$

Solutions to (2.8) are periodic functions. They are represented in the phase portrait by closed curves (circles) surrounding the origin, see Fig. 2.4.

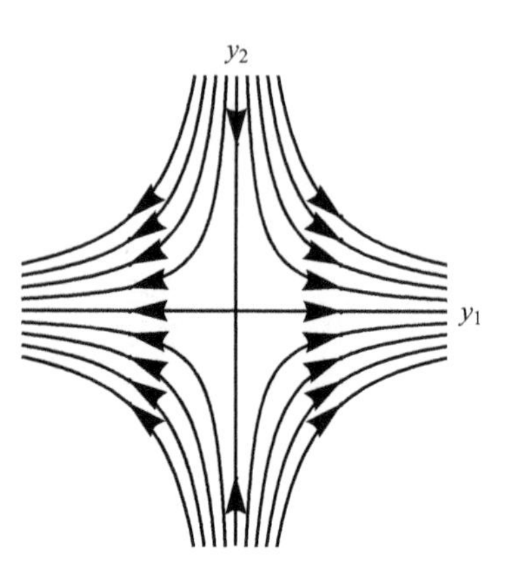

Fig. 2.3 An outlook of the phase portrait in vicinity of the saddle point.

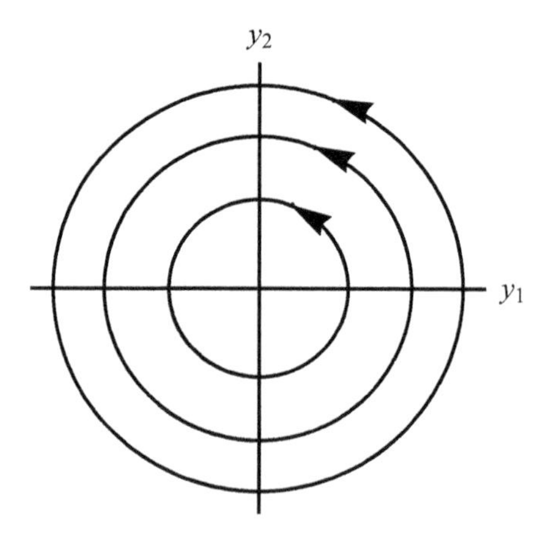

Fig. 2.4 Phase portrait of the system (2.8).

(5) The eigenvalues λ_1, λ_2 of the matrix $\hat{M}$ are complex: $\lambda_{1,2} = \alpha \pm i\omega$. In this case there exists a change of variables $(\xi, \eta) \to (y_1, y_2)$ such that in new variables the system (2.4) takes the form

$$\frac{d}{dt}\begin{pmatrix} y_1 \\ y_2 \end{pmatrix} = \begin{pmatrix} \alpha & -\omega \\ \omega & \alpha \end{pmatrix}\begin{pmatrix} y_1 \\ y_2 \end{pmatrix}. \tag{2.9}$$

The stationary point $(0, 0)$ is called *the unstable focus* when $\alpha > 0$ (see Fig. 2.5) and *the stable focus* when $\alpha < 0$.
A stable focus differs from the unstable one in its asymptotic behavior,

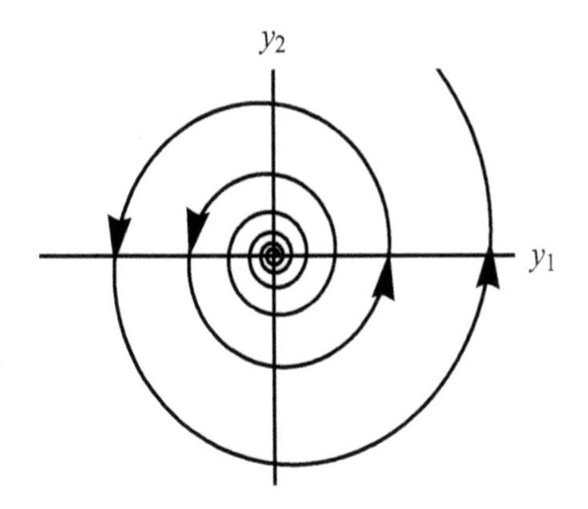

Fig. 2.5 Phase portrait of the system (2.9) in case when $\alpha > 0$.

since at $\alpha < 0$ all phase trajectories tend to the point $(0, 0)$ as $t \to +\infty$. In both cases, the movement along the trajectories is counterclockwise.

Finally, we will formulate the Grobman-Hartman theorem, which determines in which cases the local behavior of the phase trajectories of a linearized system does not change qualitatively when we take into account the rejected nonlinear terms. Let us assume that $X_0 \in \mathbb{R}^n$ is the stationary point of the system

$$\frac{dX}{dt} = F(X), \tag{2.10}$$

$\mathrm{DF}\,(X_0)$ is the Jacobi matrix of $F(\cdot)$ at the point X_0, while the system

$$\frac{d\xi}{dt} = \mathrm{DF}\,(X_0)\,\xi, \qquad \xi = X - X_0 \tag{2.11}$$

is its linearization.

Theorem 2.2. *If the matrix $DF(X_0)$ does not have zero or purely imaginary eigenvalues, then there exists a homeomorphism $h : \mathbb{R}^n \supset U \to \mathbb{R}^n$ where U is an open set containing X_0, locally taking orbits of dynamical system (2.10) to those of the linear system (2.11). The homeomorphism preserves the sense of orbits and directions of movements along the orbits.*

Corollary 2.1. *The phase portrait in the small neighborhood of the origin of the linearized system (2.11) does not qualitatively change when adding the rejected nonlinear terms for all the cases mentioned except the center.*

Let us now list the main problems solved within the framework of qualitative analysis of dynamical systems in $\mathbb{R}^2$.

(1) Finding stationary points and defining their types.
(2) Determination of attracting (repulsing) sets different from stationary points (in the case of $\mathbb{R}^2$ *the limit cycle*, i.e. nonlinear periodic solution can attract (repulse) the nearby trajectories).
(3) To determine basins of attraction (repulsion) of attracting (repulsing) sets, that is to determine the maximal set of points of the phase space which, being taken as Cauchy data, create orbits asymptotically tending to the attracting (repulsing) set as $t \to +\infty$ ($t \to -\infty$).
(4) Defining separatrices, i.e. the orbits separating disjoint basins of attraction (repulsion).
(5) Sketch the global phase portrait.

When carrying out a qualitative analysis, the following rules should be followed:

- the phase trajectories are tangent at each point to the vector field defined by the r.h.s. of the dynamical system;
- phase trajectories begin (end) at stationary points or other attractive (repulsive) sets or at infinity;
- phase trajectories can only intersect at a stationary point.

Full implementation of the indicated "program" requires additional knowledge and practical experience. However, there is a class of dynamical systems in $\mathbb{R}^2$, whose analysis is simpler. This class will be considered in the next subsection.

2.2 Conservative systems with one degree of freedom

A comprehensive qualitative description can be obtained in the case of two-dimensional Hamiltonian systems.

Definition 2.4. The dynamical system

$$\frac{dx}{dt} = P(x, y), \qquad \frac{dx}{dt} = Q(x, y) \tag{2.12}$$

is called Hamiltonian if there exists differentiable function $H(x, y)$ such that the following equations hold

$$P(x,y) = \frac{\partial H(x,y)}{\partial y}, \qquad Q(x,y) = -\frac{\partial H(x,y)}{\partial x}. \tag{2.13}$$

In case when (2.13) takes place, $H(x, y)$ is called the Hamilton function of the system (2.12), or simply Hamiltonian of (2.12).

Theorem 2.3. *The Hamilton function maintains a constant value on the solutions of the dynamical system.*

We leave the proof of this statement to the reader.

Corollary 2.2. *Each phase trajectory of a two-dimensional Hamiltonian system can be represented as a contour line*

$$H(x, y) = C$$

at certain value of the parameter $C \in \mathbb{R}$.

It turns out that a large class of mechanical (and not only mechanical) models can be represented in the form of two-dimensional Hamiltonian systems. As an example, consider the equation of motion of a material point of mass m in the field of potential forces:

$$m\frac{d^2 x}{dt^2} = F(x) = -\frac{\partial U(x)}{\partial x}. \tag{2.14}$$

This equation can be rewritten in the following equivalent form:

$$\frac{dx}{dt} = y, \qquad \frac{dy}{dt} = f(x) = -\frac{\partial V(x)}{\partial x}, \tag{2.15}$$

where $f(x) = F(x)/m$, $V(x) = U(x)/m$. In mechanics, a system of such a form is called *the system with one degree of freedom.*

Statement 2.1. The function

$$H(x, y) = y^2/2 + V(x) \tag{2.16}$$

is the Hamiltonian function of the system (2.15).

The proof is evident.

Statement 2.2. A simple stationary point of the Hamiltonian system can be either a saddle or a center.

<u>**Proof.**</u> If the point (x_0, y_0) is a stationary point of the Hamiltonian system (2.12), then

$$-Q(x, y)\Big|_{(x_0, y_0)} = \frac{\partial H(x, y)}{\partial x}\Big|_{(x_0, y_0)}$$

$$= P(x, y) \Big|_{(x_0, y_0)} = \frac{\partial H(x, y)}{\partial y} \Big|_{(x_0, y_0)} = 0.$$

Making the change of variables $\xi = x - x_0$, $\eta = y - y_0$, one can present the linearization of the Hamiltonian system as

$$\frac{d}{dt}\begin{pmatrix}\xi \\ \eta\end{pmatrix} = \begin{pmatrix} H_{yx} & H_{yy} \\ -H_{xx} & -H_{xy}\end{pmatrix}\begin{pmatrix}\xi \\ \eta\end{pmatrix},$$

where the matrix elements are the second-order partial derivatives of the Hamiltonian $H(\cdot, \cdot)$ taken at the stationary point (x_0, y_0). Hence the eigenvalues of the linearization matrix of the Hamiltonian system have the form

$$\lambda_{1,2} = \pm\sqrt{H_{xy}^2 - H_{xx} H_{yy}}$$

and now there are two possibilities. If the expression under the root is greater than zero, then the eigenvalues are real and have different signs, while if the expression under the root is negative, then the eigenvalues are purely imaginary.

Statement 2.3. All stationary points of the system (2.15) lie on the horizontal axis. Moreover, the x coordinate of the stationary point is the point at which the function $V(\cdot)$, which determines the potential energy of the system, has the local extremum. The local minimum corresponds to the center; the local maximum corresponds to the saddle.

Proof. The linearization matrix of the system (2.15) in the vicinity of a stationary point has the form

$$\begin{pmatrix} 0 & 1 \\ -V''(x_0) & 0 \end{pmatrix}.$$

Hence, $\lambda_{1,2} = \pm\sqrt{-V''(x_0)}$ and due to the fact that $V'(x_0) = 0$, the thesis becomes obvious.

The following additional properties of the Hamilton function (2.16) allow to simplify the qualitative analysis of the system (2.15):
(a)

$$H(x, -y) = H(x, y);$$

(b) the phase trajectory corresponding to the level function $y^2/2 + V(x) = C$ cannot have the coordinate x for which $V(x) > C$;
(c) the phase trajectory intersects the horizontal axis at right angle at the point $(x_0, 0)$ of phase plane for which $V(x_0) = C$, provided that $(x_0, 0)$, is not a stationary point).

Example 2.1. The pendulum.

As it was shown before, the dynamics of pendulum is described by the following equation:

$$\frac{d^2 x}{d t^2} + \omega^2 \sin[x] = 0.$$

This equation can be represented as an equivalent first-order system:

$$\frac{d x}{d t} = y, \tag{2.17}$$
$$\frac{d y}{d t} = -\omega^2 \sin[x].$$

Solving the system

$$\frac{\partial H(x, y)}{\partial y} = y, \qquad -\frac{\partial H(x, y)}{\partial x} = -\omega^2 \sin[x]$$

we obtain

$$H(x,y) = y^2/2 - \omega^2 \cos[x]. \tag{2.18}$$

Statement 2.4. The Hamilton function (2.18) possesses the following (additional) properties:

(a)

$$H(-x,y) = H(x,y);$$

(b)

$$H(x \pm 2\,k\,\pi,\, y) = H(x,y); \qquad k = 1, 2, \ldots.$$

The proof is elementary.

Corollary 2.3. *The global phase portrait can be obtained in this case by constructing a phase portrait on a set* $\mathbb{R}^2 \ni \Omega = \{(x,\, y) \in [-\pi,\, \pi] \times \mathbb{R}\}$, *and then translating by the segment of* $2\,n\,\pi$, $n = \pm 1, \pm 2, \ldots$ *along the* OX *axis.*

The stationary points of the system (2.17) are determined from the equality $y = \sin[x] = 0$. There are three stationary points in the segment $[-\pi,\, \pi]$, namely: $A = (0,0)$, $B_{\pm} = (\pm\pi,\, 0)$. Since the function $V[x] = -\omega^2 \cos[x]$, defining the potential energy of the system, has a local minimum at $x = 0$, while at points $x_{\pm} = \pm\pi$ it has local maxima, so, on virtue of the statement 2.3, the point A is a center and the points $B_{\pm}$ are saddles. It turns out that the separatrix outgoing from the saddle point $(-\pi, 0)$ and directed to the right creates one trajectory with the separatrix incoming to the saddle point $(\pi, 0)$ (the separatrix connecting two stationary points is called *a heteroclinic trajectory*). This is due to the fact that the implicit form of the trajectory starting from the point B_- is given by the formula

$$y^2/2 - \omega^2 \cos[x] = H(-\pi,\, 0) = \omega^2.$$

The orbit going upwards to the right is determined as

$$y = +\omega\sqrt{2(\cos[x] + 1)}. \tag{2.19}$$

This function grows over the segment $(-\pi,\, 0)$, reaching a maximum at the right end of the segment. The fact that the trajectory reaches B_+ at $x = \pi$ results from the symmetry of the function with respect to the reflections $x \to -x$. Due to the symmetry of the Hamilton function (2.18), the saddle separatrix of the stationary point B_+ located in the lower half-plane and pointing downwards to the left creates a heteroclinic trajectory symmetric to (2.19) with respect to the OX axis.

To conclude the qualitative analysis of the system on the set $[-\pi,\, \pi]$, it is necessary to answer the question of how the center placed at the point A behaves when the rejected nonlinear terms are taken into account. It turns out that in the case of the Hamiltonian system, the center does not change its character, what is more, all trajectories surrounding the origin and lying within the set limited by the upper and lower heteroclinics of the saddle points B_+, B_- are closed trajectories (ellipses) implicitly described by the following formula:

$$y^2/2 - \omega^2 \cos[x] = C, \qquad |C| < \omega^2. \tag{2.20}$$

In case when the modulus of the constant C in the formula (2.20) is greater than ω^2, this formula describes unbounded trajectories lying above and below the heteroclinic contour. The phase portrait of the system (2.17) shown in Fig. 2.6 is performed with the help of the command ContourPlot:

Cell 2.1.

```
"The global phase portrait of the pendulum taken at ω = 1"
   In[1]: ClearAll ["Global`*"]
   In[2]:
   V = -Cos[x];
   T = 1/2 y^2;
   H = T + V;
   ContourPlot[Evaluate[Table[H == b, {b, -2.5, 2.5, 0.5}]]], {x, -4π, 4π}, {y, -3,
3}]
```

Example 2.2. Korteweg-de Vries equation: traveling wave solution.

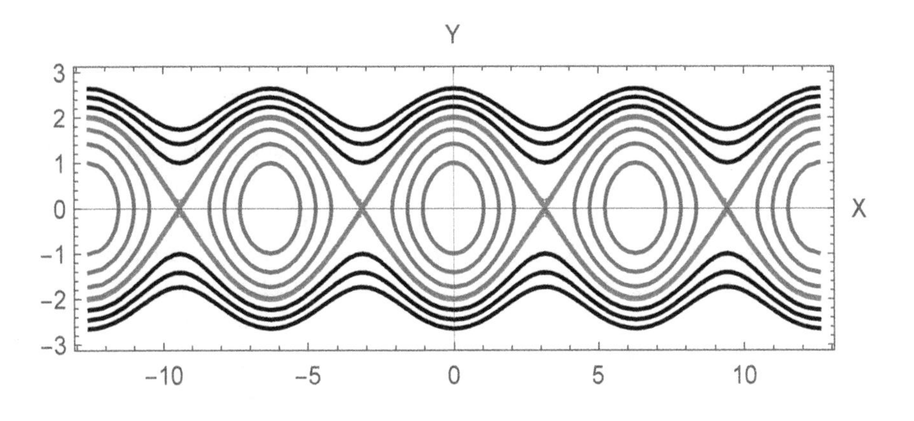

Fig. 2.6 Global phase portrait of the system (2.17).

We'll consider the equation

$$\frac{\partial u}{\partial t} + 12\, u\, \frac{\partial u}{\partial x} + \frac{\partial^3 u}{\partial x^3} = 0 \tag{2.21}$$

equivalent to that put forward by Korteweg and de Vries in 1895. This equation does not belong to the class of ordinary differential equations and, at first glance, has nothing to do with the range of problems considered in this part of the book. Indeed, a detailed analysis of this, surprising in many respects equation, is carried out in the second part. Nevertheless, in the context of the issues discussed in a given chapter, we are interested in solutions of a specific form satisfying the ordinary differential equation, namely, solutions of the traveling wave type

$$u(t,\, x) = U(\xi), \qquad \xi = x - s\, t, \tag{2.22}$$

where $0 < s$ is constant velocity of the traveling wave (TW). Substituting (2.22) into the equation (2.21), we obtain the ordinary equation of the form

$$\frac{d}{d\,\xi}\left\{-s\,U + 6\,U^2 + U''\right\} = 0,$$

which after one integration takes the form

$$-\, s\,U + 6\,U^2 + U'' = 0. \tag{2.23}$$

The constant of integration is chosen equal to zero since we are interested in analyzing solutions that vanish at infinity with their derivatives. The equation (2.23) is equivalent to the following Hamiltonian system:

$$\begin{cases} U' = W = \frac{\partial H}{\partial W}, \\ \\ W' = s\,U - 6\,U^2 = -\frac{\partial H}{\partial U}, \end{cases} \tag{2.24}$$

where

$$H = \frac{W^2}{2} + 2\,U^3 - \frac{s}{2}\,U^2 = \frac{W^2}{2} + V(U). \tag{2.25}$$

The system (2.24) has two stationary points, namely: $A = (0,\,0)$ and $B = (s/6,\,0)$. Analysis of the potential energy function $V(U)$ shown in Fig. 2.7 enables to state that A is a saddle, while B is a center. The center is surrounded by closed trajectories, which correspond to the contours on both sides limited by the edges of the potential well, see Fig. 2.7. The boundary of the area filled in the phase plane $(U,\,W)$ with periodic orbits is the *homoclinic orbit*, which is bi-asymptotic to the saddle (see Fig. 2.8). It corresponds to the level line going through the origin.

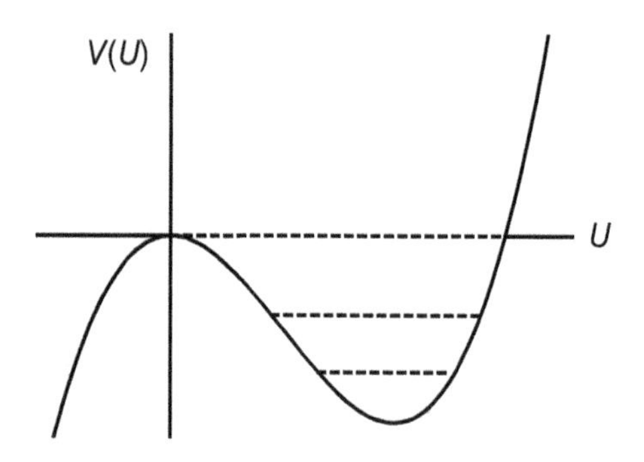

Fig. 2.7 The graph of the function $V(U)$.

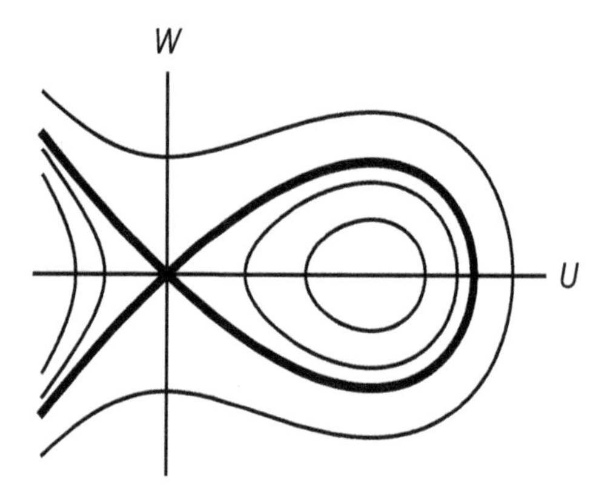

Fig. 2.8 Phase portrait of the system (2.24).

The homoclinic trajectory of the system (2.24) plays a special role as it describes the exponentially localized solution to the equation (2.21) called *soliton solution* or *soliton.* The analytical form of this solution can be found using the fact that the Hamilton function maintains a constant value on the trajectories of the dynamical system. Since the boundary of the homoclinic trajectory at $\xi \to \pm\infty$ is the stationary point (0, 0), the value of $H(U, W)$ on this trajectory covers with the value of H at the point A which is equal to zero. Hence we have equality

$$W = \frac{dU}{d\xi} = \pm\sqrt{\frac{s}{2}U^2 - 2U^3}.$$

Choosing the plus sign before the root (choice of the opposite sign leads to an analogous result due to the symmetry of the system with respect to the transformation $\xi \to -\xi$), we will obtain the differential equality

$$\frac{dU}{\sqrt{2}U\sqrt{\frac{s}{4} - U}} = d\xi,$$

which in turn implies the equality of the integrals

$$\int \frac{dU}{\sqrt{2}U\sqrt{\frac{s}{4} - U}} = \int d\xi = \xi - \xi_0.$$

The left integral can be reduced to the standard form using the substitution $\tau = \sqrt{\frac{s}{4} - U}$. By making such a change of the variable under the integral, we get:

$$\begin{aligned}\sqrt{2}\int \frac{d\tau}{\tau^2 - \frac{s}{4}} &= \frac{1}{\sqrt{s}}\int \left\{\frac{1}{\tau - \frac{\sqrt{s}}{2}} - \frac{1}{\tau + \frac{\sqrt{s}}{2}}\right\} d\tau \\ &= \frac{1}{\sqrt{s}}\log\left|\frac{\tau - \frac{\sqrt{s}}{2}}{\tau + \frac{\sqrt{s}}{2}}\right| = \frac{1}{\sqrt{s}}\log\frac{\frac{\sqrt{s}}{2} - \tau}{\tau + \frac{\sqrt{s}}{2}}.\end{aligned}$$

The last equality is related to the inequality $U = \frac{s}{4} - \tau^2 > 0$. This is how we get the equation

$$\log\left[\frac{\frac{\sqrt{s}}{2} - \sqrt{\frac{s}{4} - U}}{\sqrt{\frac{s}{4} - U} + \frac{\sqrt{s}}{2}}\right] = \sqrt{s}(\xi - \xi_0).$$

Solving it with respect to U, we obtain:

$$U(\xi) = \frac{s}{4}\operatorname{sech}^2\left[\frac{\sqrt{s}}{2}(\xi - \xi_0)\right]. \tag{2.26}$$

Problem.

Plot the phase portraits of systems with one degree of freedom for

(1) $V(x) = x^4/4 - x^2/2$;
(2) $V(x) = x^3/3 - x^2/2$;
(3) $V(x) = -x^2(x^2 - 1)$.

2.3 Dynamical systems in $\mathbb{R}^2$ having non-analytical solutions

Let us consider the system

$$K_{\pm}(m,\,n):\ \ u_t \pm (u^m)_x + (u^n)_{xxx} = 0, \qquad m,\,n \geq 2. \tag{2.27}$$

There are dynamical systems similar to Hamiltonian systems in many respects, but additionally having solutions that are not smooth everywhere [Rosenau (1997); Rosenau and Hyman (1993); Li *et al.* (1998)]. Such systems naturally arise when looking for solutions of the traveling wave type of the class of PDEs of the form:

$$K_{+}(2,\,2) = u_t + \left(u^2\right)_x + \left(u^2\right)_{xxx} = 0. \tag{2.28}$$

Inserting the ansatz $u(t,\,x) = U(z)$, $z = x - c\,t$ into (2.28), we get the ordinary differential equation

$$\left\{\left(U^2\right)'' + U^2 - c\,U\right\}' = 0. \tag{2.29}$$

After one integration w obtain the equation

$$\left(U^2\right)'' + U^2 - c\,U = a,$$

which is equivalent to the dynamical system

$$\begin{cases} U' = W, \\ W' = \frac{1}{2\,U}\left[c\,U + a - U^2 - 2\,W^2\right]. \end{cases} \tag{2.30}$$

It can be easily shown that the system (2.30) is not a Hamiltonian system, moreover, it has singularity at the point $U = 0$. There is, however, a procedure that allows us to pass to the Hamiltonian system, which determines a vector field that is in many respects identical to the vector field specified by the right sides of the system (2.30). Multiplying the left and right sides of the above system by the differentiable function $G(U)$ and going to the new independent variable τ satisfying equation $\frac{d}{d\,\tau} = G[U(z)]\,\frac{d}{d\,z}$, we get the system

$$\begin{cases} \frac{d\,U}{d\,\tau} = G(U)\,W, \\ \frac{d\,W}{d\,\tau} = \frac{G(U)}{2\,U}\left[c\,U + a - U^2 - 2\,W^2\right]. \end{cases} \tag{2.31}$$

If there exist a pair of differentiable functions $G(U)$ and $H(U,\,W)$ satisfying the system of equations

$$G(U)\,W = \frac{\partial\,H}{\partial\,W}, \quad \frac{G(U)}{2\,U}\left[c\,U + a - U^2 - 2\,W^2\right] = -\frac{\partial\,H}{\partial\,U}, \tag{2.32}$$

then the system (2.31) can be treated as Hamiltonian. Integrating the first equation of the system (2.32), we get:

$$H(U,\,W)=\frac{1}{2}G(U)\,W^2+V(U). \tag{2.33}$$

Inserting (2.33) into the second equation, we obtain:

$$\frac{1}{2}G(U)'\,W^2+V'(U)=\frac{G(U)}{2\,U}\left[U^2+2\,W^2-c\,U-a\right].$$

Equating the terms proportional to W^2, we get the equation

$$\frac{d\,G}{G}=2\,\frac{d\,U}{U},$$

whose general solution has the form $G(U)=C_1\,U^2$. The choice of constant $C_1\neq 0$ occurs to be inessential, so assuming for convenience that $C_1=2$, we obtain the equation

$$V'(U)=U\left[U^2-c\,U-a\right].$$

Integrating it, we obtain:

$$V(U)=\int U\left(U^2-a-c\,U\right)\,d\,U=\frac{1}{4}U^4-\frac{a}{2}U^2-\frac{c}{3}U^3.$$

So, under such a choice the system (2.31) is transformed into the Hamiltonian system

$$\begin{cases}\frac{d\,U}{d\,\tau}=2\,U^2\,W=H_W,\\[2ex] \frac{d\,W}{d\,\tau}=U\,\left[c\,U+a-U^2-2\,W^2\right]=-H_U\end{cases} \tag{2.34}$$

with the Hamilton function

$$H(U,\,W)=T(U,\,W)+V(U)=(U\,W)^2+\frac{1}{4}U^4-\frac{a}{2}U^2-\frac{c}{3}U^3. \tag{2.35}$$

As in the case of classical systems with one degree of freedom, the function $T=(U\,W)^2$ plays the role of kinetic energy, while the function $V(U)$ plays the role of potential energy.

We will start the analysis of the properties of the system (2.34) by examining its stationary points. Note that, contrary to what happened in the case of the KdV equation, the integration constant a cannot be removed, and the range of its values significantly affects the nature of the solutions. Let us also note that without loss of generality, we can limit ourselves to consideration of the TW solutions moving to the right, which is the case when $c>0$. This is due to the invariance of the equation (2.29) with respect to the transformation $U\to -U$, $c\to -c$.

It is easily seen, that all stationary points of the system (2.34) lie on the horizontal axis. The system always has a stationary point $A = (0,\, 0)$. The U coordinates of the remaining stationary points (if any) will satisfy the algebraic equation

$$U^2 - c\,U - a = 0,$$

having the following roots:

$$U_{\pm} = \frac{c \pm \sqrt{c^2 + 4\,a}}{2}. \tag{2.36}$$

So in the case when $c^2 + 4\,a > 0$, the system has an additional pair of stationary points $B_{\pm} = (U_{\pm},\, 0)$; in the case when $c^2 + 4\,a = 0$ there is only one additional stationary point $B_0 = (c/2,\, 0)$. However, if the expression under the root in the formula (2.36) is negative, then $A = (0,\, 0)$ is the only stationary point of the system.

If $a > 0$, then B_+ and B_- are located in the right and the left half-plane, respectively, while when $a \in (-\,c^2/4,\, 0)$, both points are located on the right half-plane. Note that the limiting case $a = 0$ is particularly important and will therefore be analyzed separately.

Despite the fact that the Hamilton function of the dynamical system (2.34) differs from the Hamilton function (2.16) characterizing the conservative system with one degree of freedom, all the statements formulated in the previous point are easily transferred to the considered case. Therefore, the nature of the stationary points is most easily determined on the basis of the analysis of the potential energy function $V(U)$. If $a > 0$, this function

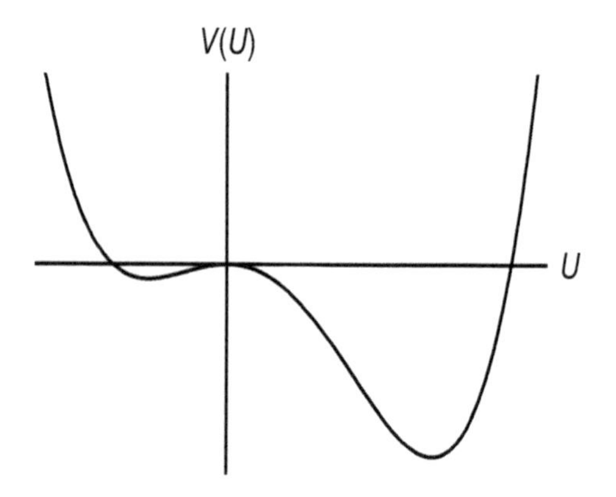

Fig. 2.9 The graph of the function (2.34) at $a > 0$.

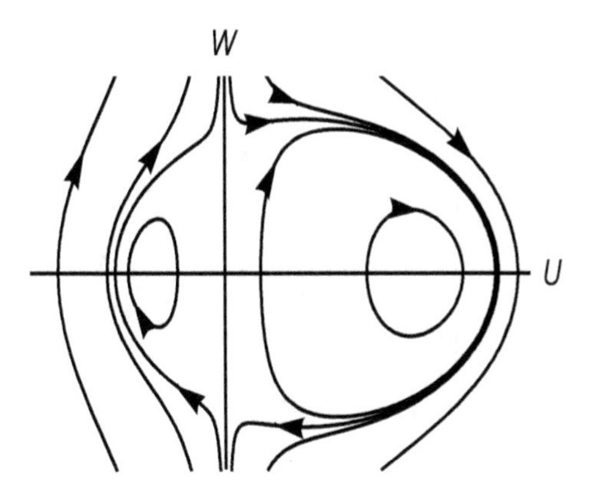

Fig. 2.10 Phase portrait of the system (2.34) at $a > 0$.

has local minima at the points $U_\pm$ (see Fig. 2.9) and hence the stationary points $(U_\pm, 0)$ are surrounded by the closed orbits, Fig. 2.10.

If $a \in (-\frac{c^2}{4}, 0)$ the potential energy has a local minimum at U_+ and a local maximum at $U_- > 0$ (Fig. 2.11). The stationary point $(U_+, 0)$ is in this case surrounded by periodic orbits, while the phase space area filled with periodic orbits is bounded by the homoclinic loop clearly seen in the Fig. 2.12.

Let us now consider the case $a = 0$. As follows from the shape of the function $V(U)$, the stationary point $(U_+, 0)$ still is the center, while

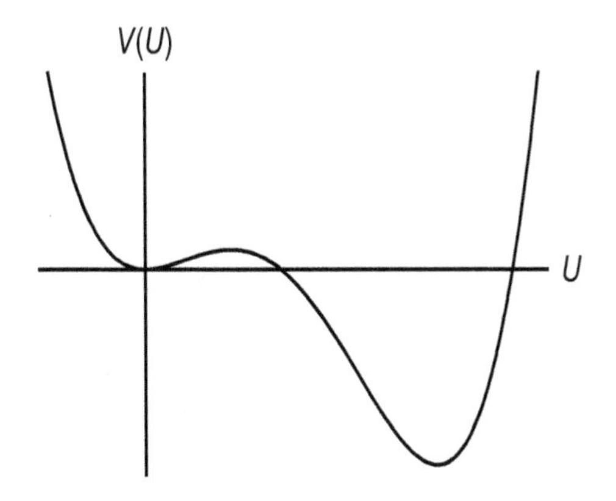

Fig. 2.11 The graph of the function (2.34) at $a \in (-\frac{c^2}{4}, 0)$.

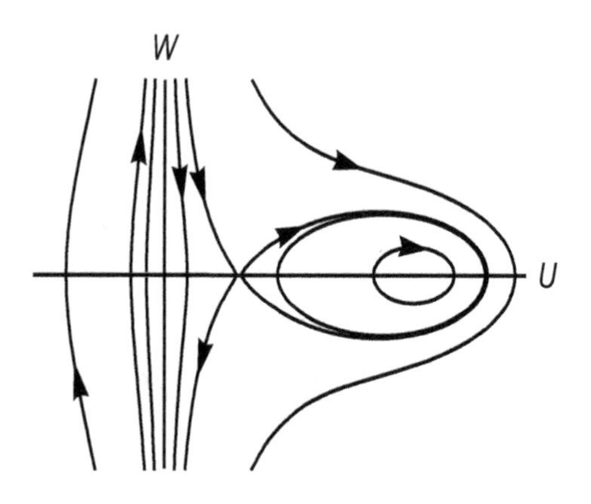

Fig. 2.12 The phase portrait of the system (2.34) at $a \in (-\frac{c^2}{4}, 0)$.

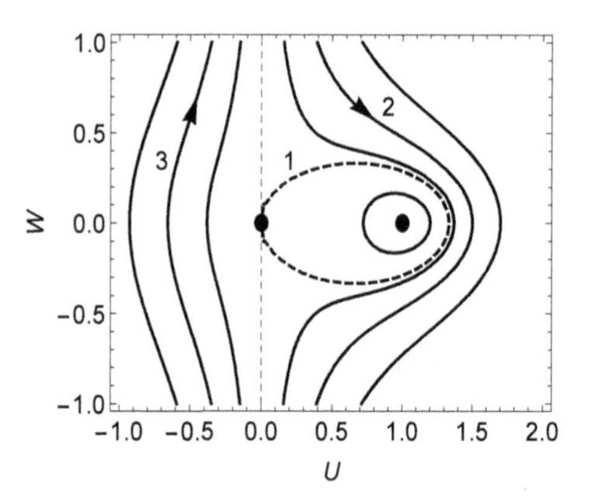

Fig. 2.13 The phase portrait of the system (2.34) at $a = 0$.

$A = (0, 0)$ becomes a compound saddle point, whose separatrices form a homoclinic loop (Fig. 2.13, dashed curve 1) seen on the phase portrait of the system (Fig. 2.13). It turns out that the "time" which is necessary to penetrate the homoclinic loop is finite. This property results from the analytical form of the solution that we can obtain using the fact that the Hamiltonian function (2.35) maintains a constant value on the trajectories of the dynamical system. Since the phase trajectory we are trying to find tends to the origin on both sides, this value is zero. Hence we have the equality

$$(U\,W)^2 + \frac{1}{4}U^4 - \frac{c}{3}U^3 = 0,$$

which can be rewritten as

$$\frac{d\,U}{d\,z} = \pm\sqrt{\left(\frac{c}{3}\right)^2 - \left(\frac{U}{2} - \frac{c}{3}\right)^2}.$$

Choosing the minus sign in the right part and then integrating the equation, we get:

$$U(z) = \frac{4\,c}{3}\cos^2\left[\frac{z - z_0}{4}\right]. \tag{2.37}$$

It follows from the above formula that the solution nullifies at the ends of the interval $[z_0 - 2\pi,\, z_0 + 2\pi]$. Analysis shows that the phase trajectory corresponding to (2.37) starts from the stationary point as $z = z_0 - 2\pi$ and returns to this point when $z = z_0 + 2\pi$. So the "time" taken to pass the homoclinic loop is equal to 4π.

Note that the solution (2.37) corresponds to the following (generalized) solution (called *compacton*) to the equation (2.28):

$$u(t,\,x) = \begin{cases} \frac{4\,c}{3}\cos^2\left[\frac{x - c\,t}{4}\right] & \text{as } |x - c\,t| < 2\,\pi, \\ 0 & \text{elsewhere.} \end{cases} \tag{2.38}$$

The profile of this solution when $c = 1$ is depicted in Fig. 2.14(a) with dashed curve. When the wave velocity decreases, e.g. $c = 0.5$ (Fig. 2.14(a), solid curve), it is evident that compacton's height decreases also but its width remains the same.

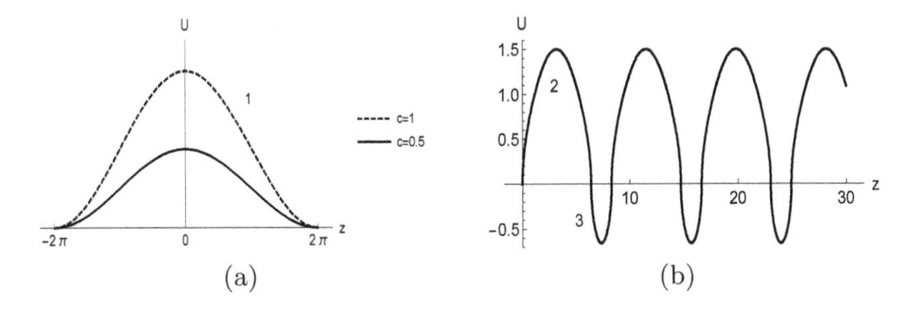

Fig. 2.14 Panel (a): Profiles of the compacton solutions; panel (b): Profile of the periodic non-analytic solution corresponding to contour composed of trajectories 2 and 3 from Fig. 2.13.

It will be shown in the second part of this book, that the solution (2.38) has a number of interesting properties similar to those of the soliton solutions satisfying the KdV equation.

Note also that system (2.34), besides a compacton, possesses a periodic nonanalytic solution (Fig. 2.14(b)) which can be composed of the phase trajectories 2 and 3 (Fig. 2.13). The matching of the phase trajectories is performed at the points (function's zeros) where vertical tangent lines exist, i.e. analyticity is violated.

Now we'll briefly discuss the properties of the TW solutions of the equation

$$K_{-}(2,\,2):\quad u_t-\left(u^2\right)_x+\left(u^2\right)_{x\,x\,x}=0. \tag{2.39}$$

Changing the sign at the second term leads to the appearance of a piecewise continuous non-analytical solutions such as *peakons*, *cuspons*, *tipons*, etc. Substituting the ansatz $u(t,\,x)=U(z),\ z=x-c\,t$, into (2.39) we get the ordinary differential equation

$$-c\,U'-\left(U^2\right)'+\left(U^2\right)'''=0,$$

which is equivalent to the system

$$\begin{cases} \frac{d\,U}{d\,z}=W, \\ \\ \frac{d\,W}{d\,z}=\frac{c\,U+U^2-\alpha(c+\alpha)-2\,W^2}{2\,U}. \end{cases} \tag{2.40}$$

The above system is obtained by posing the condition $U\to\alpha$ as $|z|\to\infty$.

Similarly as in the case of the system (2.30), the system (2.40) can be presented in the Hamiltonian form if we pass to a new independent variable τ satisfying the equation $\frac{d}{d\,\tau}=G[U(z)]\,\frac{d}{d\,z}$:

$$\frac{d\,U}{d\,\tau}=\frac{\partial H(U,\,W)}{\partial\,W},\qquad \frac{d\,W}{d\,\tau}=-\frac{\partial H(U,\,W)}{\partial\,U},$$

where

$$H(U,\,W)=(W\,U)^2+\frac{\alpha(c+\alpha)}{2}U^2-\frac{1}{4}U^4-\frac{c}{3}U^3.$$

Let us consider the trajectory of $U(z)$ which asymptotically tends to α, assuming that the constant value of the Hamilton function along this trajectory is zero. It is possible when $\alpha=-2\,c/3$. Under this condition, we are looking for solution of the form $U(z)=Ae^{r\,z}+B$. Substituting this function into the equation

$$(U'\,U)^2-\frac{c^2}{9}U^2-\frac{1}{4}U^4-\frac{c}{3}U^3=0,$$

and nullifying the coefficients at different powers of $e^{n\,r\,z}$, $n = 0,\,1,\,2$, we get that $r = \pm 1/2$, $B = \alpha = -2\,c/3$, while the constant A is arbitrary. In this case the typical phase portrait is depicted in Fig. 2.15(a).

The only bounded solution of this type satisfying an additional condition $U(0) = 0$ is that of the form

$$U = \frac{2c}{3}\left(e^{-|z|/2} - 1\right).$$

This solution is called *peakon*. Its graph is presented in Fig. 2.15(b) and corresponds to the bold lines 1 and 2 representing the separatrices of saddle fixed point. Note that the derivative of this solution has the first order discontinuity at the extreme point (the one-sided derivatives at this point attain the finite values).

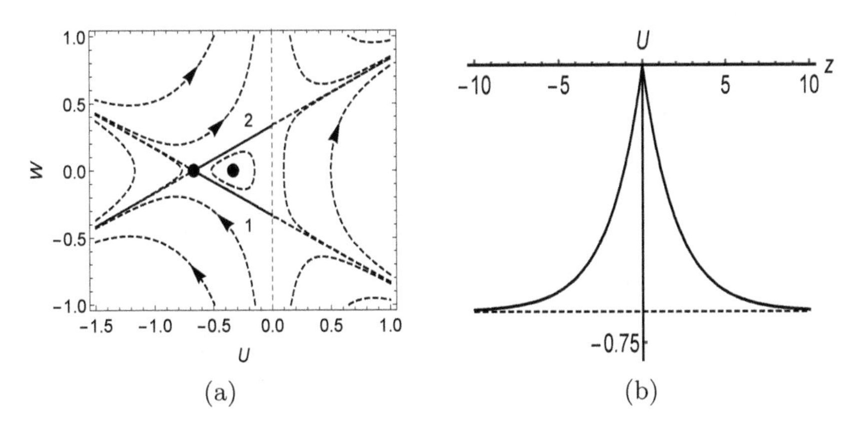

Fig. 2.15 Phase portrait (a) of the system (2.40) and peakon's profile (b) corresponding to pair bold phase trajectories from the left panel. Here the parameters $\alpha = -2c/3$ and $c = 1$.

Let us consider constructions leading to another non-classical solutions of the traveling wave type. With $\alpha > 0$, the system (2.40) has a stationary point $(\alpha,\,0)$, which is a saddle. The typical structure of the phase portrait is clear from Fig. 2.16(a) obtained at $\alpha = c = 1$. Combining two separatrices 1 and 2, we get a solution called *cuspon* (Fig. 2.16(b)). In contrast to the peakon, the one-sided derivatives of this solution at the extreme point are infinite.

The separatrices 1 and 2 together with the trajectory 3 coming from the left half-plane (see Fig. 2.16(a)) create a wave profile with two singular points. This profile is shown in Fig. 2.17(a). The solution corresponding to this trajectory is called *tipon*.

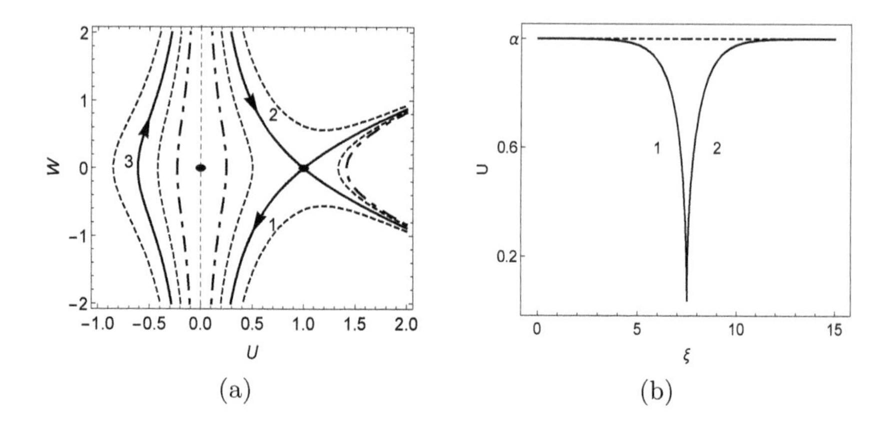

Fig. 2.16 Phase portrait (a) of the system (2.40) and cuspon profile formed by the phase trajectories 1 and 2. The parameter values are $\alpha = c = 1$.

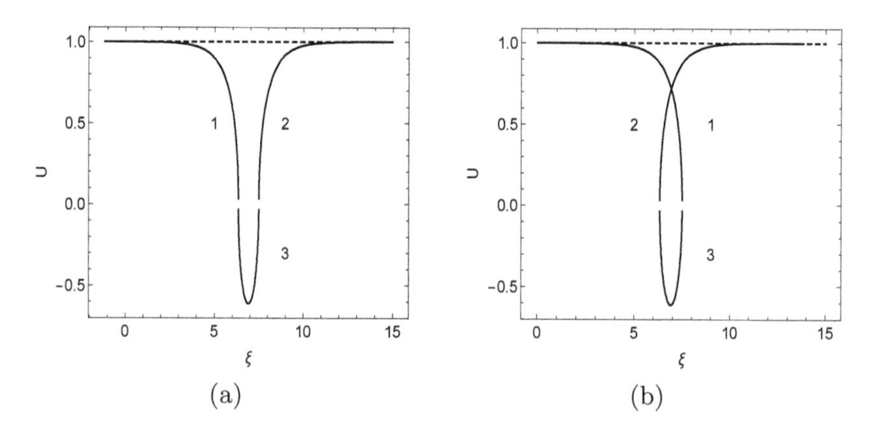

Fig. 2.17 The profiles of tipon (a) and loopon (b) formed by phase trajectories 1, 2, and 3 marked in Fig. 2.16(a). The parameter values are $\alpha = c = 1$.

It is also possible to create more complex multivalued solutions called *loopons* (Fig. 2.17(b)).

Problem.

Consider the equation

$$K(n, n) = u_t + (u^n)_x + (u^n)_{xxx} = 0, \quad n \in \mathbb{N}, \quad n \geq 2.$$

(1) Inserting the traveling wave ansatz $u(t, x) = U(\xi) = U(x - ct)$, $0 < c =$ const into the above equation, we get a third-order ODE. Integrate this ODE over the interval $(-\infty, \xi)$, using the boundary condition $U(\xi) \to 0$ as $|\xi| \to \infty$.

(2) Find the solution of the form $U(\xi) = A\cos^\gamma(B\,\xi)$, where A, γ, B are constants to be determined. The value of the index γ can be obtained by requiring that the number of different powers of the function $\cos(B\,\xi)$ in equation be minimal. Determine the parameter γ, and then find the constants A, B equating to zero the coefficients at various powers of the function $\cos(B\,\xi)$.
(3) Compare the result with the already known solution (2.38) corresponding to the case $n = 2$.

2.4 Dynamial systems in $\mathbb{R}^n$. Invariant manifolds and subspaces

In general, a linearized system has the form

$$\frac{dX}{dt} = \hat{A}X, \quad X \in \mathbb{R}^n, \tag{2.41}$$

where $\hat{A}$ is a quadratic matrix with n rows and columns. Solution of the Cauchy problem

$$\frac{dX}{dt} = \hat{A}X, \quad X(0) = X_0 \in \mathbb{R}^n \tag{2.42}$$

is given by the following formula:

$$X(t) = e^{t\hat{A}}\left[X_0\right], \tag{2.43}$$

where

$$e^{t\hat{A}} = I + \frac{t}{1!}\hat{A} + \frac{t^2}{2!}\hat{A}^2 + \frac{t^3}{3!}\hat{A}^3 + \ldots \tag{2.44}$$

(regarding the correctness of determining the mapping $(t,\,\hat{A}) \to \exp[t\,\hat{A}]$ in the above formula, see e.g. [Maurin (1973)], Chapter 9, paragraph 10). If the eigenvalues of $\hat{A}$ are real and different, the general solution to the problem (2.42) can be presented in the form of the following finite sum:

$$X(t) = \sum_{j=1}^{n} C_j\, e^{\lambda_j t} V_j, \qquad X_0 = \sum_{j=1}^{n} C_j\, V_j, \tag{2.45}$$

where $V_j \in \mathbb{R}^n$ – is the eigenvector of $\hat{A}$, corresponding to the eigenvalue λ_j. In fact, the phase flow $\phi_t\left[X\right] = \exp t\,\hat{A}\left[X\right]$ describes all solutions of the system (2.41). However, it is worth specifying certain families of solutions that play a very important role, namely the families of solutions invariant to the phase flow. The basic elements of these subspaces span the entire space $\mathbb{R}^n$. We list three families of invariant subspaces:

- (a) stable invariant subspace
$$E^s = \text{span}\{V_1, V_2, V_3, ..., V_{n_s}\}, \tag{2.46}$$
where $\{V_i\}_{i=1}^{n_s}$ are the eigenvectors of $\hat{A}$ corresponding to the eigenvalues $\{\lambda_i\}_{i=1}^{n_s}$ having the negative real parts: $Re[\lambda_i] < 0$;
- (b) unstable invariant subspace
$$E^u = \text{span}\{U_1, U_2, U_3, ..., U_{n_u}\}, \tag{2.47}$$
where $\{U_j\}_{j=1}^{n_u}$ are the eigenvectors of $\hat{A}$ corresponding to the eigenvalues $\{\lambda_j\}_{j=1}^{n_u}$ having the positive real parts: $Re[\lambda_j] > 0$;
- (c) central subspace
$$E^c = \text{span}\{W_1, W_2, W_3, ..., W_{n_c}\}, \tag{2.48}$$
where $\{W_k\}_{k=1}^{n_c}$ are the eigenvectors of $\hat{A}$ corresponding to the eigenvalues $\{\lambda_k\}_{k=1}^{n_c}$ having zero real parts: $Re[\lambda_k] = 0$.

It is easy to show that the phase flow $\phi_t[X]$ starting on the vector X belonging to one of the invariant subspaces does not leave this subspace in the course of evolution.

It turns out that in the case of nonlinear dynamical system
$$\frac{dX}{dt} = F[X], \tag{2.49}$$
whose linearization is the system (2.41) (we assume that zero is the stationary point of the system (2.49)), it is possible to specify invariant manifolds (surfaces) that are homeomorphic to the subspaces E^s and E^u. These surfaces are defined as follows:
$$W^s_{loc} = \left\{X \in U \subset \mathbb{R}^n : \phi_t[X] \underset{t\to+\infty}{\longrightarrow} 0 \bigwedge \phi_t[X] \in U\, \forall t \geq 0\right\}, \tag{2.50}$$
$$W^u_{loc} = \left\{X \in U \subset \mathbb{R}^n : \phi_t[X] \underset{t\to-\infty}{\longrightarrow} 0 \bigwedge \phi_t[X] \in U\, \forall t \leq 0\right\}. \tag{2.51}$$

The manifolds (2.50), (2.51) have the same dimensions as the subspaces E^s and E^u correspondingly and are tangent to them at the origin.

Remark. There is also an invariant surface $W^c_{loc}(0)$ tangent to the invariant subspace E^c, but unlike the previous cases, the nature of the phase flow on this manifold significantly depends on the rejected nonlinear terms (that is why an analogue of the Grobman-Hartman theorem in this case does not exist).

2.5 Discrete maps generated by the phase flows of dynamical systems

We already know how a linear system (2.41) generates a phase flow $e^{t\hat{A}}[\cdot]:\mathbb{R}^n\to\mathbb{R}^n$. We now assume that $t=\tau=\text{const}$. Then $B:=e^{\tau\hat{A}}$ is the matrix with constant coefficients that defines *a discrete dynamical system*

$$X_{n+1}=B\,X_n. \tag{2.52}$$

In a similar way, the system (2.49) generates a nonlinear mapping

$$X_{n+1}=\phi_\tau\left[X_n\right]:=G\left[X_n\right]. \tag{2.53}$$

The orbits of the linear discrete mapping are the sequences $\{X_i\}_{-\infty<i<+\infty}$, whose elements are determined by the recursive formula (2.53). Each element $X\in\mathbb{R}^n$ generates a unique orbit, provided the matrix A which defines mapping B does not have zero eigenvalues. In analogy with the continuous case, one can define invariant subspaces. If all eigenvalues of the matrix $\hat{A}$ are real and differ from each other, then in some coordinate system the discrete mapping takes the form

$$X_{n+1}=B\,X_{n+1}=\begin{pmatrix} e^{\lambda_1\tau} & 0 & \dots\dots & 0 & 0\\ 0 & e^{\lambda_2\tau} & 0\ \dots & 0 & 0\\ \dots & \dots & \dots\dots\dots & \dots & \dots\\ 0 & 0 & \dots\dots & 0 & e^{\lambda_n\tau}\end{pmatrix}\begin{pmatrix}x_1\\ x_2\\ \dots\\ x_n\end{pmatrix}$$

And now it's easy to deduce what invariant subspaces will look like in a discrete case:

- $$E^s=\text{span}\left\{V_1,\,V_2,\,\dots,V_{n_s}\right\}, \tag{2.54}$$
  where $\{V_i\}_{i=1}^{n_s}$ are the eigenvectors of the matrix B corresponding to the eigenvalues μ_{r_i} such that $|\mu_{r_i}|<1$;
- $$E^u=\text{span}\left\{U_1,\,U_2,\,\dots,U_{n_u}\right\}, \tag{2.55}$$
  where $\{U_j\}_{j=1}^{n_u}$ are the eigenvectors of the matrix B corresponding to the eigenvalues μ_{r_j} such that $|\mu_{r_j}|>1$;
- $$E^c=\text{span}\left\{W_1,\,W_2,\,\dots,W_{n_c}\right\}, \tag{2.56}$$
  where $\{W_k\}_{k=1}^{n_c}$ are the eigenvectors of the matrix B corresponding to the eigenvalues μ_{r_k} such that $|\mu_{r_k}|=1$.

The sequences (orbits) belonging to E^s and E^u have the following properties: there exists such $C > 0$ and $0 < \alpha < 1$ that for $n \geq 0$

$$|X_n| \leq C\alpha^n X_0 \quad \text{as} \quad X_0 \in E^s, \tag{2.57}$$

$$|X_{-n}| \leq C\alpha^n X_0 \quad \text{as} \quad X_0 \in E^u. \tag{2.58}$$

These inequalities apply when all eigenvalues of the matrix B (called *multipliers*) are different. In cases when B has multiple eigenvalues, the attracting (repulsive) characteristics of the mappings on the corresponding subspaces are preserved, but the estimates (2.57)–(2.58) may not take place.

Remark. In the discrete case, there is a theorem analogous to the Grobman-Hartman theorem.

2.6 Parametric resonance

This section will answer the question of when and why the lower point of equilibrium of a pendulum loses its stability and begins to make large amplitude movements under the influence of a small disturbing force. Let us consider a pendulum with the mass M suspended on a thread, the length of which changes periodically:

$$l \to l + \delta \sin[\omega t]. \tag{2.59}$$

Let us denote the angle of deviation of the thread from the vertical by $\alpha(t)$. Using the formulas (1.15), (1.16), and taking into account (2.59), we will obtain the following Lagrangian:

$$L = \frac{m}{2}\left[(l + \delta\,\sin[\omega\,t])^2\,\dot{\alpha}^2 + \delta^2\omega^2\,\cos^2[\omega\,t]\right] + m\,g\,\cos[\alpha(t)]\,\left(l + \delta\,\sin[\omega\,t]\right). \tag{2.60}$$

Calculating the variational derivative of the function (2.60) and then making the elementary transformations, we can get equation of motion for the parametric pendulum in the following form:

$$\alpha''[t] + \frac{2\alpha'[t]\delta\omega\cos[\omega t] + g\sin[\alpha[t]]}{l + \delta\sin[\omega t]} = 0. \tag{2.61}$$

Assuming that $\delta \ll 1$, $\alpha(t) \ll 1$, $|2\alpha'[t]\delta\omega| \ll |g\alpha[t]| \ll 1$, we get in the linear approximation the equation

$$\alpha''[t] = -\alpha[t]\left(\Omega^2 + \epsilon\sin[\omega t]\right), \qquad \Omega^2 = (g/l), \quad |\epsilon| \ll 1. \tag{2.62}$$

It is more convenient to write down the equation (2.62) in the form of an equivalent dynamical system

$$x_1' = x_2,$$
$$x_2' = -F[t]\,x_1, \tag{2.63}$$

where $F[t] = F[t+T] = \Omega^2 + \epsilon \sin[\omega t]$, $T = 2\pi/\omega$. The phase flow $\phi_t[\cdot] : \mathbb{R}^2 \to \mathbb{R}^2$ of the system (2.63) generates the following mapping over the period:

$$A = \phi_T[\cdot] : \qquad \mathbb{R}^2 \to \mathbb{R}^2. \tag{2.64}$$

Lemma 2.1. *The point $X_0 \in \mathbb{R}^2$ is the stationary point of the mapping (2.64) if and only if the solution of the Cauchy problem for the system (2.63) with the Cauchy data*

$$x_1[0] = X_{0\,1}, \qquad x_2[0] = X_{0\,2}$$

is periodic function with the period T.

The proof is evident.

Definition 2.5. The stationary point X_0 of the map A is stable if $\forall\ \epsilon > 0$ $\exists\, \delta > 0$:

$$|X - X_0| < \delta \Longrightarrow |A^n X - A^n X_0| < \epsilon, \quad \forall\, n \geq 0.$$

Lemma 2.2. *As a consequence of the linearity of (2.63), it appears that the map A is linear as well.*

Proof. The solution of the initial value problem

$$\frac{d\,X}{d\,t} = A[t]\,X, \quad X[0] = X_0 \tag{2.65}$$

is given by the formula

$$X[t] = \left\{ I + \int_0^t A(s_1) d\, s_1 + \int_0^t A(s_2) d\, s_2 \int_0^{s_2} A(s_1) d\, s_1 + ... \right\} [X_0]$$
$$:= \text{Tchron}\exp\left[\int_0^t A(s) d\, s \right] [X_0],$$

where $\text{Tchron}\exp\left[\int_0^t A(s) d\, s\right]$ is *the resolvent* of the linear system (2.65). We denote by "Tchron" *the chronological product* (see, e.g., [Maurin (1973)], Ch. IX). Now it is suffice to note that the operator-valued function at the r.h.s. of the above formula is linear.

Lemma 2.3. *The system (2.63) is Hamiltonian with the Hamilton function*

$$H\,(x_1,\, x_2;\, t) = \frac{x_2^2}{2} + F[t]\frac{x_1^2}{2} \tag{2.66}$$

(the correctness of the statement is proved by direct verification).

Now let us consider more general Hamiltonian system

$$q_i'(t) = H_{p_i}, \qquad p_i'(t) = -H_{q_i}, \quad i = 1, \dots, n, \tag{2.67}$$

with the Hamiltonian $H = H(p, q)$, where $(p, q) \in \mathbb{R}^{2n}$, and the phase flow

$$\phi_t : \mathbb{R}^{2n} \ni (p(0), q(0)) \Rightarrow (p(t), q(t)) \in \mathbb{R}^{2n},$$

generated by this system at the set of the Cauchy data.

Theorem 2.4 (Liouville). *Let $D_0 \subset \mathbb{R}^{2n}$ be a bounded set, and $\phi_t : \mathbb{R}^{2n} \to \mathbb{R}^{2n}$ be the phase flow of the Hamiltonian system. Then the volume of the set $D = \phi_t[D_0]$ is equivalent to the volume of the initial set D_0.*

Proof. We will use the notation

$$x'(t) = f(x), \qquad x \in \mathbb{R}^k.$$

It is obvious that we can approximate the phase flow by the formula

$$\phi_t(x) = x + t\, f(x) + O(t^2). \tag{2.68}$$

The proof comes down to showing that if $\operatorname{div} f(x) = 0$, then $\phi_t(x)$ preserves the volume, or, in other words, that

$$V(t) = \int_D d\,y = V(0) = \int_{D_0} d\,x, \quad D = \phi_t[D_0]. \tag{2.69}$$

Let us formulate the auxiliary statement:

Lemma 2.4. *The following formula takes place:*

$$\frac{d\,V(t)}{d\,t} = \int_{D_0} \operatorname{div} f(\xi)\, d\,\xi. \tag{2.70}$$

Proof. Introducing notation $y = \phi_t(x)$, we can write down the chain of equalities:

$$V(t) = \int_D d\,y = \int_{D_0} \det\left[\frac{\partial\,y}{\partial\,x}\right] d\,x = \int_{D_0} \det\left[\frac{\partial\,\phi_t(x)}{\partial\,x}\right] d\,x.$$

It appears from the formula (2.68) that

$$\left[\frac{\partial\,\phi_t(x)}{\partial\,x}\right]_{i\,j} = \delta_{i\,j} + t\,\frac{\partial\,f^i}{\partial\,x^j} + O(t^2).$$

We assert that

$$\det\left[I + t\,\frac{\partial\,f^i}{\partial\,x^j} + O(t^2)\right] = 1 + t\,\mathrm{trace}\left[\frac{\partial\,(f^1, \dots, f^n)}{\partial\,(x^1, \dots, x^n)}\right] + O(t^2)$$

$$= 1 + t \sum_{k=1}^{n} \frac{\partial f^k}{\partial x^k} + O(t^2).$$

This equality becomes evident after carefully calculating the determinant on the left-hand side. Therefore

$$V(t) = \int_{D_0} \left[I + t \, \mathrm{div} f(x) + O(t^2) \right] \, dx,$$

and hence

$$\frac{dV(t)}{dt} = \int_{D_0} \mathrm{div} f(x) \, dx.$$

In the case of Hamiltonian system (2.67)

$$f = \left(\frac{\partial H}{\partial p_1}, \dots, \frac{\partial H}{\partial p_n}; -\frac{\partial H}{\partial q_1}, \dots, -\frac{\partial H}{\partial q_n} \right)^{tr}$$

$$\mathrm{div} f = \mathrm{trace} \frac{\partial (H_p, -H_q)}{\partial (q, \ p)} = \sum_{i=1}^{n} \frac{\partial^2 H}{\partial q_i \, \partial p_i} - \sum_{i=1}^{n} \frac{\partial^2 H}{\partial p_i \, \partial q_i} = 0,$$

and this is the end of the proof.

Corollary 2.4. *Let $A : \mathbb{R}^{2n} \to \mathbb{R}^{2n}$ be the linear (or linearized) map generated by the phase flow of the Hamiltonian system. Then* $\det A = 1$.

Proof. In the case of the A mapping, Liouville's theorem results in the following chain of equalities:

$$V(t) = \int_{D} dy = \int_{D_0} \det \left[\frac{\partial y}{\partial x} \right] \, dx = \int_{D_0} \det \left[\frac{\partial (Ax)}{\partial x} \right] \, dx = \int_{D_0} \det[A] \, dx$$

$$= \det[A] \int_{D_0} dx = V(0) = \int_{D_0} dx.$$

Thus, $\det[A] = 1$.

Theorem 2.5. *Let $A : \mathbb{R}^2 \to \mathbb{R}^2$ be a volume-preserving linear mapping (that is $\det[A] = 1$). Then its stationary point X_0 is stable if and only if $|\mathrm{trace}[A]| < 2$, and unstable when $|\mathrm{trace}[A]| \geq 2$.*

Proof. Let μ_1, μ_2 be the multipliers of the map A. The multipliers satisfy the characteristic equation

$$\mu^2 - \mathrm{trace}[A]\mu + \det[A] = \mu^2 - \mathrm{trace}[A]\mu + 1 = 0.$$

And there are two options.

Case 1. If $|\text{trace}[A]| > 2$, then the multipliers are real and such that $|\mu_2| > 1 > |\mu_1|$. The stationary point X_0 is unstable in this case, because the phase trajectory running along the eigenvector $\vec{Y}_2$ corresponding to the eigenvalue μ_2 moves away from the point X_0.

Case 2. If $|\text{trace}[A]| < 2$, then $Re[\mu_{1,2}] < 1$, so the phase trajectories approach the stationary point.

Let us apply Theorem 2.5 to the system (2.63). Rescaling the time $t \to \omega t$, we can rewrite it in the form

$$x_1' = x_2, \qquad x_2' = \tilde{\omega}^2 x_1 \left[1 + \tilde{\epsilon} \sin(t)\right],$$

where $\tilde{\epsilon} = \epsilon/\Omega^2$, and $\tilde{\omega}^2 = (\Omega/\omega)^2$ is treated now as the variable parameter. We obtain the resonance frequency $\tilde{\omega}$ at which the stability of the equilibrium point $(x_1, x_2) = (0, 0)$ may be lost by constructing the map $\varphi_{2\pi} = A$ for unperturbed problem, i.e. for $\tilde{\epsilon} = 0$. We calculate the matrix A by using the fundamental solution $M[t]$, composed of the vector $(\cos[\tilde{\omega}t], -\tilde{\omega}\sin[\tilde{\omega}\, t])^{tr}$ being the solution of the initial value problem

$$x_1' = x_2, \quad x_2' = -\tilde{\omega}^2 x_1, \quad x_1[0] = 1, \ x_2[0] = 0$$

and the vector $(1/\tilde{\omega}\sin[\tilde{\omega}t], \cos[\tilde{\omega}t])^{tr}$ being the solution of the initial value problem

$$x_1' = x_2, \quad x_2' = -\tilde{\omega}^2 x_1, \quad x_1[0] = 0, \ x_2[0] = 1.$$

It is evident, that the solution of the initial value problem

$$x_1' = x_2, \quad x_2' = -\tilde{\omega}^2 x_1, \quad x_1[0] = C_1, \ x_2[0] = C_2$$

is given by the formula

$$\begin{pmatrix} x_1(t) \\ x_2(t) \end{pmatrix} = \varphi_t \begin{pmatrix} C_1 \\ C_1 \end{pmatrix} = M(t) \begin{pmatrix} C_1 \\ C_2 \end{pmatrix} = \begin{bmatrix} \cos[\tilde{\omega}t] & \frac{1}{\tilde{\omega}}\sin[\tilde{\omega}t] \\ -\tilde{\omega}\sin[\tilde{\omega}t] & \cos[\tilde{\omega}t] \end{bmatrix} \begin{pmatrix} C_1 \\ C_2 \end{pmatrix}.$$

Hence

$$\varphi_{2\pi} = M[2\pi] = \begin{bmatrix} \cos[2\pi\tilde{\omega}] & \frac{1}{\tilde{\omega}}\sin[2\pi\tilde{\omega}] \\ -\tilde{\omega}\sin[2\pi\tilde{\omega}] & \cos[2\pi\tilde{\omega}] \end{bmatrix}.$$

Thus, we see that $|\text{trace}[A]| = |\text{trace}M[2\pi]| = 2|\cos[2\pi\tilde{\omega}]| < 2$ everywhere except when

$$\tilde{\omega} \in \{k/2 : k = 0, 1, 2, ...\}. \tag{2.71}$$

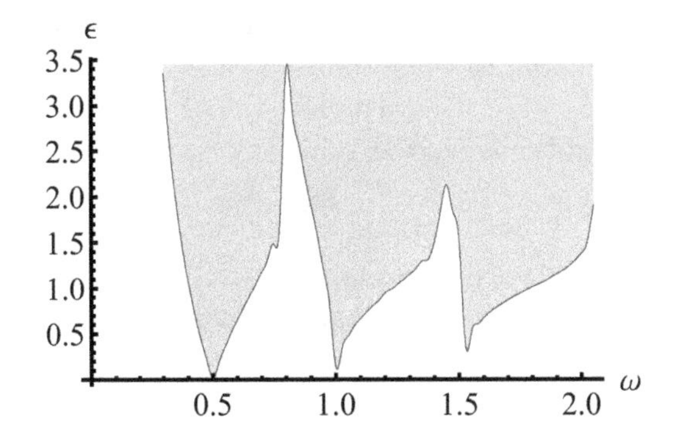

Fig. 2.18 Resonance curves in the space of parameters' ($\tilde{\omega}$, ϵ). The darkened areas above successive branches of the resonance curves in mathematical folklore are called "Arnold tongues".

So the equilibrium point is unstable when $\tilde{\omega}$ belongs to the set (2.71). At these points, the instability zones in the parametric plane (ϵ, $\tilde{\omega}$) (so called "Arnold's tongues") reach the horizontal axis (see Fig. 2.18).

Example 2.3. The numerical experiment confirming the presence of resonances with the indicated values of the ω for a nonlinear problem can be performed using the code presented in the cell 2.2. We consider the equation

$$\alpha''[t] + \omega^2 \sin[\alpha[t]]\,(1 + \epsilon\, \sin[t]) = 0, \tag{2.72}$$

which describes (in a slightly simplified version) a pendulum on a thread whose length varies periodically with the period $2\,\pi$.

Cell 2.2.

```
"Parametric resonance: dependence of solutions on the value of the parameters ω
and ϵ"
    In[1]: ClearAll["Global`*"]
    In[2]: Manipulate[
    eq1 = α'[t] == z[t] ;
    eq2 = z'[t] == -ω^2 Sin[α[t]] (1 + ϵ Sin[t]);
    rozw = NDSolve[{eq1, eq2, α[0] == 0.01, z[0] == 0}, {α[t], z[t]}, {t, 0, T}];
    cykl = Plot[α[t] /. rozw, {t, 0, T}, PlotRange → Full],
    {ω, {1/4, 1/3, 1/2, 2/3, 3/4, 1, 5/4, 3/2, 7/4, 2}}, {ϵ, {0.01, 0.1, 0.2,
0.4}},
{T, {50, 100, 200, 500, 1000, 1500, 2500}}]
```

Numerical simulations allow to determine the area in the (ω, ϵ) parameter space, corresponding to the resonance values of the parameters. This area is shown in the Fig. 2.18. For the values of the parameters (ω, ϵ) belonging to the shaded area, the solutions to the equation (2.72) grow infinitely in time.

Problem.

Carry out numerical experiments similar to those described above for the equation (2.61). Find the area of instability in the vicinity of the first three resonance values of the parameter ω and represent them in the parametric space (ω, δ). For the remaining parameters, take the values $L = 1$, $g = 10$.

2.7 Kapitza's pendulum

The model described by P.L. Kapitza [Kapitza (1951)] in the early 1950s simulates a flat pendulum whose suspension point oscillates at high frequency and low amplitude in the vertical direction (see Fig. 2.19). The physical experiments show that at a sufficiently high frequency of the vertical oscillations the upper equilibrium point becomes conditionally stable.

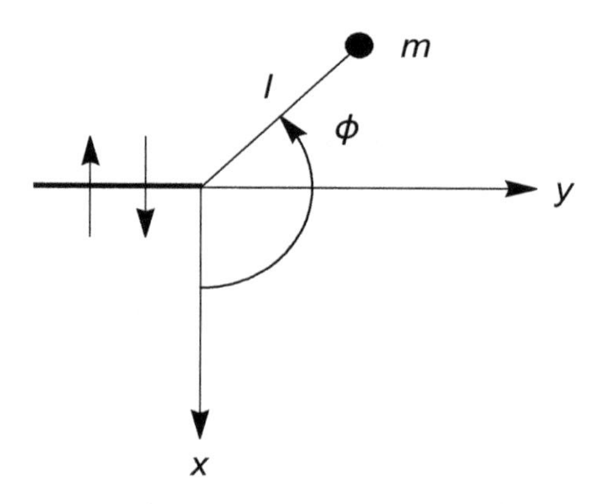

Fig. 2.19 Kapitza's pendulum.

When constructing the mathematical model, we assume that the position of the suspension point changes according to the rule

$$\zeta(t) = a\,\cos(\gamma t), \quad 0 < a \ll 1, \quad \gamma \gg 1.$$

In order to find the equations of motion, we use the Lagrange function of the system. The coordinates of the pendulum are described by the following formula:

$$x = l \cos\,\phi + a\,\cos(\gamma\,t),$$
$$y = l \sin\,\phi.$$

From this we calculate the kinetic and potential energy:

$$\begin{aligned} E_{kin} &= \frac{m}{2}\left[(x')^2 + (y')^2\right] \\ &= \frac{m}{2}\left[l^2(\phi')^2 + 2\,a\,\gamma\,l\,\phi'\sin\phi\sin(\gamma t) + a^2\gamma^2\sin^2(\gamma t)\right], \\ E_{pot} &= U(x) = -mgx = -mg[l\cos\phi + a\cos(\gamma t)]. \end{aligned}$$

Hence

$$\begin{aligned} \mathcal{L} &= E_{kin} - E_{pot} \\ &= \frac{m}{2}\left[l^2(\phi')^2 + 2\,a\,\gamma\,l\,\phi'\sin\phi\sin(\gamma t) + a^2\gamma^2\sin^2(\gamma t)\right] \\ &\quad + mg[l\cos\phi + a\cos(\gamma t)]. \end{aligned}$$

Taking the variational derivative of the function $\mathcal{L}$ and equating it to zero, we obtain:

$$\begin{aligned} 0 &= \frac{d}{dt}\frac{\partial\,\mathcal{L}}{\partial\,\phi'} - \frac{\partial\,\mathcal{L}}{\partial\,\phi} \\ &= m\,l^2\phi'' + m\,a\,\gamma\,l\,[\phi'\cos\,\phi\,\sin(\gamma\,t) + \gamma\,\sin\,\phi\,\cos(\gamma\,t)] \\ &\quad - m\,a\gamma\,l\phi'\cos\,\phi\,\sin(\gamma\,t) - m\,g\,l\,\sin\,\phi \\ &= m\,l^2\phi'' + m\,g\,l\,\sin\,\phi + m\,a\,\gamma^2\,l\,\sin\,\phi\,\cos(\gamma\,t). \end{aligned}$$

Dividing this expression by $m\,l^2$, we finally obtain:

$$\phi'' + \sin\,\phi\left[\omega^2 + \frac{a}{l}\gamma^2\cos(\gamma\,t)\right] = 0, \quad \omega = \sqrt{\frac{g}{l}}. \tag{2.73}$$

To present the equation (2.73) in a more convenient form, we introduce the scaling

$$T = \omega\,t, \qquad \Omega = \frac{\gamma}{\omega}, \qquad \beta = \frac{a}{l}.$$

After this substitution, the equation will take the form

$$\frac{d^2\,\phi}{d\,T^2} + \sin\,\phi\left[1 + \beta\,\Omega^2\,\cos(\Omega\,T)\right] = 0. \tag{2.74}$$

Next we use the substitution

$$\Omega = \frac{1}{\epsilon}, \qquad \beta = \epsilon\,\bar{\beta},$$

after which the equation takes the form

$$\frac{d^2\,\phi}{d\,T^2} + \sin\,\phi\left[1 + \frac{\bar{\beta}}{\epsilon}\,\cos\left(\frac{T}{\epsilon}\right)\right] = 0. \tag{2.75}$$

The last substitution allows for the introduction of a small parameter into the equation. Below we present a simplified version of the approach used in [Koszalka (2006)] for solving the problem.

In order to formulate the conditions which guarantee the formation of a stable point of equilibrium in the upper position of the pendulum, we use the *multiscale decomposition method* [Nayfeh (2011)]. We suppose that the function we are looking for depends on the "fast" time $\tau = T/\epsilon$ and the "slow" time T. In addition, let us express the function as a series with respect to the powers of ϵ:

$$\phi(T,\,\tau) = \phi_0(T,\,\tau) + \epsilon\,\phi_1(T,\,\tau) + O(\epsilon^2). \tag{2.76}$$

Inserting (2.76) into (2.75), using the formulas

$$\frac{d}{d\,T} = \frac{\partial}{\partial\,T} + \frac{1}{\epsilon}\frac{\partial}{\partial\,\tau}, \quad \frac{d^2}{d\,T^2} = \frac{\partial^2}{\partial\,T^2} + \frac{2}{\epsilon}\frac{\partial^2}{\partial\,T\,\partial\,\tau} + \frac{1}{\epsilon^2}\frac{\partial^2}{\partial\,\tau^2}$$

and then equating to zero the coefficients at the different powers of ϵ, we get the system

$$\begin{aligned}
O\left(\frac{1}{\epsilon^2}\right): &\quad \frac{\partial^2\phi_0}{\partial\,\tau^2} = 0,\\
O\left(\frac{1}{\epsilon}\right): &\quad 2\frac{\partial^2\phi_0}{\partial T\partial\,\tau} + \frac{\partial^2\phi_1}{\partial\,\tau^2} = -\bar{\beta}\,\cos\,\tau\,\sin\,\phi_0,\\
O\,(1): &\quad 2\frac{\partial^2\phi_1}{\partial T\partial\,\tau} + \frac{\partial^2\phi_0}{\partial\,T^2} = -\left\{\sin\,\phi_0 + \bar{\beta}\,\phi_1\,\cos\,\phi_0\right\}.
\end{aligned}$$

Our goal is to find a particular solution to the above system, describing movements with bounded amplitudes. We also can assign arbitrary values to the initial conditions. Integrating twice the equation proportional to $O(1/\epsilon^2)$, we get the general solution

$$\phi_0 = \tau\,F(T) + G(T)$$

dependent on two arbitrary functions. Note, however, that the first term at the r.h.s. grows indefinitely as τ increases. In order to maintain the boundedness of solution, we put $F(T) = 0$. With this condition, the second equation takes the form

$$\frac{\partial^2\phi_1}{\partial\,\tau^2} = -\bar{\beta}\,\cos\,\tau\,\sin\,\phi_0.$$

Integrating it with respect to τ, we get an equality

$$\frac{\partial \phi_1}{\partial \tau} = -\bar{\beta} \sin \tau \sin \phi_0 + h(T).$$

We should also assume that $h(T) \equiv 0$, for otherwise we will get an expression proportional to τ after the next integration. Integrating the above equation with respect to τ under such a supposition, and using the initial condition $\phi_1(0) = 0$, we finally get the solution

$$\phi_1 = \bar{\beta} \sin \phi_0 (\cos \tau - 1). \tag{2.77}$$

The last equation now takes the form

$$\frac{d^2 \phi_0}{d T^2} = 2\bar{\beta}\dot{\phi}_0 \sin \tau \cos \phi_0 - \sin \phi_0 \left[1 + \bar{\beta}^2 \cos \phi_0 \cos \tau (\cos \tau - 1)\right]. \tag{2.78}$$

Next, we will apply the procedure of averaging over the fast variable τ:

$$\bar{\Psi}(T) = \frac{1}{R} \int_\tau^{\tau+R} \Psi(T, \xi)\, d\xi, \qquad R \gg 2\pi.$$

It is easy to see that the following formulas take place:

$$\frac{1}{R} \int_\tau^{\tau+R} \sin \xi\, d\xi \approx 0, \quad \frac{1}{R} \int_\tau^{\tau+R} \cos \xi\, d\xi \approx 0, \quad \frac{1}{R} \int_\tau^{\tau+R} \cos^2 \xi\, d\xi \approx \frac{1}{2}.$$

After the averaging, we get the equation

$$\frac{d^2 \bar{\phi}_0}{d T^2} + \sin \bar{\phi}_0 \left(1 + \frac{\bar{\beta}^2}{2} \cos \bar{\phi}_0\right) = 0. \tag{2.79}$$

This equation is equivalent to the Hamiltonian system

$$\begin{cases} \dfrac{d \bar{\phi}_0}{d T} = \bar{\varphi}_0 = \dfrac{\partial H_{ef}}{\partial \bar{\varphi}_0}, \\ \dfrac{d \bar{\varphi}_0}{d T} = -\sin \bar{\phi}_0 \left[1 + \dfrac{(\beta \Omega)^2}{2} \cos \bar{\phi}_0\right] = -\dfrac{\partial H_{ef}}{\partial \bar{\phi}_0}, \end{cases} \tag{2.80}$$

where

$$H_{ef} = \frac{\bar{\varphi}_0^2}{2} + U_{ef}, \qquad U_{ef} = \frac{(\beta \Omega)^2}{4} \sin^2 \bar{\phi}_0 - \cos \bar{\phi}_0.$$

It is evident that

$$-\sin \bar{\phi}_0 \left[1 + \frac{(\beta \Omega)^2}{2} \cos \bar{\phi}_0\right]\Bigg|_{\bar{\phi}_0 = \pm\pi} = 0,$$

hence the effective potential of U_{ef} will have local extrema in the points $\pm\pi$. In order that these extrema be local minima, it is sufficient that the following condition holds:

$$0 < \frac{\partial^2 U_{ef}}{\partial \bar{\phi}_0^2} = \left(\cos \bar{\phi}_0 + \frac{(\beta \Omega)^2}{2} \cos(2\bar{\phi}_0)\right)\Bigg|_{\bar{\phi}_0 = \pm\pi} = \frac{(\beta \Omega)^2}{2} - 1,$$

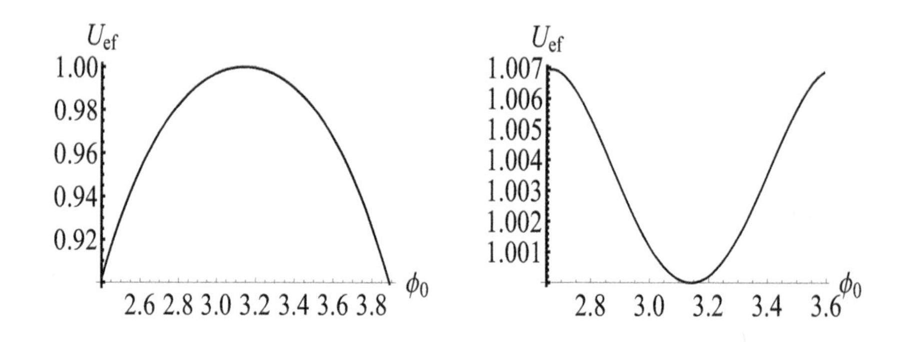

Fig. 2.20 The graphs of the function $U_{ef}(\bar{\phi}_0)$ in the vicinity of the point $\bar{\phi}_0 = \pi$ obtained at the values $\beta = 0.1$, $\Omega = 12$ for which the condition (2.81) is not true (left panel), and for $\beta = 0.125$, $\Omega = 12$ for which it is fulfilled (right panel).

or

$$\frac{(\beta\,\Omega)^2}{2} > 1. \tag{2.81}$$

Figure 2.20 shows the graphs of the function $U_{ef}(\bar{\phi}_0)$ for different parameter values.

Thus, the upper equilibrium point $\bar{\phi}_0 = \pi$ is a stable stationary point (the center) when the frequency Ω is large enough to satisfy the condition (2.81). Note that with a fixed value of Ω, the stationary point is stable when the parameter β is small enough, which we assumed at the beginning of this section. One can see on the right panel of Fig. 2.20 that the potential well that arises when the product $\beta\,\Omega$ slightly higher than $\sqrt{2}$ is shallow, which in practice means stability under the small deviations of the pendulum from the upper equilibrium position. Animation demonstrating the movement of the presented dynamical system can be found in the file PM1_4.nb.

2.8 Modification of the problem posed by Kapitza

Let us consider a mass m suspended on a weightless rod of length L in the gravitational field. The suspension point is assumed to perform rapid low-amplitude oscillations in the horizontal direction. As before, we will assume that the OX axis is directed vertically downward, and the OY axis is directed to the right in the horizontal direction, see Fig. 2.21.

In the selected reference frame, the coordinates of the mass m will be given by the following relations:

$$x = L\cos\phi, \qquad y = L\,\sin\phi + a\,\sin\gamma\,t,$$

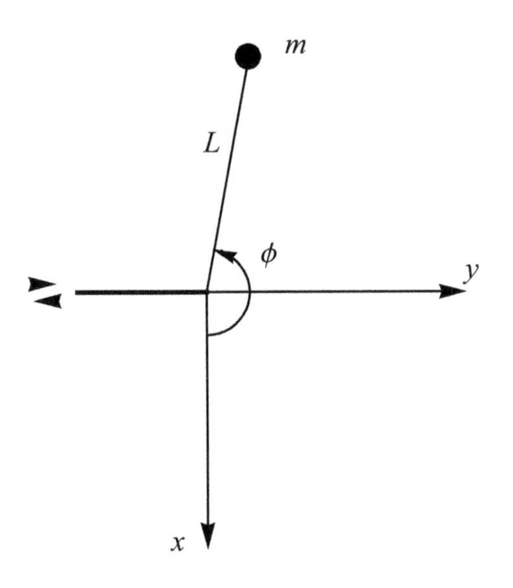

Fig. 2.21 Modified Kapitza's pendulum.

where ϕ is the angle between the vertical direction and the instantaneous direction of the rod on which the mass m is fixed. The kinetic energy in the case under consideration is expressed by the formula

$$E_{kin} = \frac{m}{2}\left(\dot{x}^2 + \dot{y}^2\right) = \frac{m}{2}\left[L^2\dot{\phi}^2 + a^2\gamma^2\cos^2\gamma\, t + 2\, a\, L\, \gamma\dot{\phi}\cos\phi\, \cos\gamma\, t\right],$$

while the potential energy is as follows:

$$E_{pot} = -m\, g\, x = -m\, g\, L\cos\phi.$$

So equating to zero the variational derivative of the function $L = E_{kin} - E_{pot}$ we obtain, after some algebraic manipulations, the equation

$$\ddot{\phi} + \omega^2\sin\phi - \frac{a\,\gamma^2}{L}\cos\phi\sin\gamma\, t = 0, \tag{2.82}$$

where $\omega = \sqrt{g/L}$.

We apply to the equation (2.82) two scaling transformations coinciding with that applied to the equation (2.73). The transformed equation takes the following form:

$$\frac{d^2\,\phi}{d\,T^2} + \sin\,\phi - \frac{\bar{\beta}}{\epsilon}\,\cos\phi\,\sin\,\tau = 0, \tag{2.83}$$

where $T = \omega t$, $\tau = T\Omega = T/\epsilon$, $\bar{\beta} = \beta\Omega = \beta/\epsilon$. As in the case of the Kapitza's pendulum, we assume that $\phi = \phi(T, \tau)$ and present this function as a series with respect to ϵ:

$$\phi(T, \tau) = \phi_0(T, \tau) + \epsilon\, \phi_1(T, \tau) + O(\epsilon^2). \tag{2.84}$$

Inserting (2.84) into the formula (2.83), changing d/dT with $\partial/\partial T + 1/\epsilon\, \partial/\partial \tau$, and equating the coefficients at the same powers of ϵ, we obtain the system

$$\begin{aligned}
O\left(\frac{1}{\epsilon^2}\right): &\quad \frac{\partial^2 \phi_0}{\partial \tau^2} = 0,\\
O\left(\frac{1}{\epsilon}\right): &\quad 2\frac{\partial^2 \phi_0}{\partial T \partial \tau} + \frac{\partial^2 \phi_1}{\partial \tau^2} = \bar{\beta}\, \sin\, \tau\, \cos\, \phi_0,\\
O\,(1): &\quad 2\frac{\partial^2 \phi_1}{\partial T \partial \tau} + \frac{\partial^2 \phi_0}{\partial T^2} = -\sin\, \phi_0 \left[1 + \bar{\beta}\, \phi_1\, \sin \tau\right].
\end{aligned}$$

Integrating twice the equation proportional to $O\left(\frac{1}{\epsilon^2}\right)$ and dropping out the secular term (i.e. the term proportional to τ) we obtain that ϕ_0 is an arbitrary function of the "slow" variable T. Next, integrating twice the equation proportional to $O\left(\frac{1}{\epsilon}\right)$ and dropping out the secular term, we obtain the solution

$$\phi_1 = -\bar{\beta} \sin\, \tau \cos\, \phi_0,$$

satisfying zero initial conditions as $\tau = 0$. Inserting this equation into the equation proportional to $O\,(1)$, we obtain the equation, which after the application of the procedure of averaging over the "fast" variable τ will finally take the form

$$\frac{d^2\, \bar{\phi}_0}{d\,T^2} + \sin\, \bar{\phi}_0 \left(1 - \frac{\bar{\beta}^2}{2}\, \cos\, \bar{\phi}_0\right).$$

The above equation is equivalent to the Hamiltonian system

$$\begin{cases}
\frac{d\,\bar{\phi}_0}{d\,T} = \bar{\psi}_0 = \frac{\partial\, H_{ef}}{\partial\, \bar{\psi}_0},\\[2ex]
\frac{d\,\bar{\psi}_0}{d\,T} = -\sin\, \bar{\phi}_0 \left(1 - \frac{\bar{\beta}^2}{2}\, \cos\, \bar{\phi}_0\right) = -\frac{\partial\, H_{ef}}{\partial\, \bar{\phi}_0},
\end{cases} \tag{2.85}$$

where

$$H_{ef} = \frac{\bar{\psi}_0^2}{2} + V_{ef}(\bar{\phi}_0), \qquad V_{ef}(\bar{\phi}_0) = \frac{(\beta\Omega)^2}{8} \cos 2\, \bar{\phi}_0 - \cos\, \bar{\phi}_0. \tag{2.86}$$

Similarly to (2.80), the system (2.85) has the stationary points $A = (0,\, 0)$ and $B_{\pm} = (\pm\pi,\, 0)$. Besides, there is an extra pair of stationary points

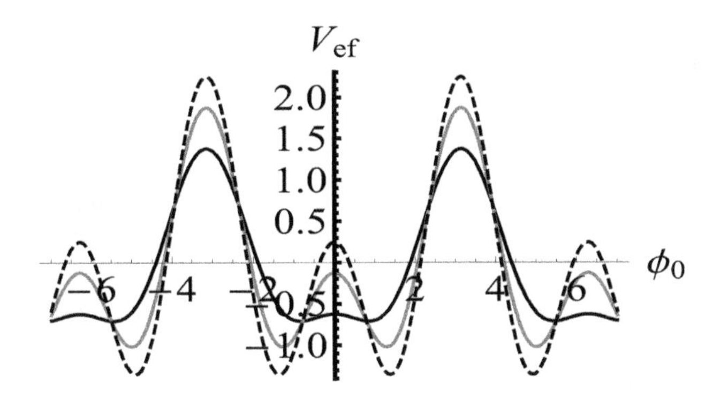

Fig. 2.22 Dependence of the effective potential V_{ef} from the formula (2.86) on the parameters' values: black line corresponds to $(\beta\Omega)^2 = 3$; gray line corresponds to $(\beta\Omega)^2 = 7$; dashed line corresponds to $(\beta\Omega)^2 = 10$.

$C_{\pm} = \pm\arccos\frac{2}{(\beta\Omega)^2}$ in between, provided that the inequality $\frac{(\beta\Omega)^2}{2} > 1$ is fulfilled. Assuming that this condition is satisfied, we obtain the following estimates for the signs of the second derivative of the function $V_{eff}(\bar{\phi}_0)$ at the stationary points:

$$\frac{d^2\, V_{ef}}{d\, \bar{\phi}_0^2}|_{\bar{\phi}_0=\pm\pi} = -\left(1 + \frac{(\beta\Omega)^2}{2}\right) < 0,$$

$$\frac{d^2\, V_{ef}}{d\, \bar{\phi}_0^2}|_{\bar{\phi}_0=0} = 1 - \frac{(\beta\Omega)^2}{2} < 0,$$

$$\frac{d^2\, V_{ef}}{d\, \bar{\phi}_0^2}|_{\pm\bar{\phi}_0^*=\pm\arccos\frac{2}{(\beta\Omega)^2}} = \frac{2}{(\beta\Omega)^2}\sin^2\bar{\phi}_0^* > 0.$$

This means that when the inequality $\frac{(\beta\Omega)^2}{2} > 1$ is satisfied, the stationary points A and $B_{\pm}$ become unstable, while a pair of stable stationary points appears between them.

Numerical experiments show that the modified Kapitza's model has one nuance. Despite the fact that new stable stationary points do appear immediately when the value of $(\beta\Omega)^2$ exceeds two, however, the observation of the effect in numerical experiment is difficult because at first the potential well in the vicinity of the angle $\phi_0^* = \arccos\frac{2}{(\beta\Omega)^2}$ is shallow (see Fig. 2.22) and it is not easy to keep the mass m in it. However, the effect is clearly manifested as the angle $\bar{\phi}_0^* = \arccos\frac{2}{(\beta\Omega)^2}$ approaches $\pi/2$ from below.

Problem.

Consider a mass m suspended on a weightless rod of length L in the gravitational field. Assume that the suspension point of the pendulum performs fast oscillations of small amplitude along a straight line directed at an angle α to the horizontal axis.

(1) Derive the equation of motion of the pendulum.
(2) Repeating the reasoning involved in the previous subsections, derive and analyze the Hamiltonian system describing the angle of deviation of the pendulum from the vertical direction, averaged over the "fast" time.
(3) Consider separately the case when $\alpha \in (0, \pi/4)$, and when $\alpha \in (\pi/4, \pi/2)$. Analyze for each of the cases the appearance of additional stationary points and derive the conditions under which the effective potential has local minima at these points.
(4) Perform numerical experiments to confirm the results of theoretical studies.

Chapter 3

Models describing nonlinear oscillations

3.1 Introduction

Periodic processes can without exaggeration be called the most common natural phenomena. The change of day and night, periodically repeating seasons, sounds and colors of the world — all this is related to periodicity. Researches on nonlinear oscillations in mechanical and electrical systems are being conducted form many decades due to great practical demand. Recently, interest in such research has also been manifested by representatives of chemical and biological sciences, ecologists and financiers, as the universality of periodic phenomena is by no means limited to natural phenomena. Periodic solutions supported by two-dimensional dynamical systems are usually associated with some regularity (constant period of oscillation, stability of solution, etc.). Nevertheless, even in this case, substantially nonlinear solutions cannot be described in terms of elementary functions or series, therefore qualitative and numerical methods are widely used in this field of knowledge. In the multidimensional case, the situation becomes even more complicated, because changing the parameters of the system can result in appearance of multi-periodic and chaotic solutions, and these two types of vibration can most often be distinguished only by using special methods. We start this chapter with the presentation of simple models. In the course of the further presentation, we will introduce new tools necessary for the study of multidimensional and chaotic models.

3.2 Predator-prey model

It was observed at the beginning of the 20th century, that the populations of large fish (predators) and small fish (prey) in the Adriatic Sea change periodically. More or less at the same time, a similar effect was noticed when processing empirical data on the population of lynx and hares in the forests of Canada. The mechanism of the observed phenomena is quite common. Intuitively, it is rather understandable that an increase in food resources favors an increase in the population of predators. The appearance of an excessive number of predators leads, in turn, to a decrease in the prey population, which after some time causes hunger among predators and a reduction in their number, which after a while contributes to an increase in the prey population. Consequently, the ecosystem returns to its initial stage. One of the simplest models used to describe the above ecosystem is the Lottka-Volterra model. Let us denote by N_B the size of the predator population (B means "big fish") and by N_L the population of victims (L means "little fish"). The population dynamics in this model is described by the system of equations

$$\dot{N}_B = N_B\,(\gamma\,N_L - \beta), \qquad \dot{N}_L = N_B\,(\alpha - \gamma\,N_B). \tag{3.1}$$

It is assumed that all coefficients are non-negative. Below we present the numerical scheme aimed at solving the above system.

Cell 3.1.

```
"Lottka-Volterra model"
In[1]: ClearAll["Global'*"];
In[2]:
    {α, β, γ} = {2, 1, 0.1};
    eq1a = x'[t] - (α - γ y[t]) x[t];
    eq2a = y'[t] - (-β + γ x[t]) y[t];
    sol = NDSolve[{eq1a == 0, eq2a == 0, x[0] == 100, y[0] == 2}, {x[t], y[t]},
{t, 0, 250}];
    U = x[t] /. sol;
    V = y[t] /. sol;
    Plot[{U, V}, {t, 0, 25}, PlotRange → Full, PlotStyle → {{Black}, Dash-
ing[{.02}]}, {Red, Thick}},
    PlotLabel → "solid: big fish, dashed: little fish"]
```

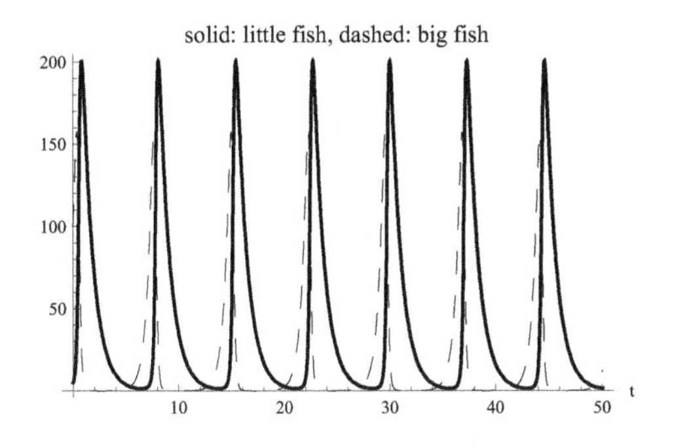

As one can see in the attached drawing, the system (3.1) demonstrates the periodic solutions for $\alpha > 0$, $\beta > 0$, $\gamma > 0$. Let us note that the maximum values of individual populations are separated by a certain period of time. The reader can convince by direct verification that the periodic nature of solutions is observed in this model in wide range of parameters' values.

3.3 Van der Pol's equation

Another model we present here describes oscillations in an electrical circuit with non-linear resistance. This model assumes that electrical energy is dissipated at high amplitudes and generated at low amplitudes. Equation, introduced by Van der Pol in the 1920s to describe such an effect, has the form

$$\ddot{x} - \epsilon(1 - x^2)\dot{x} + \omega^2 x = 0, \tag{3.2}$$

where $\omega > 0$, $\epsilon > 0$. The equation (3.2) was introduced to model of an electric circuit with a triode valve, the conductive properties of which change with the change of current intensity. It is more convenient to present Eq. (3.2) in the form of the following dynamical system:

$$\dot{x} = y, \qquad \dot{y} = \epsilon(1 - x^2)\, y - \omega^2 x. \tag{3.3}$$

Let us consider the numerical solutions of the system (3.3). The numeric value of the constant $\omega > 0$ is irrelevant, so we put it equal to one. The procedure of solving the system together with the graphical presentation of the Cauchy problem solutions is presented below. Numerical simulations indicate that the solution, regardless of the selection of (non-zero) initial data, tends to a non-linear periodic solution called *limit cycle.*

Cell 3.2.

```
"Van der Pol's model"
In[1]: ClearAll["Global'*"];
In[2]: eps = 1;
    eq1a = x'[t] - y[t];
    eq2a = y'[t]- eps (1 - x[t]^2) y[t]+x[t];
    sol = NDSolve[{eq1a == 0, eq2a == 0, x[0] == 0.6, y[0] == 0}, {x[t], y[t]},
{t, 0, 250}];
    U = x[t] /. sol;
    V = y[t] /. sol;
    Plot[U, {t, 0, 50}, PlotRange → Full]
```

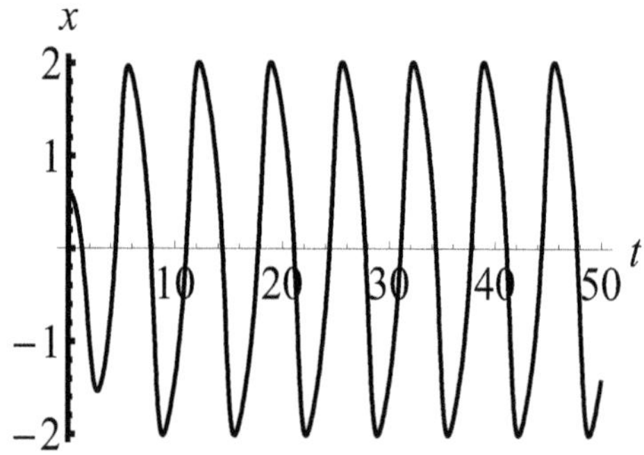

Remark. Note that the Van der Pol's model is unique in that the limit cycle appears in it for any values of the parameter $\epsilon > 0$.

3.4 Conditions for the existence of periodic trajectories

Due to the large role of periodic solutions in the models of natural and technical phenomena, we will provide criteria for the existence (or non-existence) of closed orbits (cycles) in specific areas of the phase spaces of dynamical systems. Let us consider the dynamical system

$$\dot{x} = P(x, y), \qquad \dot{y} = Q(x, y). \tag{3.4}$$

We assume that the functions P and Q are differentiable in some open set $U \subset \mathbb{R}^2$.

Theorem 3.1 (Bendixon's criterion). *If the expression*

$$\frac{\partial P}{\partial x} + \frac{\partial Q}{\partial y} \neq 0$$

does not change the sign in the set Ω ($\Omega \subset U \subset \mathbb{R}^2$), then periodic solutions inside this region do not exist.

Proof. (Ad absurdum). Suppose that some periodic solution γ lying entirely in the set Ω does exist. Denote by D the interior of the set bounded by γ. The following equalities hold:

$$0 \neq \iint_D \left(\frac{\partial P}{\partial x} + \frac{\partial Q}{\partial y} \right) dx\, dy = \oint_\gamma (P\, dy - Q\, dx) .$$

The system (3.4) can also be presented in the form

$$dt = \frac{dx}{P} = \frac{dy}{Q}.$$

It appears from this presentation that $P\, dy - Q\, dx \equiv 0$ on solutions of (3.4). Hence we get the contradiction.

Now we will give without proof the Poincare-Bendixon theorem, facilitating the study of the existence of periodic trajectories in a given area.

Theorem 3.2. *Let $\Gamma_+ = \{x[t],\, y[t]\}_{t \geq t_0}$, be the half-trajectory of the system (3.4), lying entirely in the bounded connected set D. Let us suppose in addition that there are no stationary points in D. Then one of the two options holds: either Γ_+ is a periodic orbit, or D has at least one periodic orbit. In the latter case, Γ_+ tends to the periodic orbit lying inside the set D.*

Corollary 3.1. *Suppose that D is a bounded connected set which does not contain stationary points. If the vector field $\{P(x,\, y),\, Q(x,\, y)\}$ is directed inside D at each point of its boundary ∂D, then at least one periodic orbit exists in this set.*

Problem.

1. Apply the Bendixon criterion for the following vector fields:

(a)

$$P = -x + 4y; \quad Q = -x - y^3; \quad \Omega \text{ is a ring } r_1 < \sqrt{x^2 + y^2} < r_2,$$

(b)

$$P = -2\, x\, e^{x^2+y^2}; \quad Q = -2\, y\, e^{x^2+y^2}; \ \Omega = R^2.$$

2. Using the Poincare-Bendixon theorem, study whether the system

$$\dot{x} = -x\left(x^2 + y^2 - 2x - 3\right) + y, \quad \dot{y} = -y\left(x^2 + y^2 - 2x - 3\right) - x$$

has periodic solutions in the set $1/2 < \sqrt{x^2+y^2} < 4$. Verify the answer obtained by visualizing the vector field inside the square $x \in [-2, 4]$, $y \in [-3, 3]$.

3. Using the Poincare-Bendixon theorem, show that the system

$$\dot{x} = x + y - x\left(x^2 + 2y^2\right), \quad \dot{y} = -x + y - y\left(x^2 + 2y^2\right)$$

has the periodic solutions inside the set $1/2 < \sqrt{x^2+y^2} < 2$. Validate the result obtained visualizing the vector field inside the set $x \in [-2,\ 2]$, $y \in [-1.5,\ 1.5]$.

Remark. The correctness of the answers obtained while solving the problems 2, 3 can be checked (as recommended) by constructing phase portraits in the vicinity of the examined areas using the StreamPlot command as is shown below:

"Checking the problem 2"

```
StreamPlot [{-(x (x^2 + y^2 - 2 x - 3) - y), -(y (x^2 + y^2 - 2 x - 3) + x)}, {x, -2, 4}, {y, -3, 3}]
```

"Checking the problem 3"

```
StreamPlot[{ + y - x (x^2 + 2 y^2), -x + y - y (x^2 + 2 y^2)}, {x, -2, 2}, {y, -1.5, 1.5}]
```

3.5 Andronov-Hopf bifurcation

3.5.1 *Introduction*

Let us consider the dynamical system

$$\frac{dX}{dt} = F_\mu(X), \qquad X \in \mathbb{R}^n, \quad \mu \in \mathbb{R}^k. \tag{3.5}$$

The qualitative nature of the vector field and the types of the stationary points depend as a rule on the parameter values. The changes that occur when the parameters exceed some critical values are called *bifurcations*, while the parameters' values at which these changes occur are called *bifurcation values.* If the changes of the vector field take place in a small neighborhood of a stationary point and, in order to "unfold" bifurcation, it is sufficient to make small changes of the parameters' values in the neighborhood of their bifurcation values, then we say that we are dealing with the *local bifurcation.* If the dimension of the phase space is small ($n = 1, 2$), then it is convenient to extend the phase space by including a set of parameters.

Example 3.1. Let us consider the system

$$\frac{dx}{dt} = x\left(\mu - x^2\right), \qquad x \in \mathbb{R}^1, \quad \mu \in \mathbb{R}^1.$$

If $\mu < 0$, then the system has one stationary point $x_0 = 0$, which is the stable node. If $\mu > 0$, then the point $x_0 = 0$ changes its type, turning into the source; simultaneously a pair of stable nodes appear, having the coordinates $x_{1,2} = \pm\sqrt{\mu}$. The dynamical system thus undergoes the bifurcation called *pitchfork bifurcation.* The situation is illustrated in Fig. 3.1. In this

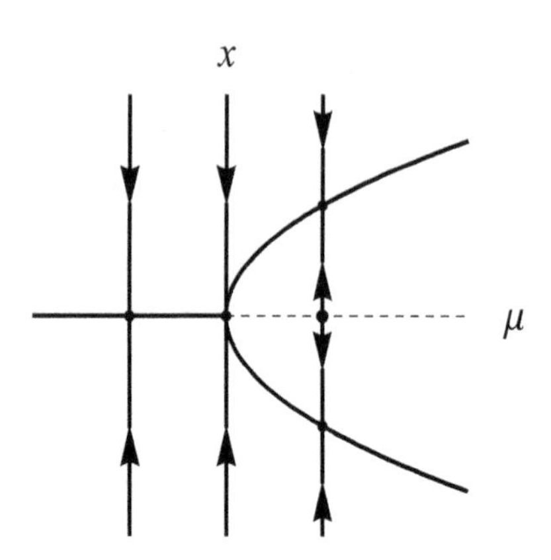

Fig. 3.1 Scheme of the pitchfork bifurcation.

figure, the parameter values are plotted on the horizontal axis, while the corresponding phase portraits are plotted on the vertical lines.

A sign that the bifurcation can take place near the stationary point X_0 of the system (3.5) at μ_0, is the presence among the eigenvalues of the linearization matrix $D\,f_{\mu_0}(X_0)$ one or more eigenvalues with zero real part. In the simplest case, the linearization matrix has one zero eigenvalue or a pair of purely imaginary eigenvalues. In the first case, the linearized system can be reduced by the linear transformation $X \to Y$ to the form

$$Y' = \begin{pmatrix} 0 & 0 \\ 0 & A \end{pmatrix} Y, \tag{3.6}$$

where A is quadratic matrix with $(n-1)$ rows and columns such that all of its eigenvalues have nonzero real parts. In the second case the system can be presented in the form

$$Y' = \begin{pmatrix} \begin{pmatrix} 0 & -\omega \\ \omega & 0 \end{pmatrix} & 0 \\ 0 & B \end{pmatrix} Y, \tag{3.7}$$

where B is quadratic matrix with $(n-2)$ rows and columns, whose eigenvalues have nozero real parts. Before we proceed to the analysis of the Hopf bifurcation in a system whose linear part comes down to the form (3.7), we need to develop two techniques that are necessary to perform such an analysis.

3.5.2 *Central manifold*

The study of local bifurcation comes down to the analysis of the central manifold W^c_{loc}, because small changes in the system cannot change the nature of vector fields on invariant manifolds $W^{s,u}_{loc}$. As is well-known, there is always possible to orthogonalize the E^s, E^u, and E^c invariant subspaces, using, for example, the Gram-Schmidt method. "Orthogonalization" of the corresponding tangent manifolds by linear transformations, in general, is not possible. On the other hand, it is possible to perform orthogonalization in the vicinity of zero using the technique described below, which allows to "project" the vector field onto the central manifold.

The algorithm is as follows. Suppose that the system under study is reduced to the form

$$\begin{cases} \dfrac{dx}{dt} = Bx + f(x,y), \quad x \in \mathbb{R}^n, \quad y \in \mathbb{R}^m, \\ \dfrac{dy}{dt} = Cy + g(x,y), \end{cases} \tag{3.8}$$

where B, C are square matrices of dimensions n and m, correspondingly; all the eigenvalues of the matrix B have zero real parts, while all the eigenvalues of the matrix C have negative real parts. The central manifold is presented in the form of the mapping $h : \mathbb{R}^n \supset U \to \mathbb{R}^m$:

$$W^c_{loc} = \{(x,\, y) : \; y = h(x); \; h(0) = 0, \;\; Dh(0) = 0\}. \tag{3.9}$$

The map $h(x)$ can be found approximately using the equalities

$$\frac{dy}{dt} = Dh(x)\frac{dx}{dt} = Dh(x)\left[Bx + f\left(x, h(x)\right)\right] = Ch(x) + g\left(x, h(x)\right),$$

or, which is the same,

$$Dh(x)\left[Bx + f\left(x, h(x)\right)\right] - Ch(x) - g\left(x, h(x)\right) = 0. \tag{3.10}$$

Example 3.2. As an example, let us consider the system

$$\begin{pmatrix} x \\ y \\ z \end{pmatrix}' = \begin{pmatrix} 0 & -1 & 0 \\ 1 & 0 & 0 \\ 0 & 0 & -1 \end{pmatrix} \begin{pmatrix} x \\ y \\ z \end{pmatrix} + \begin{pmatrix} xz \\ yz \\ z^2 - x^2 - y^2 \end{pmatrix}, \tag{3.11}$$

where

$$B = \begin{pmatrix} 0 & -1 \\ 1 & 0 \end{pmatrix}, \qquad C = -1,$$

$f(x, y, z) = (xz, yz)^{tr}$, $g(x, y, z) = z^2 - x^2 - y^2$. This system can be written in the following form:

$$\begin{pmatrix} x \\ y \end{pmatrix}' = \begin{pmatrix} 0 & -1 \\ 1 & 0 \end{pmatrix} \begin{pmatrix} x \\ y \end{pmatrix} + \begin{pmatrix} x\,z \\ y\,z \end{pmatrix}, \tag{3.12}$$

$$z' = -z + z^2 - x^2 - y^2. \tag{3.13}$$

Let

$$h(x, y) = a\,x^2 + b\,x\,y + c\,y^2 + d\,x^3 + f\,x^2\,y + gx\,y^2 + k\,y^3 + O(x^4,\, y^4).$$

Then the equation (3.10) now takes the form

$$(h_x,\, h_y)\left\{\begin{pmatrix} 0 & -1 \\ 1 & 0 \end{pmatrix}\begin{pmatrix} x \\ y \end{pmatrix} + \begin{pmatrix} x\,h(x,\, y) \\ y\,h(x,\, y) \end{pmatrix}\right\} - \{h(x,\, y) - x^2 - y^2\} = 0.$$

Using the *Mathematica* package, we get the following approximate solution (see file PM1_5.nb):

$$h(x,\, y) = -(x^2 + y^2) + O(x^4,\, y^4).$$

Thus up to $O(x^4,\, y^4)$, the projection of the system onto the central manifold has the form

$$\begin{pmatrix} x \\ y \end{pmatrix}' = \begin{pmatrix} 0 & -1 \\ 1 & 0 \end{pmatrix} \begin{pmatrix} x \\ y \end{pmatrix} - (x^2 + y^2)\begin{pmatrix} x \\ y \end{pmatrix}. \tag{3.14}$$

Problem.

Consider the following dynamical system:

$$\begin{pmatrix} x_1 \\ x_2 \\ x_3 \end{pmatrix}' = \begin{pmatrix} 0 & -1 & -1 \\ 1 & 0 & 0 \\ 0 & 0 & -1 \end{pmatrix} \begin{pmatrix} x_1 \\ x_2 \\ x_3 \end{pmatrix} + \begin{pmatrix} 0 \\ 0 \\ x_3\,x_1 - \alpha(x_1^2 + x_2^2) \end{pmatrix}$$

$$= \hat{M}\begin{pmatrix} x_1 \\ x_2 \\ x_3 \end{pmatrix} + \text{nonl. terms.} \tag{3.15}$$

Note that the matrix $\hat{M}$ has a pair of pure imaginary eigenvalues $\lambda_{1,2} = \pm\, i$ and one negative eigenvalue $\lambda_3 = -1$.

(1) Using the substitution

$$\begin{pmatrix} x_1 \\ x_2 \\ x_3 \end{pmatrix} = (R, -T, W)\begin{pmatrix} x \\ y \\ z \end{pmatrix},$$

where $V = R+iT$ is the eigenvector of the matrix $\hat{M}$, corresponding to the eigenvalue $\lambda_1 = i$, while W is the eigenvector of $\hat{M}$ corresponding to the eigenvalue $\lambda_3 = -1$, reduce the system to the canonical form

$$\begin{pmatrix} x \\ y \end{pmatrix}' = \begin{pmatrix} 0 & -1 \\ 1 & 0 \end{pmatrix} \begin{pmatrix} x \\ y \end{pmatrix} + F(x, y, z) = P(x, y, z),$$

$$z' = -z + G(x, y, z) = Q(x, y, z).$$

(2) Find out the function

$$h(x, y) = \sum_{a+b=2} M_{a,b} x^a y^b + \sum_{m+n=3} N_{m,n} x^m y^n + O(|x|^4 + |y|^4)$$

by solving the equation

$$(h_x, h_y) \cdot P(x, y, h(x, y)) - Q(x, y, h(x, y)) = 0$$

up to $O(|x|^4 + |y|^4)$.

(3) Project the system onto the central manifold.

3.5.3 *Perturbations of the center*

When classifying simple stationary points in $\mathbb{R}^2$, we drew attention to the fact that the vicinity of the stationary point is not structurally stable in the case when the linearization matrix of the system has eigenvalues with zero real part. The only stationary point of this kind is the center, which can be reduced by the linear change of variables to the following form:

$$\frac{d}{dt} \begin{pmatrix} y_1 \\ y_2 \end{pmatrix} = \begin{pmatrix} 0 & -\omega \\ \omega & 0 \end{pmatrix} \begin{pmatrix} y_1 \\ y_2 \end{pmatrix}. \tag{3.16}$$

The qualitative properties of the vector field can be changed in this case both by the disturbance of the linear part and by taking into account the nonlinear terms rejected during the linearization. For example, the system

$$\frac{d}{dt} \begin{pmatrix} y_1 \\ y_2 \end{pmatrix} = \begin{pmatrix} \epsilon & -\omega \\ \omega & \epsilon \end{pmatrix} \begin{pmatrix} y_1 \\ y_2 \end{pmatrix},$$

differing from (3.16) by the factor ϵ changes its properties at arbitrarily small values of the disturbing parameter. All solutions of this system asymptotically tend to zero (if $\epsilon < 0$), or to infinity (if $\epsilon > 0$). The nature of solutions can also be changed by adding rejected nonlinear terms. Such an effect is almost always observed (except for Hamiltonian systems). For example, it can be shown that the non-linear system

$$\frac{d}{dt} \begin{pmatrix} y_1 \\ y_2 \end{pmatrix} = \begin{pmatrix} 0 & -\omega \\ \omega & 0 \end{pmatrix} \begin{pmatrix} y_1 \\ y_2 \end{pmatrix} - (y_1^2 + y_2^2) \begin{pmatrix} y_1 \\ y_2 \end{pmatrix} \tag{3.17}$$

in polar coordinates $y_1 = \rho \cos\theta$, $y_2 = \rho \sin\theta$ takes the form

$$\frac{d\rho}{dt} = -\rho^3, \qquad \frac{d\theta}{dt} = \omega.$$

Due to the fact that in the polar representation the variables have separated, it is possible to find an exact solution to the system in question. Assuming in addition that $y_1(0) = [2\,t_0]^{-1/2}$, $y_2(0) = 0$ we obtain the following solution:

$$y_1 = \rho(t)\cos[\omega\,t], \qquad y_2 = \rho(t)\sin[\omega\,t], \qquad \rho(t) = \frac{1}{\sqrt{2(t+t_0)}}. \tag{3.18}$$

Equation (3.18) describes the solution asymptotically tending to zero as $t \to +\infty$. An interesting situation arises when the disturbance of the linear part and the addition of the rejected nonlinear part act as if in "out-of-phase". In the considered example of a linearly disturbed system (3.16) it happens when $\epsilon > 0$. As a result of action into the system of two opposite factors, a nonlinear periodic solution called *limit cycle* can appear. Since in the vast majority of cases periodic solutions cannot be described analytically, the conditions assuring the emergence of the limit cycle are a very valuable tools enabling to capture the periodic solution "in embryo". The following subsections are devoted to a brief overview of the birth of a limit cycle as a result of the Andronov-Hopf bifurcation.

3.5.4 *Reduction of the dynamical system to the canonical form*

Let us consider two-dimensional system

$$\begin{pmatrix} x_1 \\ x_2 \end{pmatrix}' = F_\mu(x_1, x_2). \tag{3.19}$$

We assume that (3.19) has a stationary point at the origin and at $\mu = 0$ the linearization matrix $DF_0[0] = A$ has the pair of purely imaginary eigenvalues $\lambda_{1,2} = \pm i\omega$, while at $|\mu| \ll 1$ the pair of complex eigenvalues $\lambda = \mu \pm i\omega(\mu)$. We assume in addition that $\omega = O(1)$ and that $\omega(\mu)$ in a differentiable way depends on the parameter μ, and hence $\omega(\mu) = \omega + O(|\mu|)$.

<u>**Remark.**</u> If any of the above assumptions is not fulfilled, for example, the stationary point has non-zero coordinates (x_{10}, x_{20}), then using the change of variables $\bar{x}_1 = x_1 - x_{10}$, $\bar{x}_2 = x_2 - x_{20}$ one can put the stationary point into the origin. The same is true for μ: if the linearization matrix

has a pair of purely imaginary eigenvalues at $\mu = \mu_0 \neq 0$, then the above assumptions can be met by passing to the parameter $\bar{\mu} = \mu - \mu_0$.

Below we will describe the procedure enabling to pass to the canonical coordinates in which the linearization matrix has a specific form. We begin with the system

$$\begin{pmatrix} x_1 \\ x_2 \end{pmatrix}' = A \begin{pmatrix} x_1 \\ x_2 \end{pmatrix} + \begin{bmatrix} f_1(x_1, x_2) \\ f_2(x_1, x_2) \end{bmatrix}, \tag{3.20}$$

where $A = D\,F_\mu[0]$. As above, we assume that the matrix A has a pair of the complex eigenvalues $\lambda_{1,2} = \mu \pm i\,\omega$, and that $V_\pm = R \pm i\,I$, $R, I \in \mathbb{R}^2$ are the eigenvectors corresponding to them. From this we have the equality

$$A(R + iI) = (\mu + i\,\omega)(R + iI),$$

which can also be presented in the following way:

$$A\,R = -\omega\,I + \mu\,R, \qquad A\,I = \omega\,R + \mu\,I. \tag{3.21}$$

Let us define the matrices $P = (R, -I)$ and P^{-1}, where

$$P^{-1} = \begin{pmatrix} R^{-1} \\ -I^{-1} \end{pmatrix},$$

and the vectors $R^{-1} = (a_1, a_2)$, $I^{-1} = (b_1, b_2)$ satisfy the conditions

$$R^{-1} \cdot R = 1, \quad R^{-1} \cdot I = 0, \quad I^{-1} \cdot R = 0, \quad I^{-1} \cdot I = 1. \tag{3.22}$$

Using the above formulas, we can easily prove the correctness of the following statement:

Lemma 3.1. *The change of variables*

$$\begin{pmatrix} x \\ y \end{pmatrix} = P^{-1} \begin{pmatrix} x_1 \\ x_2 \end{pmatrix}$$

reduces the matrix A to the following form:

$$\tilde{A} = \begin{pmatrix} \mu & -\omega \\ \omega & \mu \end{pmatrix}.$$

Proof. Acting on the system (3.20) with the operator P^{-1} from the left, we get:

$$\begin{pmatrix} x \\ y \end{pmatrix}' = P^{-1} \begin{pmatrix} x_1 \\ x_2 \end{pmatrix}' = P^{-1}\,A\,P \begin{pmatrix} x \\ y \end{pmatrix} + P^{-1} \begin{bmatrix} f_1(P(x, y)^{tr}) \\ f_2((P(x, y)^{tr}) \end{bmatrix},$$

where $(\cdot,\,\cdot)^{tr}$ means the transposition. Let us consider now the linearization matrix. Using the formulas (3.21), (3.22) we obtain:

$$P^{-1}\,A\,P = P^{-1}\,A\,(R,\,-I) = \begin{pmatrix} R^{-1} \\ -I^{-1} \end{pmatrix}(-\omega I + \mu R,\,-\omega R - \mu I) = \begin{pmatrix} \mu & -\omega \\ \omega & \mu \end{pmatrix}.$$

So the system (3.20) attains in new variables the following form:

$$\begin{pmatrix} x \\ y \end{pmatrix}' = \begin{pmatrix} \mu & -\omega \\ \omega & \mu \end{pmatrix}\begin{pmatrix} x \\ y \end{pmatrix} + \begin{bmatrix} f(x,\,y) \\ g(x,\,y) \end{bmatrix}, \tag{3.23}$$

where

$$f(x,\,y) = \sum_{a+b=2} f^2_{a\,b} x^a\,y^b + \sum_{m+n=3} f^3_{m\,n} x^m\,y^n + \ldots,$$

$$g(x,\,y) = \sum_{a+b=2} g^2_{a\,b} x^a\,y^b + \sum_{m+n=3} g^3_{m\,n} x^m\,y^n + \ldots.$$

3.5.5 *Normal form. Criterion of the appearance of the limit cycle*

The crucial step in the analysis of the Andronov-Hopf bifurcation is construction of the *normal form* corresponding to the system (3.23). Having the normal form, one can formulate the conditions leading to the limit cycle formation and to test its stability. When constructing the normal form, it is instructive to pass to the complex variable $z = x + i\,y$. This transition allows us to present the system (3.23) as

$$\begin{cases} z' = \lambda z + F(z,\bar{z}), \\ \bar{z}' = \bar{\lambda}\,\bar{z} + \bar{F}(\bar{z},\,z), \end{cases} \tag{3.24}$$

where

$$F(z,\bar{z}) = f\left(\frac{z+\bar{z}}{2},\,\frac{z-\bar{z}}{2\,i}\right) + i\,g\left(\frac{z+\bar{z}}{2},\,\frac{z-\bar{z}}{2\,i}\right). \tag{3.25}$$

We are interested in the first two terms of the decomposition of $F(z,\bar{z})$ into the Taylor series, and therefore we will present it in the following form:

$$F(z,\bar{z}) = \sum_{a+b=2} F^2_{a\,b} z^a\,\bar{z}^b + \sum_{m+n=3} F^3_{m\,n} z^m\,\bar{z}^n + O(|z|^4).$$

Let us make a change of variables

$$\begin{pmatrix} z \\ \bar{z} \end{pmatrix} = \begin{pmatrix} w \\ \bar{w} \end{pmatrix} + P_2(w,\,\bar{w}), \tag{3.26}$$

where

$$P_2(w,\,\bar{w}) = \begin{pmatrix} \sum\limits_{a+b=2} P^2_{a\,b} w^a\,\bar{w}^b \\ \sum\limits_{a+b=2} \bar{P^2_{a\,b}}\,\bar{w}^a\,w^b \end{pmatrix}.$$

The coefficients $P^2_{a\,b}$ are arbitrary for now complex numbers. Inserting this formula into the equation (3.24), we get:

$$[I_2 + D\,P_2]\begin{pmatrix} w \\ \bar{w} \end{pmatrix}' = \Lambda\left(\begin{pmatrix} w \\ \bar{w} \end{pmatrix} + P_2\right) + F_2[w,\,\bar{w}] + O(|w|^3), \tag{3.27}$$

where I_2 is two-dimensional unit matrix,

$$F_2(w,\,\bar{w}) = \begin{pmatrix} \sum\limits_{a+b=2} F^2_{a\,b} w^a\,\bar{w}^b \\ \sum\limits_{a+b=2} \bar{F^2_{a\,b}}\,\bar{w}^a\,w^b \end{pmatrix},$$

$$\Lambda = \begin{pmatrix} \mu + i\,\omega & 0 \\ 0 & \mu - i\,\omega \end{pmatrix},$$

$$[D\,P_2] = \begin{bmatrix} \sum_{a+b=2} a\,P^2_{a\,b} w^{a-1}\,\bar{w}^b & \sum_{a+b=2} b\,P^2_{a\,b} w^a\,\bar{w}^{b-1} \\ \sum_{a+b=2} b\,\bar{P}^2_{a\,b}\bar{w}^a\,w^{b-1} & \sum_{a+b=2} a\,\bar{P}^2_{a\,b}\bar{w}^{a-1}\,w^b \end{bmatrix}.$$

From now on, we will treat the transformation (3.26) as asymptotic, and drop out the terms of the order $O(|w|^3)$. Multiplying the equation (3.27) by the operator

$$I_2 - D\,P_2 = (I_2 + D\,P_2)^{-1} + O(|w|^2)$$

from the left, we get within the given precision equation

$$\begin{pmatrix} w \\ \bar{w} \end{pmatrix}' = \Lambda\begin{pmatrix} w \\ \bar{w} \end{pmatrix} + R_2 + O(|w|^3), \tag{3.28}$$

where

$$R^1_2 = \sum_{a+b=2} \left[(\lambda - a\lambda - b\bar{\lambda})\,P^2_{a\,b} + F^2_{a\,b}\right] w^a\,\bar{w}^b, \tag{3.29}$$

$R^2_2 = \bar{R^1_2}$. Let us assume that $\lambda = i\,\omega$. This assumption will not affect the final result, as the parameter μ is independent and therefore we can choose it arbitrarily small. In new variables, the coefficient of the monomial $w^a\,\bar{w}^b$ will disappear if we make the following choice:

$$P^2_{a\,b} = -\frac{F^2_{a\,b}}{i\,\omega(1-a+b)}. \tag{3.30}$$

The formula (3.30) will be well defined if $1-a+b\neq 0$. Monomials $w^a\,\bar{w}^b$ for which the formula (3.30) contains singularity are called *resonant monomials.* The presence of second-order resonant monomials can be analyzed by solving the system of algebraic equations

$$\begin{aligned} 1-a+b&=0,\\ a+b&=2. \end{aligned}$$

The above system has the unique solution $a = 3/2$, $b = 1/2$, so it is not satisfied by any pair of numbers $a,\ b\in \mathbb{N}\bigcup 0$. It follows that all monomials of degree 2 are removable and with appropriate selection of constants $P^2_{a\,b}$ the system (3.28) can be reduced to the form

$$\begin{aligned} w' &= i\,\omega\,w + \sum_{m+n=3} R^3_{m\,n} w^m \bar{w}^n + O(|w|^4),\\ \bar{w}' &= -i\,\omega\,\bar{w} + \sum_{m+n=3} \bar{R}^3_{m\,n} \bar{w}^m w^n + O(|w|^4). \end{aligned} \tag{3.31}$$

Note that the coefficients $R^3_{m\,n}$ after the asymptotic transformations are not identical with the corresponding coefficients of the equation (3.28) hidden in the term $O(|w|^3)$.

In order to remove non-resonant monomials of the third order, we use the substitution

$$w = v + P^3 \equiv v + \sum_{m+n=3} P^3_{m\,n} v^m\,\bar{v}^n.$$

Inserting it into the formula (3.31) and multiplying the result obtained by the operator $I_2 - DP_3 \approx (I_2 + DP_3)^{-1}$ from the left, we get the system

$$\begin{pmatrix} v \\ \bar{v} \end{pmatrix}' = \Lambda \begin{pmatrix} v \\ \bar{v} \end{pmatrix} + S_3 + O(|v|^4), \tag{3.32}$$

where

$$S^1_3 = \sum_{m+n=3} \left[i\,\omega\,(1-m+n)\,P^3_{m\,n} + R^3_{m\,n}\right] v^m\,\bar{v}^n, \tag{3.33}$$

$S^2_3 = \bar{S^1_3}$. It turns out that most of the third order monomials can also be removed by selecting $P^3_{m\,n}$ in the form

$$P^3_{m\,n} = -\frac{R^3_{m\,n}}{i\,\omega\,(1-m+n)}.$$

In order to check the presence of resonance monomials, we consider the system of equations

$$1 - m + n = 0,$$
$$m + n = 3.$$

The unique solution to this system takes the form $m = 2$, $n = 1$. These numbers belong to the set $\mathbb{N} \bigcup 0$, and therefore the monomial $v^2 \bar{v}$ cannot be removed. So, with proper choice of the coefficients $P^3_{m\,n}$ the first solution of the system (with the re-consideration of the disturbing parameter μ) can be presented as

$$v' = (\mu + i\,\omega)\,v + (a + i\,b)|v|^2\,v + O(|v|^4). \tag{3.34}$$

The variable $v(t)$ can be presented in the trigonometric form as

$$v[t] = \rho(t)e^{i\phi(t)}.$$

Inserting this formula into (3.34) we get:

$$\rho'\,e^{i\phi(t)} + i\,\rho\,\phi'(t)\,e^{i\phi(t)} = (\mu + i\,\omega)\,\rho\,e^{i\phi(t)} + (a + i\,b)\,\rho^3\,e^{i\phi(t)} + O(\rho^4).$$

By writing the appropriate equations for the real and imaginary parts, we get the dynamical system

$$\begin{aligned} \rho' &= \rho(\mu + a\,\rho^2) + O(|\rho|^4), \\ \phi' &= \omega + O(|\rho|^2). \end{aligned} \tag{3.35}$$

The parameter a, appearing in the equations (3.34) and (3.35), is the real part of so-called the *first Floquet index* $C_1 = a + i\,b$. It plays a leading role in studying the appearing of the limit cycle and determining its stability. Let us consider the equation

$$\rho' = \rho(\mu + a\,\rho^2).$$

Regardless of the values of the parameters, it always has the stationary point $\rho_0 = 0$. Besides, it also has the stationary point $\rho_1 = \sqrt{-\mu a}$ provided that $-\mu a > 0$. And here we have two options:

(1) $a < 0, \quad \mu > 0$. In this case the stationary point ρ_0 is unstable, while ρ_1 is stable. The trivial solution $\rho(t) = \rho_1 = \text{const} > 0$, together with the approximate solution

$$\phi = (t - t_0)\,\omega$$

of the second equation of system (3.35), present on the complex plane $(Re(v),\ Im(v))$ the periodic solution. Due to the stability of the stationary point ρ_1, all nearby trajectories in the course of evolution are attracted to the periodic solution described above, which is what characterizes its stability (see Fig. 3.2).

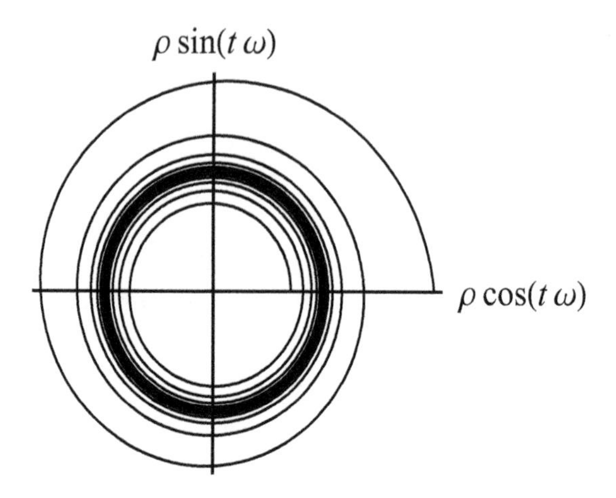

Fig. 3.2 Stable limit cycle in the system (3.35), obtained with $a = 1$, $\mu = -0.1$, $\omega = 1$ after the passage to physical variables.

(2) $a > 0$, $\mu < 0$. In this case the stationary point ρ_0 is stable, while the stationary point ρ_1 is unstable. So the unstable limit cycle corresponds to the solution $\rho(t) = \rho_1 > 0$, $\phi = (t - t_0)\,\omega$ in the physical variables $(Re(v),\ Im(v))$.

It is shown below how the parameter a can be calculated. In the representation (3.23) it can be done using the formula:[1]

$$\begin{aligned} a = Re\, C_1 \; &= \frac{1}{16}\left[f_{xxx} + f_{xyy} + g_{xxy} + g_{yyy}\right] \\ &+ \frac{1}{16\,\omega}\left[f_{xy}\left(f_{xx} + f_{yy}\right) - g_{xy}\left(g_{xx} + g_{yy}\right) - f_{xx}\, g_{xx} + f_{yy}\, g_{yy}\right], \end{aligned} \tag{3.36}$$

where

$$f_{xx} = \frac{\partial^2 f}{\partial x^2}(0,\,0), \quad f_{xy} = \frac{\partial^2 f}{\partial x\,\partial y}(0,\,0) \quad \text{and etc.}$$

Finally, we will present the fundamental theorem formulated by A. Andronov in 1937 and independently by E. Hopf in 1942.

Theorem 3.3. *Let $\dot{x} = f_\mu(x)$ be the dynamical system in $\mathbb{R}^n$, depending on the parameter $\mu \in \mathbb{R}$. Assume that for all the values μ belonging to some vicinity of $\mu_0 = 0$ there exists a stationary point $p \in \mathbb{R}^n$ such, that*

[1] Details of the derivation of this formula can be found in the appendix to Chapter 3.4 of the book [Guckenheimer and Holmes (1987)].

(1) $D_x\left[f_0(x)\right]\Big|_{x=p}$ *has a pair of purely imaginary eigenvalues, while the remaining eigenvalues have negative real parts.*

(2) The eigenvalues $\lambda(\mu)$, $\bar{\lambda}(\mu)$, *which are purely imaginary at* $\mu = 0$, *are differentiable with respect to* μ *and, besides,*

$$\frac{d}{d\,\mu}\,Re\left[\lambda(\mu)\right]\Big|_{\mu=0} = d \neq 0.$$

Then there exist the coordinates at which the central manifold up to $O[(x^2+y^2)^2]$ *has the following Taylor decomposition*

$$\begin{pmatrix} x \\ y \end{pmatrix}' = \begin{pmatrix} \mu & -\omega \\ \omega & \mu \end{pmatrix}\begin{pmatrix} x \\ y \end{pmatrix} + (x^2+y^2)\begin{pmatrix} a & -b \\ b & a \end{pmatrix}\begin{pmatrix} x \\ y \end{pmatrix}.$$

If $a \neq 0$, *then the system at* $|\mu| \ll 1$ *possesses the limit cycle which is stable as* $a < 0$ *and unstable when* $a > 0$.

3.5.6 *Examples of appearance of limit cycles*

Below we present two relatively easy examples illustrating the techniques enabling to investigate the emergence of the limit cycles. A technically more complicated example with comprehensive studies of a three-dimensional system can be found in the file PM1_6.nb.

Example 3.3. Let us consider the system

$$\begin{aligned} \dot{x} &= \epsilon\,x - y + \sigma\,y^2 + \beta\,x^2\,y, \\ \dot{y} &= x - \gamma\,y^2 + \delta\,x\,y - y^3. \end{aligned} \tag{3.37}$$

If $\epsilon = 0$, then the stationary point (0, 0) is a center. We want to determine the real part of the first Floquet index, and to analyze the conditions of the Andronov-Hopf bifurcation formation and the dependence of the stability of the limit cycle on the parameters σ, β, γ and δ.

The eigenvalues of the linearization matrix

$$A = \begin{pmatrix} \epsilon & -1 \\ 1 & 0 \end{pmatrix}$$

of the system (3.37) take the form

$$\lambda_{1,\,2} = \frac{\epsilon}{2} \pm i\,\sqrt{1-\epsilon^2/4} = \frac{\epsilon}{2} \pm i\,\left[1+O(\epsilon^2)\right].$$

Now we solve the vector equation

$$A(R+i\,J) = \left\{\frac{\epsilon}{2} + i[1+O(\epsilon^2)]\right\}\,[R+i\,J],$$

or the equivalent system

$$A\begin{pmatrix}1\\ \alpha\end{pmatrix}=\begin{pmatrix}\epsilon-\alpha\\ 1\end{pmatrix}=\frac{\epsilon}{2}\begin{pmatrix}1\\ \alpha\end{pmatrix}-\begin{pmatrix}j_1\\ j_2\end{pmatrix}$$

$$A\begin{pmatrix}j_1\\ j_2\end{pmatrix}=\begin{pmatrix}1\\ \alpha\end{pmatrix}+\frac{\epsilon}{2}\begin{pmatrix}j_1\\ j_2\end{pmatrix}$$

assuming that $R=(1,\,\alpha)^{tr}$, $J=(j_1,\,j_2)^{tr}$ (note that we drop all the terms of the order $O(\epsilon^2)$) and higher during the calculations). From the first equation we get

$$(j_1,\,j_2)^{tr}=(\alpha-\epsilon/2,\,\alpha\,\epsilon/2-1)^{tr}.$$

Inserting it into the second equation, we obtain that $\alpha=0$. So

$$P=\begin{pmatrix}1 & \epsilon/2\\ 0 & 1\end{pmatrix}=\begin{pmatrix}1 & 0\\ 0 & 1\end{pmatrix}+\Lambda(\epsilon).$$

With this transformation, the nonlinear part up to $O(|\epsilon|)$ remains unchanged, and therefore the formula (3.36) can be calculated directly from the system (3.37). Calculating the appropriate derivatives and substituting the result into the formula (3.36) we get

$$16\,a=2\,\gamma(\delta-2\,\sigma)-6.$$

Thus, the stable limit cycle appears at $\epsilon>0$ in the case when the inequality

$$\gamma(\delta-2\,\sigma)<3$$

takes place. Let us note, that the parameter β does not influence the stability of the limit cycle. The result obtained is confirmed by the numerical experiment given below.

Cell 3.3.

```
"Appearance of the limit cycle in the dynamical system (3.37)"
In[1]: ClearAll["Global`*"]; Clear[σ, β, δ];
In[2]:
  {σ, δ} = {1, 1};
  Manipulate [eq1a = x'[t]==ϵ x[t]-y[t]+σ y[t]^2+β x[t]^2 y[t];
  eq2a = y'[t]==x[t]-γ y[t]^2+δ x[t] y[t]-y[t]^3;
  sol = NDSolve[{eq1a, eq2a, x[0] == a0, y[0] == 0}, {x[t], y[t]}, {t, 0, 250}];
  ParametricPlot[{x[t], y[t]}/.sol, {t, 0, T}, PlotRange → Full],
  {γ, {-1, 0.5, 1.5}},{β, {-1, 0.5, 1.5}},{ϵ, -0.1, 0.6},
  {T, {50, 100, 150, 200, 250}}, {a0, {0.05, 0.1, 0.2, 0.3, 0.4}}]
```

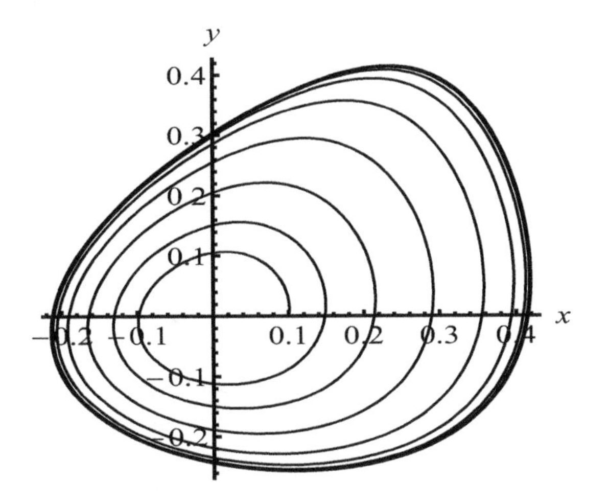

Example 3.4. Let us show that with a certain relationship between the parameters the limit cycle appears in the system

$$\begin{pmatrix} x \\ y \end{pmatrix}' = \begin{pmatrix} 0 & -1 \\ b & \mu \end{pmatrix}\begin{pmatrix} x \\ y \end{pmatrix} - \begin{pmatrix} 0 \\ x\,y^2 \end{pmatrix} = \hat{M}\begin{pmatrix} x \\ y \end{pmatrix} - \begin{pmatrix} 0 \\ x\,y^2 \end{pmatrix}, \quad b > 0. \tag{3.38}$$

The system possesses a stationary point at the origin, which is a center when $\mu = 0$. We know that the linear transformation that brings the system to the canonical form has the form $P = (R, -J)$, where $R + i\,J$ is the eigenvector of the matrix $\hat{M}$:

$$\begin{pmatrix} 0 & -1 \\ b & \mu \end{pmatrix}(R + i\,J) = \lambda\,(R + i\,J),$$

where $\lambda = \frac{\mu}{2} + i\sqrt{b} + O(\mu^2)$ is the eigenvalue of the matrix $\hat{M}$. Assuming that $R = (1,\, \alpha)^{tr}$, we get, up to $O(\mu^2)$, the following system defining $J = (j_1,\, j_2)^{tr}$:

$$\begin{pmatrix} 0 & -1 \\ b & \mu \end{pmatrix}\begin{pmatrix} 1 \\ \alpha \end{pmatrix} = \begin{pmatrix} -\alpha \\ b + \mu\alpha \end{pmatrix} = \frac{\mu}{2}\begin{pmatrix} 1 \\ \alpha \end{pmatrix} - \sqrt{b}\frac{\mu}{2}\begin{pmatrix} j_1 \\ j_2 \end{pmatrix},$$

$$\begin{pmatrix} 0 & -1 \\ b & \mu \end{pmatrix}\begin{pmatrix} j_1 \\ j_2 \end{pmatrix} = \sqrt{b}\begin{pmatrix} 1 \\ \alpha \end{pmatrix} + \frac{\mu}{2}\begin{pmatrix} j_1 \\ j_2 \end{pmatrix}.$$

Solving the first equation we obtain

$$j_1 = (\mu/2 + \alpha)/\sqrt{b}, \qquad j_2 = -\sqrt{b} - \mu\,\alpha/(2\sqrt{b}).$$

It appears from the second equation that $\alpha = 0$. So

$$P = (R,\, -J) = \begin{pmatrix} 1 & 0 \\ 0 & \sqrt{b} \end{pmatrix} + \begin{pmatrix} 0 & -\frac{\mu}{2\sqrt{b}} \\ 0 & 0 \end{pmatrix} = A + T(\mu).$$

With such a precision, we can write down what follows:

$$P^{-1} = A^{-1} - A^{-1}\,T(\mu)\,A^{-1}, \quad \text{where } A^{-1} = \begin{pmatrix} 1 & 0 \\ 0 & 1/\sqrt{b} \end{pmatrix}.$$

Let us note that $A^{-1}T = T$, hence $P^{-1} = (I - T)A^{-1}$. Multiplying P^{-1} by P from the right, we get:

$$P^{-1}\,P = (I - T)A^{-1}(A + T) = I - T^2 = I + O(\mu^2).$$

In fact, on the right-hand side we get the unit operator, because $T^2 = 0$. Multiplying (3.38) by P^{-1} from the left and using the substitution

$$\begin{pmatrix} y_1 \\ y_2 \end{pmatrix} = P^{-1} \begin{pmatrix} x \\ y \end{pmatrix}$$

we obtain:

$$\begin{pmatrix} y_1 \\ y_2 \end{pmatrix}' = \begin{pmatrix} \mu/2 & -\sqrt{b} \\ \sqrt{b} & \mu/2 \end{pmatrix} \begin{pmatrix} y_1 \\ y_2 \end{pmatrix} - \begin{pmatrix} 0 \\ \sqrt{b}\,y_1^2 y_2 + O(|\mu|) \end{pmatrix} + O(\mu^2).$$

Employing the formula (3.36), we finally get:

$$Re\,C_1 = -\sqrt{b}/8 + O(|\mu|).$$

So the stable limit cycle appears in the system (3.38) under the condition $b > 0$ and $0 < \mu \ll 1$, and this is confirmed by the results of the numerical experiment presented below.

Cell 3.4.

```
"Appearance of the limit cycle in the system (3.38)"
In[1]: ClearAll["Global`*"];
In[2]: Manipulate[
    eq1a = x'[t]==-y[t];
    eq2a = y'[t]==b x[t]+μ y[t]^2 - x[t]^2 y[t];;
    sol = NDSolve[{eq1a, eq2a, x[0] == a0, y[0] == 0}, {x[t], y[t]}, {t, 0, 250}];
    ParametricPlot[{x[t], y[t]}/.sol, {t, 0, T}, PlotRange → Full, AspectRatio → k
],
    {μ, {-0.01, 2}},{b, {0.5, 1.5, 2.5,3.5}},
    {T, {50, 100, 150, 200, 250}}, {a0, {0.05, 0.1, 0.2, 0.3, 0.4}}, {k, {0.5, 1, 1.5,
2}}]
```

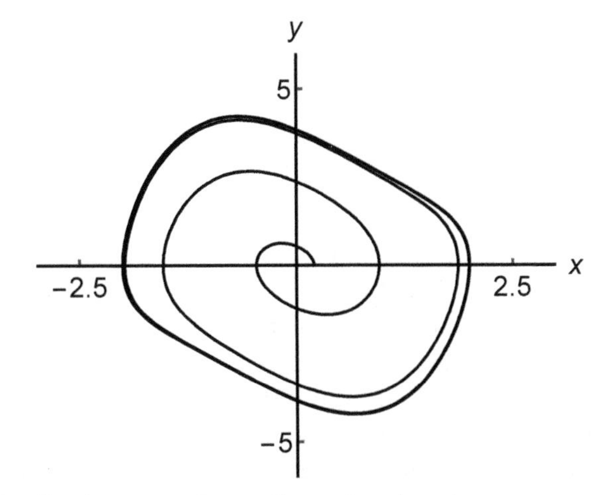

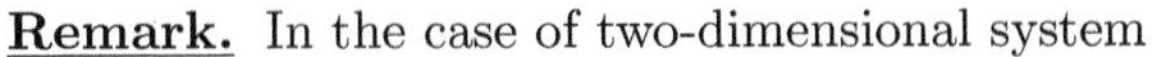

Remark. In the case of two-dimensional system

$$\begin{pmatrix} x_1 \\ x_2 \end{pmatrix}' = \hat{M}(\mu) \begin{pmatrix} x_1 \\ x_2 \end{pmatrix} + \begin{pmatrix} f_1(x_1,\, x_2) \\ f_2(x_1,\, x_2) \end{pmatrix}$$

whose linearization matrix $\hat{M}(\mu)|_{\mu=0}$ has a pair of purely imaginary eigenvalues $\lambda_{1,\,2} = \pm i\omega$, a passage to the canonical representation is possible using the matrix

$$P = (R,\, -J), \quad \text{where} \quad R = \begin{pmatrix} 1 \\ 0 \end{pmatrix}, \quad J = -\frac{1}{\omega}\, \hat{M}(0)\, R.$$

Problem.

Consider the following system:

$$\begin{pmatrix} x_1 \\ x_2 \\ x_3 \end{pmatrix}' = \begin{pmatrix} 0 & -1 & -1 \\ 1 & \mu & 0 \\ 0 & 0 & -1 \end{pmatrix} \begin{pmatrix} x_1 \\ x_2 \\ x_3 \end{pmatrix} + \begin{pmatrix} 0 \\ 0 \\ x_3\, x_1 - \alpha(x_1^2 + x_2^2) \end{pmatrix}. \tag{3.39}$$

This system differs from the system (3.15) analyzed in Section 3.5.2 only by the presence of the parameter μ (which is assumed to be small) in the linear part.

(1) Using the results of the analysis of system (3.15), and the instructions supplied with it, find the projection of the system (3.39) on the center manifold.

(2) Calculate the real part of the first Floquet index up to $O(\mu)$. Based on this result, formulate the conditions under which a stable limit cycle appears in the system (3.39).

(3) Confirm the results of calculations with a numerical experiment. Choosing μ as the control parameter, observe how the periodic mode changes with increasing μ.

3.5.7 *The homoclinic bifurcation*

The appearance of the limit cycle can be described using the theory of normal forms. This theory allows for significant simplification of the nonlinear terms in the right side of the dynamical system in a small but finite neighborhood of a stationary point (center). With the change of the bifurcation parameter, which leads to an increase in the size of the limit cycle, this theory becomes useless, and the further evolution of the periodic trajectory depends on global factors, and in particular on the nature of the nearby stationary points. The most common scenario for the evolution of the periodic trajectory located near the saddle point is as follows. As the parameter changes, the limit cycle becomes larger, approaching the saddle point, and at a certain critical value of the parameter *the homoclinic bifurcation* occurs at which the stable and unstable saddle separatrices form a single trajectory, called the homoclinic loop. This orbit is not structurally stable, therefore the small changes of the parameter usually cause that one of the trajectories coming from the unstable focus enters the saddle, while all other trajectories, including the outcoming separatrix of the saddle, abandon the area where the periodic orbit was previously located. This scenario is realized, for example, in the system

$$\begin{aligned} x'[t] &= -y[t]; \\ y'[t] &= x[t] + \mu\, y[t] + x[t]^2 + x[t]y[t]. \end{aligned} \tag{3.40}$$

The system (3.40) possesses two stationary points: $A(0,\,0)$ and $B(-1,\,0)$. The point $A(0,\,0)$ at $\mu = 0$ is a center, while the stationary point $B(-1,\,0)$ is a saddle at arbitrary values of the parameter μ. Applying the theory of normal forms, and, in particular, the formula (3.36), it can be seen that the limit cycle, appearing in the system when $0 < \mu \ll 1$, is stable.

The further evolution of the limit cycle can be traced using numerical simulations. If we want to present the separatrices of the saddle point the $B(-1,\,0)$ on the global phase portrait of the system (3.40), then we should find characteristic directions along which the separatrices enter (exit) to (from) the stationary point. Using the change of variables $(z_1,\,z_2) = (x + 1,\,y)$ and dropping out the nonlinear terms, we get the system

$$\begin{pmatrix} z_1 \\ z_2 \end{pmatrix}' = \begin{pmatrix} 0 & -1 \\ -1 & \mu \end{pmatrix} \begin{pmatrix} z_1 \\ z_2 \end{pmatrix} = \hat{M} \begin{pmatrix} z_1 \\ z_2 \end{pmatrix}.$$

The eigenvalues of the matrix $\hat{M}$ are expressed by the formula

$$\lambda_{\pm} = \frac{\mu \pm \sqrt{\mu^2 + 4}}{2}.$$

These values correspond to the eigenvectors $B_{\pm} = (1, -\lambda_{\pm})^{tr}$. The phase portraits presented in Fig. 3.3, are obtained by means of the numerical simulations based on the standard tools of the *Mathematica* package.

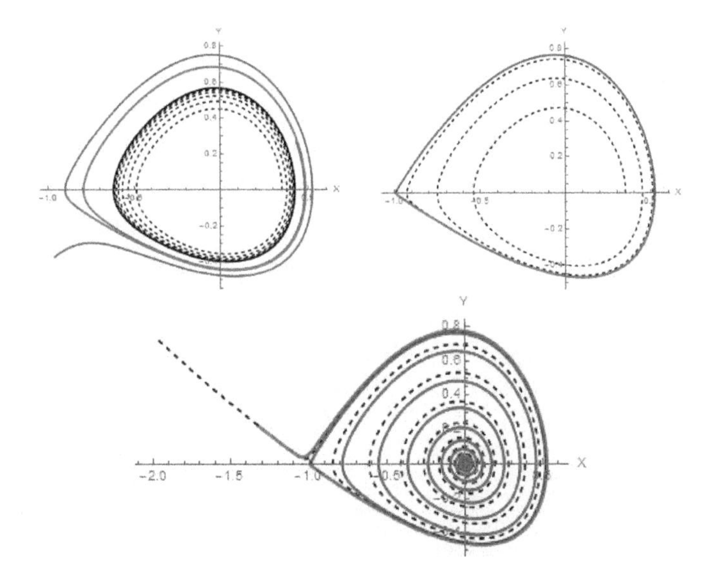

Fig. 3.3 Phase portraits of the system (3.40) obtained respectively: with $\mu = 0.07$ (left panel); $\mu = 0.1355$ (middle panel) and $\mu = 0.14$ (right panel). Blue color in the last two figures shows the separatrices entering the saddle $B(-1, 0)$; red color — outgoing separatrices, and the dashed line — trajectories coming out of the focus $A(0, 0)$.

Chapter 4

Oscillations in non-autonomous and multidimensional systems

4.1 Introduction

Oscillations in multidimensional dynamical systems, as well as in two-dimensional non-autonomous systems, which can be reduced to the three-dimensional dynamical systems by change of parametrization, run in a very different way from what we could observe in the case of autonomous two-dimensional systems. Therefore, for their research advanced tools are employed, such as the Poincaré sections technique or study of the Lyapunov index. It is quite unexpected fact that bifurcation phenomena occurring in discrete dynamical systems of dimension one proceed in a manner similar to bifurcations in multidimensional systems. At the same time, discrete systems are much easier to study, moreover, in contrast to multidimensional dynamical systems, such studies can be carried out using the analytical methods. Therefore, at the end of this chapter, we present the results of a study of one of the most common scenarios for the transition from periodic to chaotic oscillations using as a model logistic map.

4.2 Bifurcations in the Duffing equation with the periodic inhomogeneous part

A modified Duffing equation we'll deal with takes the following form:

$$\ddot{x} + 2\,\gamma\,\dot{x} - x + x^3 = F\,\cos[\omega\,t]. \tag{4.1}$$

It describes the deviation of the steel bar from the equilibrium position in the magnetic field (see Fig. 4.1). Two permanent magnets serve as a source of the magnetic field, the north poles of which are below the bar, to the left and to the right of its axis of symmetry. In addition, the system includes a source of an alternating magnetic field, being delivered by the iron frame

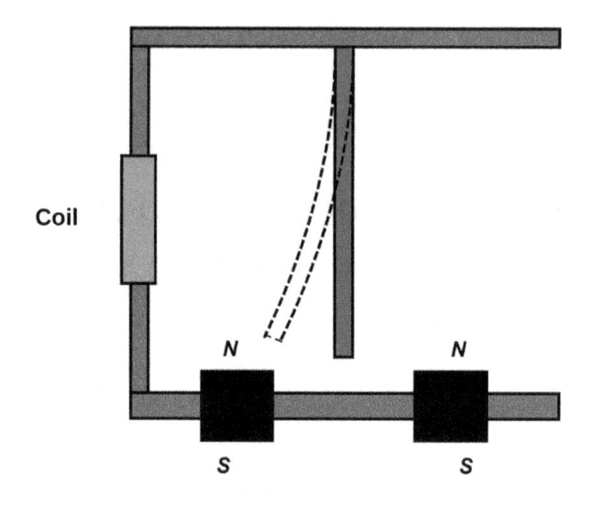

Fig. 4.1 Steel bar in the magnetic field.

to which the bar is attached. This source is active only when a current flowing in the coil surrounding the bar is periodically changing with the frequency ω.

With the external source turned off ($F = 0$) and lack of friction ($\gamma = 0$), the bar oscillates around one of the two equilibrium positions created by the presence of the constant magnetic field. Under such assumptions (4.1) becomes equivalent to the Hamiltonian system

$$\dot{x} = -y = -H_y \qquad \dot{y} = x^3 - x = H_x, \tag{4.2}$$

where

$$H(x, y) = \frac{y^2}{2} + \frac{x^4}{4} - \frac{x^2}{2}. \tag{4.3}$$

The location and nature of the stationary points of the system (4.2) is determined by the potential energy $U(x) = \frac{x^2}{4}(x^2 - 2)$, see Fig. 4.2.

It is seen that the system has a saddle point A with the coordinates $(0, 0)$ and a pair of the stationary points $B_{\pm}$ with the coordinates $(\pm 1, 0)$, which are centers. In the phase portrait shown in Fig. 4.3 a symmetrical pair of homoclinic solutions can be seen, separating oscillations of relatively small amplitudes from the periodic movements during which the bar passes through all equilibrium points.

Remark. If the friction and the external force are non-zero, the equation (4.1) becomes equivalent to a three-dimensional autonomous dynamical

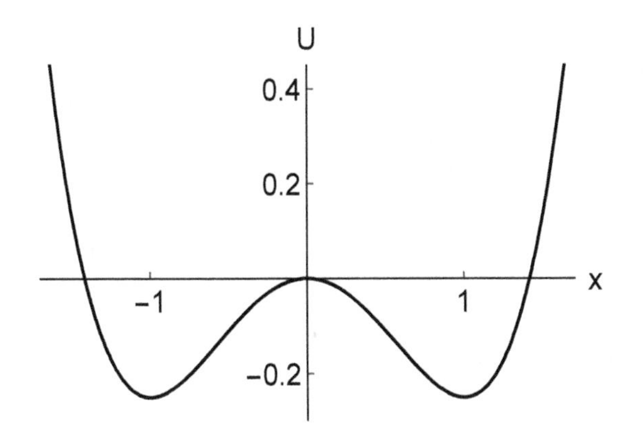

Fig. 4.2 The graph of the function $U(x)$.

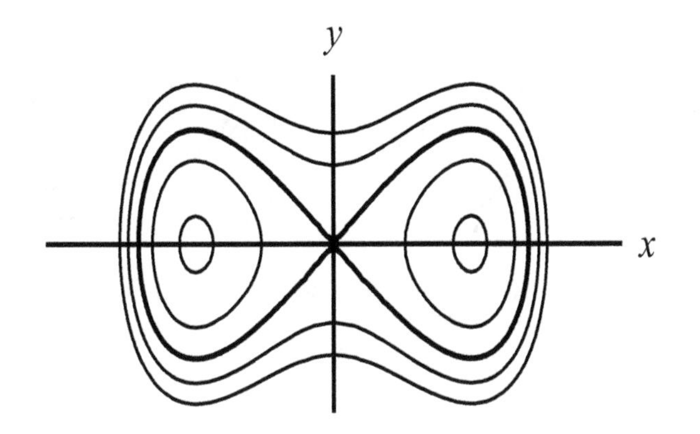

Fig. 4.3 Phase portrait of the system (4.3).

system

$$\begin{aligned}\frac{dx}{d\tau} &= -y,\\ \frac{dy}{d\tau} &= x^3 - x - 2\gamma y - F\cos[\tau],\\ \frac{dz}{d\tau} &= \omega\end{aligned} \tag{4.4}$$

attained from (4.1) in a standard way by formal introduction of new independent variable $\tau = \omega t$.

Throughout the numerical simulations of the system (4.4) we used the constant parameter values $\gamma = 0.25$, $\omega = 1$ and fixed initial data $x[0] = 0.09$, $y[0] = 0$, $z[0] = 0$. Changing the range of variability of F within the segment $[0, 1]$, we observed various periodic and chaotic oscillations in the system. The results obtained are illustrated in Figs. 4.4–4.10. When $F = 0$ the system is autonomous. The phase trajectory in the course of evolution approaches the stationary point $B_+ = (1, 0)$, which at $\gamma > 0$ becomes a stable focus (see Fig. 4.4, left panel).

As F belongs to the segment $(0, 0.34)$, a stable limit cycle appears in the vicinity of the stationary point B_+ (see Fig. 4.4).

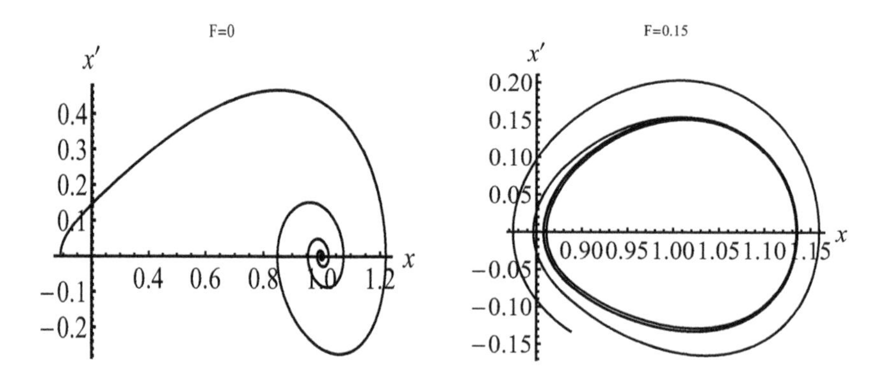

Fig. 4.4 Projections of phase trajectories of the system (4.4) on the (x, y) plane. Left panel: $F = 0$; right panel: $F = 0.15$.

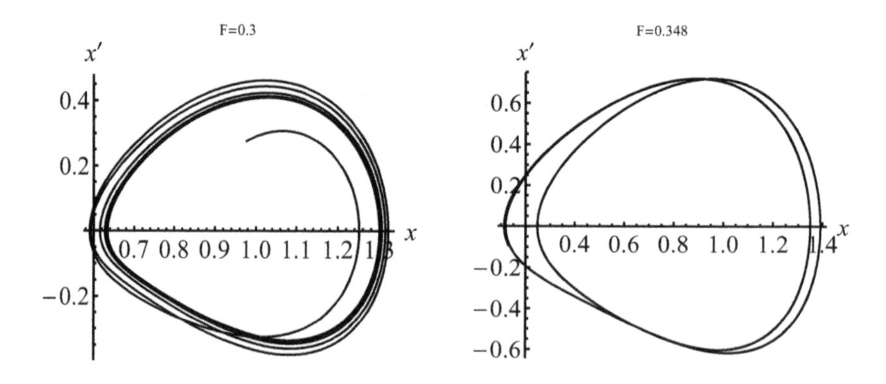

Fig. 4.5 Projections of phase trajectories of the system (4.4) on the (x, y) plane. Left panel: $F = 0.3$; right panel: $F = 0.348$.

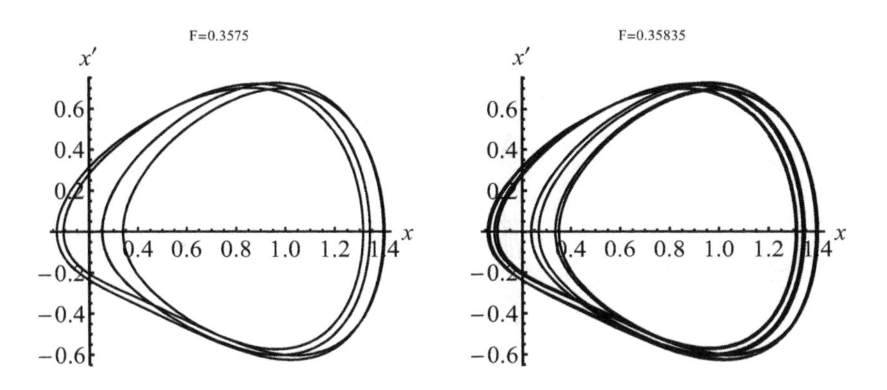

Fig. 4.6 Projections of phase trajectories of the system (4.4) on the $(x,\ y)$ plane. Left panel: $F = 0.3575$; right panel: $F = 0.35835$.

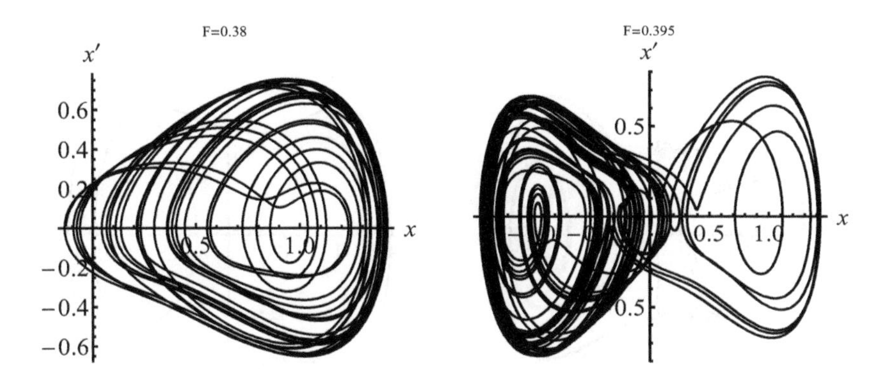

Fig. 4.7 Projections of phase trajectories of the system (4.4) on the $(x,\ y)$ plane. Left panel: $F = 0.38$; right panel: $F = 0.395$.

As the parameter F grows, the size of the cycle increases and its left edge approaches the origin (left panel of Fig. 4.5 shows the cycle corresponding to the value of $F = 0.3$). At $F \approx 0.3405$, there are signs of a period doubling. At $F = 0.348$, we observe a cycle with the period $\approx 2T$ (Fig. 4.5, right panel).

At $F \approx 0.3488$ another period doubling is observed. The phase portrait of $4T$ cycle is presented on the left panel of Fig. 4.6. With a further increase of the parameter F, another period doubling is observed. On the right panel of Fig. 4.6, most probably, the period $8T$ cycle is observed at $F = 0.35835$, although it cannot be clearly stated on the basis of visual observations.

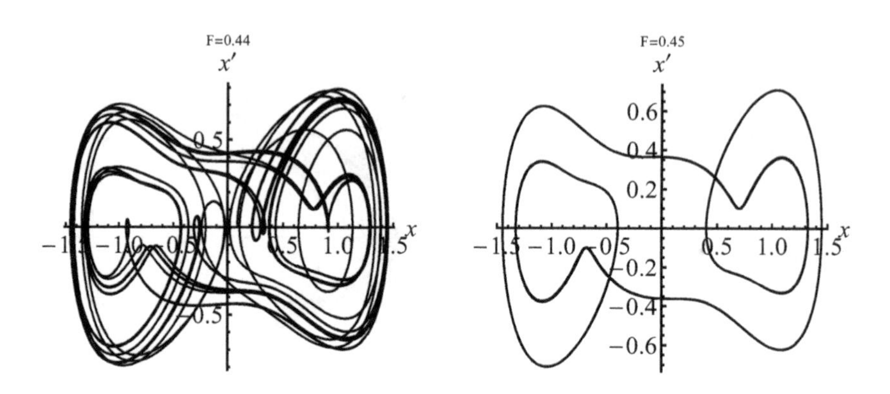

Fig. 4.8 Projections of phase trajectories of the system (4.4) on the $(x,\ y)$ plane. Left panel: $F = 0.44$; right panel: $F = 0.45$.

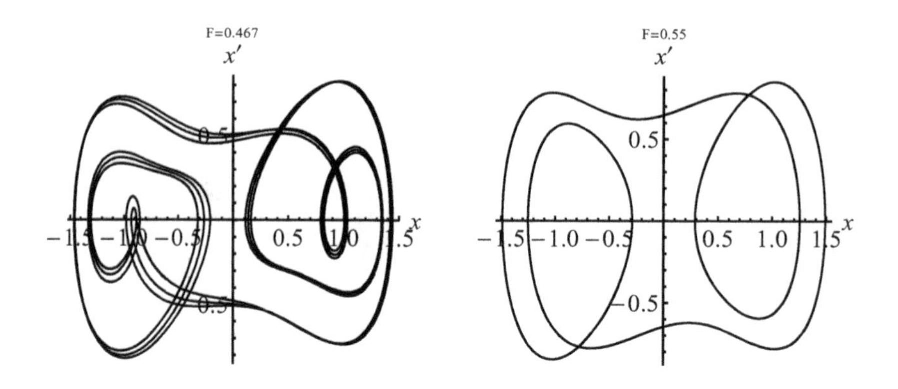

Fig. 4.9 Projections of phase trajectories of the system (4.4) on the $(x,\ y)$ plane. Left panel: $F = 0.467$; right panel: $F = 0.55$.

At $F = 0.38$, oscillations are observed that look as chaotic (see Fig. 4.7, left panel). At $F = 0.395$, the chaotic vibrations fill the area containing all three stationary points of the undisturbed system (Fig. 4.7, right panel). At $F = 0.44$, signs of trajectory regularization are observed, and the phase portrait suggests that the system implements multi-period vibrations, covering all three stationary points (Fig. 4.8, left panel). At $F = 0.45$, one can clearly see that the system demonstrates large periodic oscillations (Fig. 4.8, right panel).

Above this parameter value, period doubling of large-amplitude oscillations' takes place. Increasing the amplitude, we still observe the

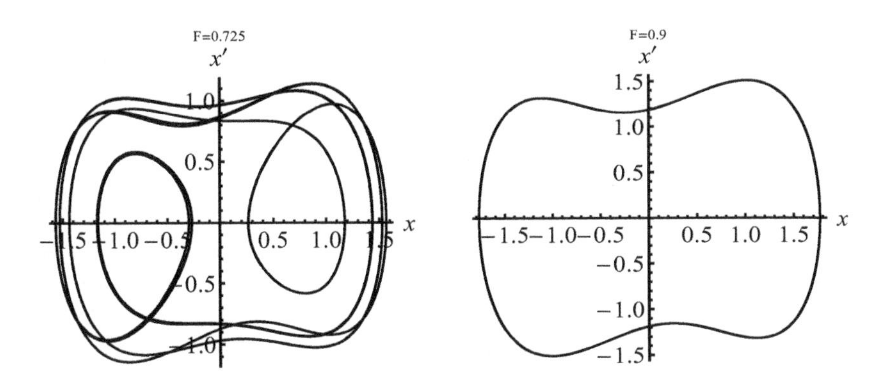

Fig. 4.10 Projections of phase trajectories of the system (4.4) on the $(x,\ y)$ plane. Left panel: $F = 0.725$; right panel: $F = 0.9$.

multi-periodic oscillations (Figs. 4.9–4.10, left panels). In the right panel of Fig. 4.10 a one-period solution involving all three stationary points is shown at $F = 0.9$.

In summary, the equation (4.1) with the periodic perturbation and the amplitude F belonging to the segment $(0,\ 1)$ demonstrates a very complicated behavior. To analyze this behavior (at least partially), it becomes necessary to use new research tools.

4.3 Methods of investigations of chaotic solutions

4.3.1 *Lyapunov index*

Visualization of complex solutions (e.g. multi-periodic and chaotic) gives an insight into the behavior of the trajectories and allows one to observe changes in the system with the change of the parameters' values, but it does not answer the question whether we are dealing with chaotic solutions or solutions with a period of mT, with big enough $m \in \mathbb{N}$. The strict characteristics of complicated phase trajectory would give the *Lyapunov index*. Let us start by mentioning that chaotic trajectories are extremely sensitive to the choice of initial conditions. An instructive example illustrating this sensitivity will be the solution of the initial value problem using numerical methods of varying accuracy. This is illustrated by the example below (see PM1_7.nb file).

Let us solve the Cauchy's problem for the Rössler system

$$\begin{cases} \dot{x} = -(y+z), & x(0) = 4.0, \\ \dot{y} = x + 0.2\,y, & y(0) = 0, \\ \dot{z} = 0.2 + z(x-c), & z(0) = 0 \end{cases} \tag{4.5}$$

with different accuracy (rounding error).

In the first series of simulations we use the options AccuracyGoal $\to$ 6 and PrecisionGoal $\to$ 6, while in another series we use the options AccuracyGoal $\to$ 10 and PrecisionGoal $\to$ 10. The results of numerical experiments, performed in the file PM1_7.nb are shown in Fig. 4.11. It is seen on this figure that after some time the solutions begin to differ more and more from each other.

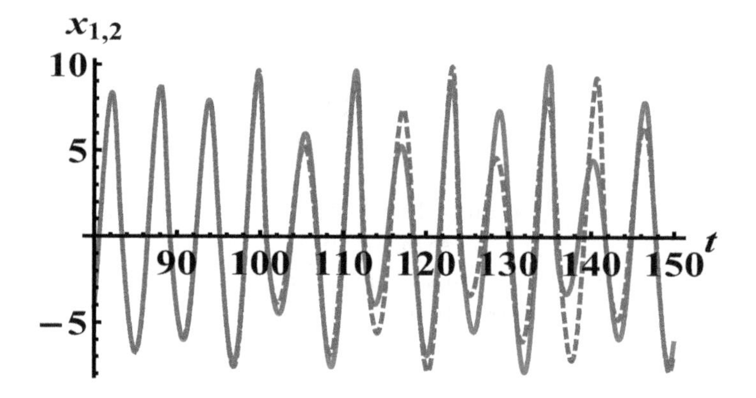

Fig. 4.11 Comparison of numerical simulations of the initial value problem (4.5) carried out with different precision (rounding error): the red curve corresponds to the series in which the options AccuracyGoal $\to$ 6, PrecisionGoal $\to$ 6 were employed; the blue dashed curve corresponds to the simulation in which the options AccuracyGoal $\to$ 10, PrecisionGoal $\to$ 10 were used.

If to present graphically the differences $\Delta = x_1[t] - x_2[t]$, where $i \in \{1, 2\}$ stands for the number of the experiment, then it can be seen that the difference, which in the range $t \in (0, 50)$ is of the order $10^{-3} \div 10^{-4}$ (Fig. 4.12, left panel), begins to rapidly increase on the segment $t \in (50, 100)$ then the trajectories are getting further apart (Fig. 4.12, right panel), which starts to be clearly visible on the graph presented, starting from $t \approx 115$.

The above effect is rather a general feature of systems demonstrating chaotic behavior. If to compare a selected trajectory with the trajectories

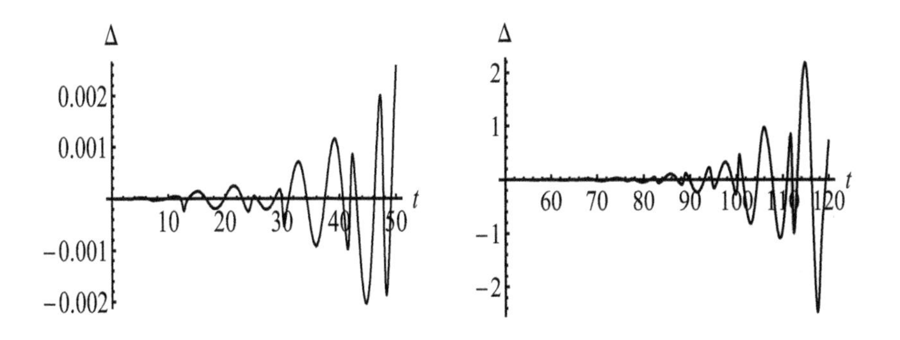

Fig. 4.12 Graphs of the function $\Delta = x_1[t] - x_2[t]$, where $i \in \{1, 2\}$ marks the number of the numerical experiment. Left panel: $t \in (0, 50)$; right panel: $t \in (50, 120)$.

in its vicinity, then one can see that the average distance between them increases exponentially: $|\Delta x| \sim exp[\lambda t]$, $\lambda > 0$, $t > 0$.

The constant λ, which characterizes the degree of divergence of the trajectories, is called *Lyapunov's index (exponent)*. The value of the λ can be determined in a numerical experiment, which will be described below. First, however, we should give a definition of this quantity. We will continue to use the Rössler system, demonstrating the chaotic behavior as a model system. Let us choose the point $p_0 = (x_0, y_0, z_0)$ as the initial data and consider the phase trajectory

$$p[t, p_0] = [x(t; x_0, y_0, z_0), y(t; x_0, y_0, z_0), z(t; x_0, y_0, z_0)]$$

passing through this point when $t = 0$ ($p[0, p_0] = p_0$). Then we consider a set of small deviations $d_{0\,i} = \{(\Delta x_{0\,i}, \Delta y_{0\,i}, \Delta z_{0\,i})\}_{i \in I}$ from the initial data.

Definition 4.1. Lyapunov index is the quantity

$$\lambda[p_0] = \lim_{\substack{d_{0i} \to 0 \\ t \to \infty}} \frac{1}{t} \frac{1}{N} \sum_{i=1}^{N} \log \left[\frac{|p(t, p_0 + d_{0i}) - p(t, p_0)|}{|d_{0i}|} \right], \tag{4.6}$$

where the number N is assumed to be sufficiently large ($N \gg 1$), and the infinitesimal deviations d_{0i} surround evenly the point p_0.

Using the above definition, we will provide the algorithm for search of the Lyapunov index. We'll still use the Rössler's system as a test case.

Cell 4.1.

"Determining the Lyapunov index for the phase trajectory of the Rössler system $x' = -(y + z)$, $y' = x + 0.2y$, $z' = 0.2 + z(x - 4.99)$ passing at $t = 0$ through the point $(x_0, y_0, z_0) = (4, 0, 0)$"

```
In[1]: Clear["Global'*"]
In[2]: {a,b,c}={0.2,0.2,4.99};
    z1={x[t],y[t],z[t]}/.NDSolve[{x'[t]==- (y[t]+z[t]), y'[t]==x[t]+a y[t], z'[t] ==
b + z[t] (x[t] - c),
    x[0] == 4, y[0] == 0, z[0] == 0}, {x[t], y[t], z[t]}, {t, 0, 100}, MaxSteps →
500 000][[1]];
    z0=Table[{4+10^-7 (2 Random[ ] - 1), 10^-7 (2 Random[ ]-1), 10^-7 (2 Random[
]-1), {m,1,60}];
    Δz[t_]=Table[Sqrt[({x[t], y[t], z[t]}-z1).({x[t],y[t],z[t]}-z1)]/.
    NDSolve[{x'[t]==- (y[t]+z[t]), y'[t]==x[t]+a y[t], z'[t] == b + z[t] (x[t] - c),
    x[0] == z0[[m,1]], y[0] == z0[[m,2]], z[0] == z0[[m,3]]}, {x[t], y[t], z[t]}, {t, 0,
100},
    MaxSteps → 500 000][[1]], {m,1,60}];
    λ [t_]=1/60 Sum[Log[Δz[t][[n]]/Δz[0][[n]]],{n, 1, 60}]/t;
    Plot[λ[t],{t,0,100}[, PlotRange →{{0,100},{-0.1, 0.3}}, AxesLabel → {"t",
"λ(t)"}]
Out[2]:
```

λ(t)
0.3
0.2
0.1
0.0
-0.1
20 40 60 80 100 t

Remark. The applied algorithm for calculating the Lyapunov index, the implementation of which in the case of the Rössler system is presented in the file PM1_7.nb, does not allow the exact calculation of the Lyapunov index. The problem is that the distance d_0 cannot be less than the precision of the numeric method used. If, on the other hand, the distance d_0 is too large, then the value of $\lambda[z_0]$ will be burdened with an error due to the fact that the trajectories move apart exponentially quickly only when they are close enough to each other. Nevertheless, the method suggested is suitable for practical purposes. The calculations carried out with it show that the function sought tends to a certain positive value. From this we can conclude, that the phase trajectories of the Rössler system demonstrate chaotic features.

In general, the change of the distance between trajectories depends on their relative position. Therefore, it would be more correct to talk about the set of Lyapunov's indices. The number of indices is equal to the number of degrees of freedom in the direction perpendicular to the trajectory, i.e. $N - 1$, where N is the dimension of the dynamical system. The formula (4.6) specifies only one index, called *the largest Lyapunov index.* This is explained below basing on the Poincaré mapping generated by the phase flow Φ_t of the dynamical system in the vicinity of the periodic orbit. The construction of the mapping is as follows. Let us assume that the periodic orbit γ crosses the plane Σ at some point p (see Fig. 4.13).

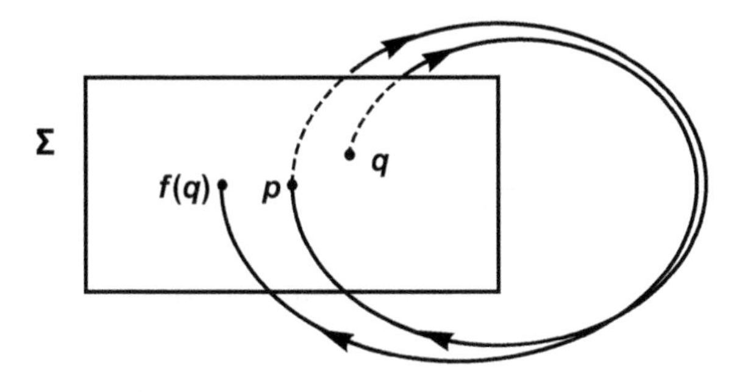

Fig. 4.13 Poincaré section $P : \mathbb{R} \ni U_p \to \Sigma$.

To each point $q \in U_p \subset \Sigma$, where $\Sigma \supset U_p$ is a sufficiently small open set containing p, corresponds the point is $f(q) = \Phi_\tau(q) \in \Sigma$, where τ is the time of the first intersection of the phase flow $\Phi_t(q)$ with the Poincaré section Σ. It is obvious that the point p at which the periodic orbit γ intersects Σ plane is the stationary point of the Poincaré map. Since the phase flow in the small neighborhood of the point p is linear, the trajectories of the discrete map will behave like phase trajectories of planar dynamical system in vicinity of the stationary point. This can be visualized by linking the points of successive iterations $f^{(n)}(q)$ of the Poincaré map $f : U_p \Rightarrow \Sigma$. Depending on the nature of the phase flow $\Phi_t(q)$ (for which the Lyapunov index (4.6) is positive), the trajectories around p will have the character of saddle, source or unstable focus. Using this analogy, we will provide arguments explaining the operation of the algorithm based on the formula (4.6). We will conduct our reasoning assuming that the point p is a saddle point. With the appropriate selection of variables, the Poincaré map in a small neighborhood of the stationary point $p \in \Sigma$ takes the form

$$\begin{pmatrix} x_{n+1} \\ y_{n+1} \end{pmatrix} = \begin{pmatrix} \lambda_1 & 0 \\ 0 & \lambda_2 \end{pmatrix} \begin{pmatrix} x_n \\ y_n \end{pmatrix}, \qquad \lambda_1 > 1 > \lambda_2 > 0.$$

In these variables, the stationary point p has the coordinates (0, 0). Consider the orbit starting at $q = (x_0,\, y_0) \in \Sigma$. The distance between the orbit $\{f^{(n)}(q)\}_{n=1}^{\infty}$ and the origin will be changed in the course of iterations in accordance with the formula

$$d_n = \sqrt{x_0^2\,\lambda_1^{2n} + y_0^2\lambda_2^{2n}} = \lambda_1^n\sqrt{x_0^2 + y_0^2\left(\frac{\lambda_2}{\lambda_1}\right)^{2n}} = \lambda_1^n\left[|x_0| + O\left(\left|\left(\frac{\lambda_2}{\lambda_1}\right)^n\right|\right)\right].$$

Let us estimate the discrete analog of the formula (4.6) for a single phase trajectory:

$$\begin{aligned} L[q] &= \lim_{n\to\infty} \tfrac{1}{n} \ln\left(\tfrac{d_n}{d_0}\right) = \lim_{n\to\infty} \tfrac{1}{n}\ln\left[\frac{\lambda_1^n\left(|x_0| + O\left(\left|\left(\frac{\lambda_2}{\lambda_1}\right)^n\right|\right)\right)}{\sqrt{x_0^2+y_0^2}}\right] \\ &= \lim_{n\to\infty} \ln\left[\lambda_1\left(\frac{|x_0|}{\sqrt{x_0^2+y_0^2}} + O\left(\left|\left(\tfrac{\lambda_2}{\lambda_1}\right)^n\right|\right)^{1/n}\right)\right] = \ln[\lambda_1] > 0. \end{aligned}$$

So, for almost all initial deviations $L(q) = \ln\lambda_1$. The exceptions are the initial deviations located along the separate directions $q_\pm = (0,\, \pm d)$. For such deviations

$$L[q] = \lim_{n\to\infty} \ln\left(\frac{d_n}{d_0}\right) = \lim_{n\to\infty} \frac{1}{n} \ln\left[\frac{\lambda_2^n\,|y_0|}{|y_0|}\right] = \ln\,\lambda_2.$$

It is assumed in the algorithm used for studying stability of the periodic trajectory γ that $|d_{0i}| \ll 1$ (for the reasons given above, the deviations cannot be too small). In the given coordinate frame p coincides with the origin and the following formula applies:

$$\lambda[p] = \lim_{n\to\infty} \frac{1}{n}\,\frac{1}{N} \sum_{i=1}^{N} \ln\left[\frac{d_{ni}}{d_{0i}}\right],$$

where $d_{ni} = \sqrt{x_{0i}^2\lambda_1^{2n} + y_{0i}^2\lambda_2^{2n}}$, $d_{0i} = \sqrt{x_{0i}^2 + y_{0i}^2}$. In the case when the deviations of the form $q_\pm^i = (0,\, \pm d_{0i})$ are met in a random sample exactly j times $(0 \le j \ll N)$, the formula can be written as follows:

$$\begin{aligned} \lambda[p] &= \lim_{n\to\infty} \tfrac{1}{n}\,\tfrac{1}{N} \textstyle\sum_{i=1}^{N} \ln\left[\tfrac{d_{ni}}{d_{0i}}\right] = \tfrac{1}{N} \textstyle\sum_{i=1}^{N} \lim_{n\to\infty} \tfrac{1}{n}\ln\left[\tfrac{d_{ni}}{d_{0i}}\right] \\ &= \tfrac{1}{N}\left(\textstyle\sum_{i=1}^{N-j} \lim_{n\to\infty} \ln\left[\lambda_1\left(\frac{|x_0|}{\sqrt{x_0^2+y_0^2}} + O\left[\left(\tfrac{\lambda_2}{\lambda_1}\right)^n\right]\right)^{1/n}\right] + j \lim_{n\to\infty} \tfrac{1}{n}\ln\tfrac{\lambda_2^n|y_0|}{|y_0|}\right) \\ &= \tfrac{1}{N}\left\{(N-j)\ln\lambda_1 + j\ln\lambda_2\right\}, \end{aligned}$$

which in the case when $N = 40$ (N is the sample size used in calculations) gives for $\lambda[p]$ value which is very close to $\ln[\lambda_1] > 0$.

Problem.

- Show that employment of the above formula in the case when the stationary point of the Poincaré map behaves as a non-degenerate source, i.e. the eigenvalues (multipliers) of the linearization matrix of the Poincaré map satisfy the inequalities $1 < \lambda_2 < \lambda_1$) we'll obtain the following result: $\lambda[p] \approx \ln[\lambda_1] > 0$.
- What will be the result in case when $0 < \lambda_2 < \lambda_1 < 1$?

4.3.2 *Rössler system and the Lorenz map*

Our next goal is to understand the mechanism underlying the period doubling cascade. The standard three-dimensional system in which such a scenario is realized is the Rössler system (4.5). The period-doubling cascade can be observed in the numerical experiment in which the parameter c varies within the interval $[2, 5]$, see Fig. 4.14.

An interesting information concerning the period doubling bifurcations in the Rössler system can give us the study of the Poincaré map. The phase flow $\Phi_t(\cdot)$ of the dynamical system containing the periodic trajectory γ, generates a map of a plane Σ transversal to γ into itself (Fig. 4.15, left panel). If γ is a stable orbit, then the points of successive intersections of the orbit passing through the selected point q in the initial moment of time approach the fixed point $p \in \Sigma$, creating an analog of a stable node (or a stable focus if the multipliers are complex). At certain value of the parameter c, the period T cycle loses stability, but does not disappear. Simultaneously a stable cycle with the period $\approx 2T$ appears in its vicinity. In the Poincaré plane Σ, the period doubling corresponds to the pitchfork bifurcation (Fig. 4.15, right panel). Further increase of the parameter's c values causes the instability of the period $2T$ cycle, which does not disappear but merely becomes unstable. As before, in a small vicinity of the $2T$ cycle that has lost the stability, a stable cycle of period $4T$ appears. The points of intersection of the plane Σ with the period $2T$ orbit are subjected to the pitchfork bifurcation. This process repeats again and again, with progressively smaller changes of the parameter c at which successive period-doubling bifurcations occur. Ultimately, in a finite region of the phase space simultaneously exist a countable set of unstable orbits with the periods $2^n T$ ($n = 1, 2, \dots$) and the unique stable orbit with the period $2^\infty T$ (chaotic orbit).

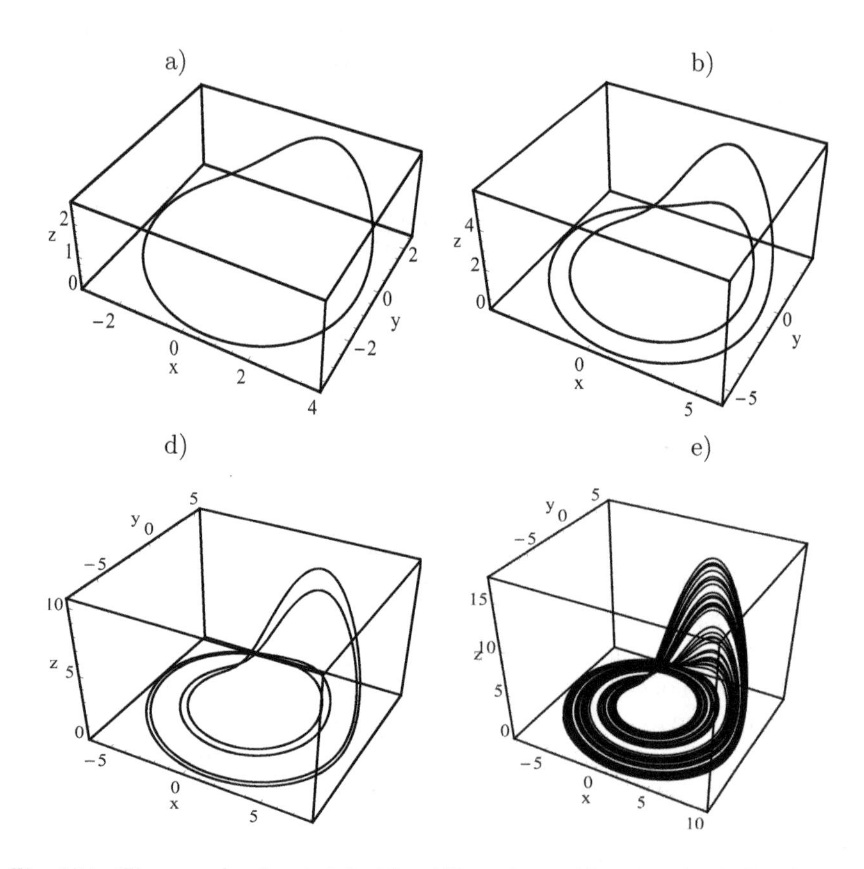

Fig. 4.14 The cascade of period-doubling bifurcations taking place in the Rössler system, when the parameter c varies within the interval [2, 5].

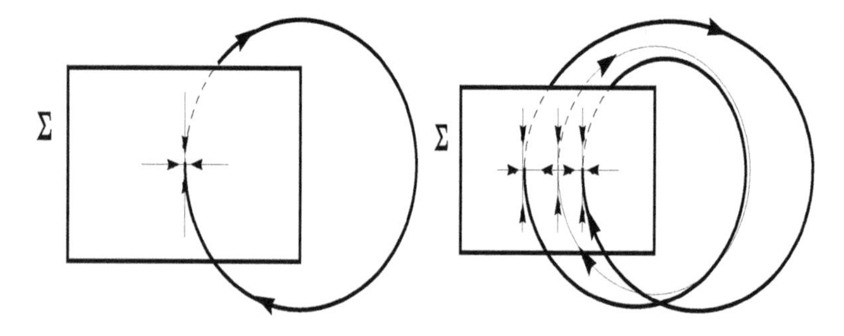

Fig. 4.15 The pitchfork bifurcation induced on the Poincaré section Σ by the phase flow of a dynamical system subjected to the period doubling bifurcation.

We've presented the scenario of the cascade of period-doubling bifurcations without addressing the question of the genesis of this phenomenon. Unfortunately, it is not possible to obtain a simple answer to this question by analyzing a three-dimensional dynamical system. However, we can take advantage of *the principle of shortening information* [Haken (1983)], which in a given context boils down to the fact that information about the nature of solutions to a three-dimensional system is contained in each of the component of the vector-functions $[x(t),\, y(t),\, z(t)]$, for example in $x(t)$. Getting to this information was proposed by E. Lorenz, who is one of the founders of the theory of *deterministic chaos*. E. Lorenz put forward a hypothesis that $n+1$-th local maximum of the function $x(t)$ is connected by functional dependence with the n-th local maximum of this function.

Let's denote the value of the n-th local maximum of the function $x(t)$ by the index x_n (see Fig. 4.17). The result of presenting the value of x_{n+1} as a function of x_n for a large number of iterations is shown in Fig. 4.17 (this procedure is carried out in the PM1_8.nb file). One can see that in a given particular case the next local maxima lie on a curve resembling an overturned parabola.

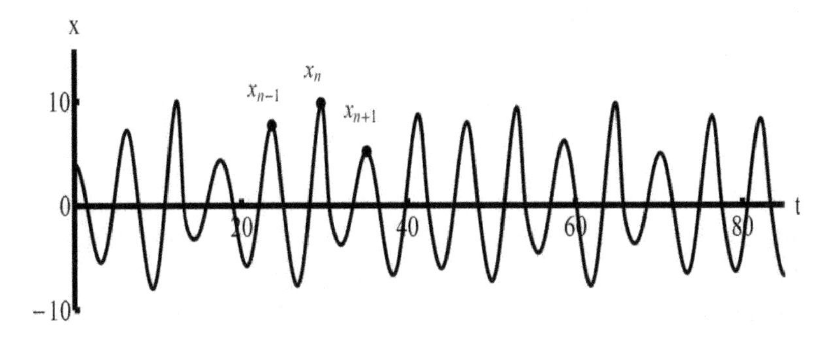

Fig. 4.16 Evolution of the x variable of the Rössler system. Successive local maxima of the $x(t)$ are marked with dots.

Definition 4.2. Function $y = f(x)$, mapping the segment $I = [a, b] \subset \mathbb{R}$ into itself is called *unimodular function*, if $f(a) = f(b)$, and there exists a point $x_0 \in (a, b)$ such that $f'\Big|_{[a,x_0)} > 0$ and $f'\Big|_{(x_0,\, b]} < 0$.

Using translations $x \to x - a$, $y \to y - b$ and scaling $x \to \bar{x} = \alpha x$, $y \to \bar{y} = \beta y$, $\alpha > 0$, $\beta > 0$, any unimodular function can be converted to the form $\bar{y} = \bar{f}[\bar{x}]$, where $\bar{f}$ is the unimodular function mapping the segment

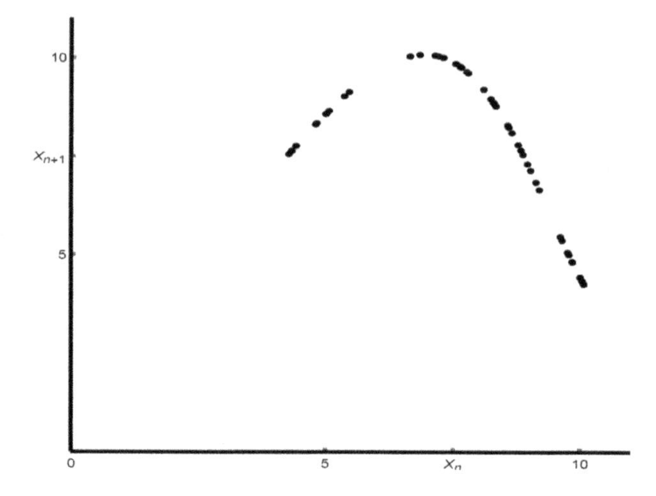

Fig. 4.17 Representation of the map $x_{n+1} = f(x_n)$, where x_n denotes n-th local maximum of the x coordinate of the Rössler system.

[0, 1] into itself, such that $\bar{f}[0] = \bar{f}[1] = 0$. As shown in [Metropolis *et al.* (1973)], each unimodular function $x_{n+1} = f[x_n]$, $f(0) = f(1) = 0$, mapping the segment $[0, 1]$ into itself and satisfying the condition

$$S[f] = \frac{f'''}{f'} - 3/2 \left(\frac{f''}{f'} \right)^2 < 0 \tag{4.7}$$

($S[\cdot]$ is called the *Schwartz derivatives*) demonstrates the period doubling cascade. Moreover, the bifurcation phenomena do not depend on the particular choice of the unimodular function possessing the above features. The simplest representative of a given class, convenient for analysis, is the logistic mapping $x_{n+1} = r\, x_n\, (1 - x_n)$, which was briefly discussed at the beginning of the first chapter. The properties of this map, revealing the mechanism of the period doubling bifurcation, will be analyzed in the following subsection.

4.3.3 *Logistic map*

It is easy to see that the logistic map

$$x_{n+1} = r\, x_n\, (1 - x_n) \tag{4.8}$$

transforms the segment [0, 1] into itself when $r \in (0, 4]$. So we'll assume further on that $0 < r \leq 4$. To begin with, let us locate stationary points and analyze their stability. The stationary point is defined as a solution to the equation

$$x_* = r\, x_* \,(1 - x_*). \tag{4.9}$$

Solution $x_*^0 = 0$ of the equation (4.9) defines the stationary point existing at arbitrary value of the parameter r. The second solution, given by the formula $x_*^1 = (r-1)/r$, exists when $r \in (1,\, 4]$. The stability type of the stationary point x_* of a map $x_{n+1} = f(x_n)$ depends on whether the points belonging to a sufficiently small vicinity of this point are approaching or moving away from it. This is defined in terms of the linearization of the function f at the stationary point x_*:

$$|f[x] - x_*| = |D\, f[x_*]\,(x - x_*)| + O\left(|x - x_*|^2\right) \approx r\,|1 - 2\,x_*|\;|x - x_*|\,.$$

For the logistic map $D\,f[0] = r$, hence the stationary point $x_*^0 = 0$ is stable when $r \in (0,1)$. The loss of the stability of this point occurs simultaneously with the appearance of the stationary point x_*^1. Linearizing the logistic map at this point we get

$$|f[x] - x_*^1| = |2 - r|\;|x - x_*^1| + O\left(|x - x_*^1|^2\right)$$

so the stationary point x_*^1 is stable if the inequality $|2 - r| < 1$ takes place, and this is so when $r \in (1,3)$.

The orbit of the point $x_0 \in (0,\, 1)$, which is understood as the sequence of points $\{x_0, x_1, ..., x_n, ...\}$, where $x_k = f^k[x_0] = f^{k-1}[f[x_0]] = f^{k-2}[f^2[x_0]]$, etc., can be presented graphically in the form of so-called Lameray diagrams (or steps). In the first step of constructing the diagram, the point x_0 is connected by the vertical segment with its image on the parabola $f[x] = rx(1-x)$. This point, in turn, is connected with the diagonal by the horizontal line. Since the x coordinate of the point on the diagonal coincides with $x_1 = f[x_0]$, we will get the image of the point $f^2[x_0]$ by connecting it with the parabola, using the vertical segment. Repeating this procedure many times, we can recreate the orbit of x_0. The implementation of this procedure for $r = 1.6$, $r = 2.5$ and $r = 3.1$ is shown in Fig. 4.18.

We have shown some typical Lameray diagrams that occur when $r \in (1,\, 3+\delta]$, $0 < \delta \ll 1$. In this set of the parameter values, the map (4.8) demonstrates following types of behavior: the orbit follows monotone to the stationary point x_*^1, when $r \in (1,\, 2)$; at $r \in (2,3)$ iterations approaching the stationary point x_*^1, form a spiral; for $r > 3$, when both x_*^0 and x_*^1 become unstable, iterations tend to a periodic orbit. Thus, we are dealing here with an analog of the Andronov-Hopf bifurcation. When the parameter r is increased further on, a period doubling cascade is observed leading to a chaotic mapping where almost all orbits move apart from each other.

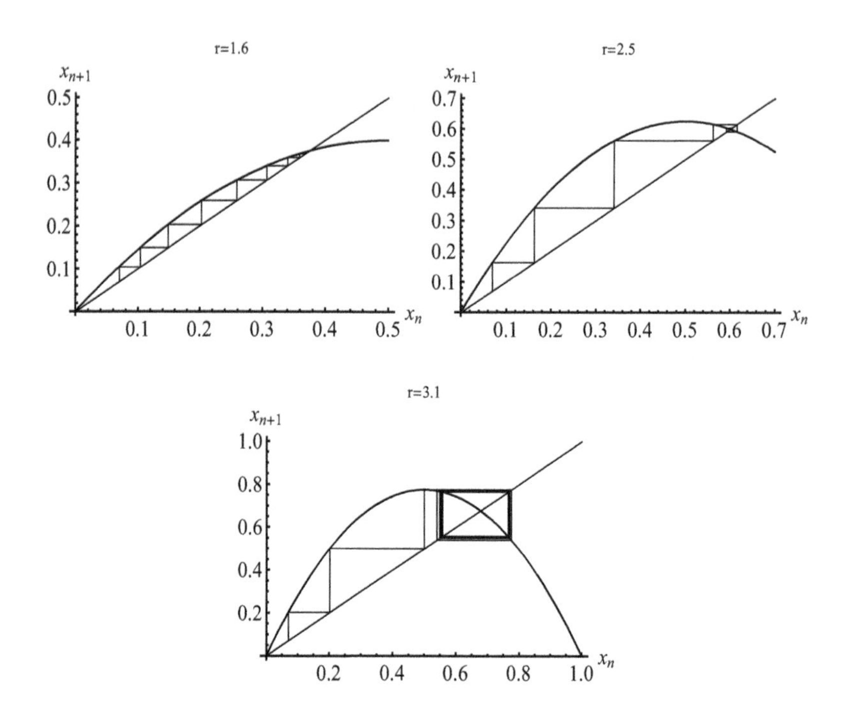

Fig. 4.18 Lameray diagrams obtained for the logistic mapping at different values of the parameter r.

4.3.4 *Period doubling bifurcations in the logistic map*

The behavior of Lameray diagrams at values of r slightly exceeding $r_1 = 3$ can be analyzed by considering simultaneously with $f[x_n]$ subsequent iterations and examining the properties of the maps $f^k[x_n] = f^{k-1}[f[x_n]] = \ldots,\ k = 2, 4, 8, \ldots, 2^n, \ldots$.

Lemma 4.1. *The function $f^2[x]$ possesses local extrema at the points $x_0 = 1/2$ and $x_{1,2} = f^{-1}[1/2]$.*

Proof. The necessary condition for the existence of local extremum takes the form

$$\frac{d}{dx} f^2(x) = f'[f(x)]\, f'(x) = r[1 - 2f(x)]r(1 - 2x) = 0.$$

The last (factorized) equation nullifies at the points $x_0 = 1/2$ and $x_{1,2} = f^{-1}[1/2]$. In order to check the fulfillment of the sufficient condition, and to determine the types of local extrema, we calculate the second derivative of $f^2[x]$:

$$\frac{d^2}{d\,x^2} f^2[x] = \frac{d}{d\,x}\, r^2[1 - 2f(x)]\,(1 - 2x)$$
$$= -2\,r^3(1 - 2\,x)^2 - 2\,r^2[1 - 2\,r\,x(1 - x)].$$

Substituting the coordinates of the appropriate points into the above formula, we get:

$$\frac{d^2}{d\,x^2} f^2[x] \;\Big|_{x=1/2} = r^2\,(2 - r) > 0;$$
$$\frac{d^2}{d\,x^2} f^2[x] \;\Big|_{x=f^{-1}[1/2]} = -2\,r^3\,(1 - 2\,f^{-1}[1/2])^2 < 0.$$

Thus, the function $f^2[x]$ has the local minimum at x_0, and a pair of local maxima at the points $x_{1,2}$.

Lemma 4.2. *If* $f(x_*) = 0$, *then* $f^n(x_*) = 0, n = 2, 3, \ldots.$

The proof is evident.

Lemma 4.3. *If the stationary point* x_* *is stable with respect to* f *then it is also stable with respect to the maps* f^n, $n = 2, 3, \ldots.$

Proof. Assume that $|f'(x_*)| < 1$. Then

$$\left|(f^2)'(x_*)\right| = \left|f[f(x_*)]'\right| = \left|f'[f(x_*)]\,f'(x_*)\right| = \left[f'(x_*)\right]^2 < 1.$$

For higher iterations the proof is analogous.

Lemma 4.4. *The point* x_*^1, *which is a stable stationary point of the* f^2 *map when* $r < r_1 = 3$, *becomes unstable and undergoes the pitchfork bifurcation as* $r > r_1$. *As a result, two stable stationary points* $\bar{x}_1$ *and* $\bar{x}_2$ *of the map* f^2 *appear* $(\bar{x}_1 < x_*^1 < \bar{x}_2)$ *and the following equations hold:*

$$f[\bar{x}_1] = \bar{x}_2, \qquad f[\bar{x}_2] = \bar{x}_1. \tag{4.10}$$

The pair $\{\bar{x}_1, \bar{x}_2\}$ *forms the periodic orbit attracting the nearby trajectories as long as* $r_1 = 3 < r < r_2 = 1 + \sqrt{6} \approx 3.449$.

Sketch of the proof. A simple way to find additional stationary points of $f^2[x]$ is to discard those roots that this map shares with $f(x)$. To do this, let us consider the quotient equation

$$0 = \frac{f^2(x) - x}{f(x) - x}. \tag{4.11}$$

The right side can be found by using the command PolynomialQuotient and PolynomialRemainder in the *Mathematica* package. The last command enables to check whether the polynomials are completely divisible.

Cell 4.2.

```
In[1]:
    P = r^2 (1 - x) (1 - r x (1 - x)) - 1;
    Q = r (1 - x) - 1;
In[3]: PolynomialQuotient[P, Q, x]
Out[3]: 1+r+(-r-r^2) x+r^2 x^2
In[4]: PolynomialRemainder[P, Q, x]
Out[4]: 0
```

Thus, after cutting off already known roots we get:

$$1 + r - r(r+1)x + r^2x^2 = 0.$$

The roots of this algebraic equation are given by the formula

$$\bar{x}_{1,\,2} = \frac{r + 1 \pm \sqrt{(r+1)(r-3)}}{2\,r}. \tag{4.12}$$

Let us analyze the expression (4.12). First of all, we note that it does not have real roots when $r < 3$. At $r = r_1 = 3$ it has the only solution $\bar{x}_{1,\,2} = x_* = 2/3$. So the additional stationary points $\bar{x}_{1,\,2}$ appear as a result of "branching" of the non-zero stationary point of the map $f^2[x]$.

Let us show the validity of the formula (4.10). It follows from the formula $f^2[\bar{x}_1] = \bar{x}_1$ that

$$f^2[f(\bar{x}_1)] = f[f^2(\bar{x}_1)] = f(\bar{x}_1).$$

So $f(\bar{x}_1)$ is a stationary point of the map f^2, and therefore it belongs to the set $\{0,\, x_*^1,\, \bar{x}_1,\, \bar{x}_2\}$. It remains to determine what this point is and we will do it by eliminating wrong options.

- The equality $f(\bar{x}_1) = 0$ should be rejected, because $f^{-1}(0) \in \{0, 1\}$;
- The equality $f(\bar{x}_1) = x_*^1$ does not take place because it implies what follows:

  $$f^2(\bar{x}_1) = \bar{x}_1 = f(x_*^1) = x_*^1.$$

  From this we conclude that $\bar{x}_1$ coincides with x_*^1, which is not true for $r > 3$;
- $f(\bar{x}_1) \neq \bar{x}_1$, because $\bar{x}_1$ is not a stationary point of the map f.

So there remains the option $f(\bar{x}_1) = \bar{x}_2$. For the point $\bar{x}_1$ the proof is analogous.

In order to show the stability of the stationary points $\bar{x}_{1\,2}$, we should consider the derivative of $f^2[x]'(x)$ at these points. Since $f^2[x]' = f'[f[x]] \cdot f'[x]$, it is instructive to count the derivatives of each component separately. Using the formula (4.12) we get:

$$f'(\bar{x}_{1,2}) = r(1 - 2\,x_{1,2}) = r\left[1 - \frac{r+1 \pm \sqrt{(r+1)(r-3)}}{r}\right] = -1 \mp \sqrt{(r+1)(r-3)}.$$

Next,

$$f'[f[\bar{x}_{1,2}]] = r(1 - 2f[\bar{x}_{1,2}]) = r(1 - 2\hat{x}_{2,1}) = r\left(1 - 2\frac{r+1 \mp \sqrt{(r+1)(r-3)}}{2r}\right)$$
$$= -1 \pm \sqrt{(r+1)(r-3)}.$$

Multiplying these two terms we get the formula

$$(f^2)'(\bar{x}_{1,2}) = 1 - (r+1)(r-3).$$

At $r_1 = 3$ the expression $(f^2)'(\bar{x}_{1,2})$ is equal to unit. It decreases monotonically as $r > 3$ grows, attaining the value -1 when $r = r_2 = 1 + \sqrt{6} \approx 3.449$. At $r > r_2$ the stationary points $\bar{x}_{1,2}$ become unstable; each of them undergoes the pitchfork bifurcation resulting in appearance of the stable cycle of the period 4. The points $\{\bar{x}_{1\,1}, \bar{x}_{1\,2}, \bar{x}_{2\,1}, \bar{x}_{2\,2}\}$, forming the period 4 orbit, are the stationary points of the map $f^4[x]$, which plays the same role in relation to $f^2[x]$ as the mapping $f^2[x]$ does in relation to $f[x]$.

The period doubling bifurcation is repeated an infinite many times, while the difference $\Delta_n = r_{n+1} - r_n$ between the successive bifurcation values becomes smaller and smaller. Finally, as r exceeds the value $r_\infty \approx 3.5699$ the iterations become chaotic. The chaotic nature of the iterations is manifested in that the adjacent trajectories are rapidly moving away from each other. As in the continuous case, the mutual "attraction" or "repulsion" of the trajectories in the vicinity of a point $x_0 \in (0,\,1)$ can be characterized by the Lyapunov index $\lambda(x_0)$, which is defined by the formula

$$\epsilon e^{N\,\lambda(x_0)} \approx \left|f^N(x_0 + \epsilon) - f^N(x_0)\right|.$$

In the limiting case $\epsilon \to 0$, $N \to \infty$ the formula

$$\lambda(x_0) = \lim_{N\to\infty}\lim_{\epsilon\to 0}\frac{1}{N}\log\left|\frac{f^N(x_0+\epsilon) - f^N(x_0)}{\epsilon}\right| = \lim_{N\to\infty}\frac{1}{N}\log\left|\frac{d\,f^n(x_0)}{d\,x_0}\right| \tag{4.13}$$

gives the exact value of $\lambda(x_0)$. Dependence of the Lyapunov index on the value of the parameter r is presented in Fig. 4.19.

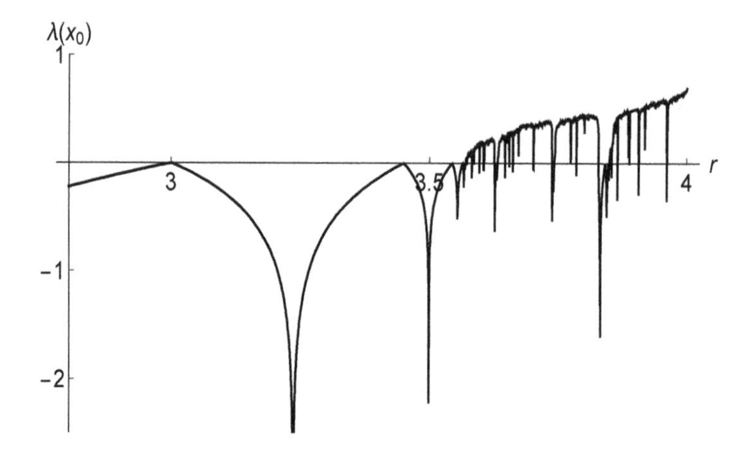

Fig. 4.19 Dependence of the Lyapunov index on the parameter r.

Construction of the bifurcation diagram illustrating the period doubling cascade in the logistic map and the parallel computations of the Lyapunov index can be found in the file PM1_9.nb. Note that the Lyapunov index approaches the horizontal axis from below at those points where the logistic

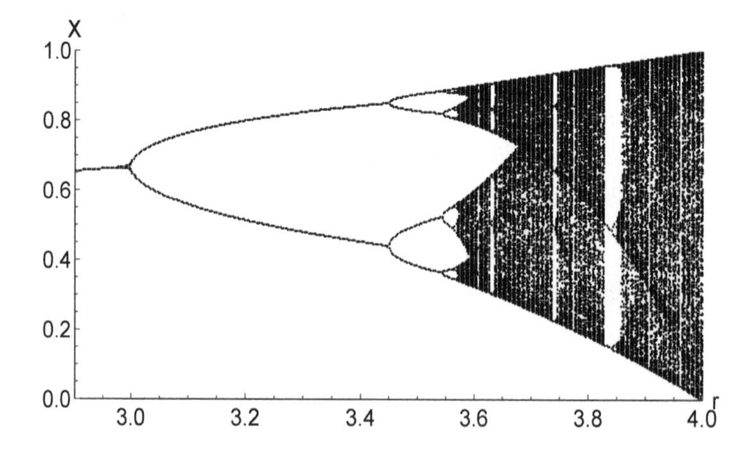

Fig. 4.20 Bifurcation diagram of the logistic map. Each vertical section of the presented graph can be thought of as a one-dimensional analog of the Poincaré section. The sections show iterations of the $f^n[x_0]$ map after removal of the first N iterations ($N \gg 1$). The diagram illustrates the period doubling scenario leading to the appearance of the chaotic orbits in the logistic mapping occurring at $r > r_\infty \approx 3.5699$.

map undergoes the period doubling bifurcations. At $r > r_\infty \approx 3.5699$ the graph moves to the upper half-plane, demonstrating at the same time the oscillatory features. Those values of the parameter r at which the Lyapunov index returns to the lower half-plane correspond to the *windows of periodicity*. Many of them are exposed in Fig. 4.20.

The results presented show that the logistic map demonstrates the period-doubling cascade. It is related to the universal property of superposition of unimodular functions that have a negative Schwartz derivative (4.7). The application of the "principle of shortening information" allows us to approach the understanding of the mechanism of period doubling bifurcations observed in the Rössler system by examining the related Lorenz map, or, more precisely, the logistic map belonging to the same class and exhibiting the same universal characteristics.

Problem.

(1) Explore the period doubling bifurcations of the map $x_{n+1} = r\, x_n(1-x_n^3)$ using the Lameray diagrams form, $r \in (0,\, 2.1)$.
(2) Based on the algorithm presented in the file PM1_9.nb, build the bifurcation diagram and the diagram showing the dependence of the associated Lyapunov index on the parameter $r \in (0,\, 2.1)$.

PART 2

Models described in terms of partial differential equations

Chapter 5

Models based on the concept of fields

5.1 Introductory remarks

In this part we consider the models of processes taking place in continual media. According to the hypothesis of *continuum*, a certain area $\Omega \subset \mathbb{R}^n$ is continuously filled with matter. The model is assumed to describe processes characterized by the changes of quantities $\{u_i(t,x)\}_{i \in I}$, $t \in \mathbb{R}$, $x \in \Omega$ called *the fields*, which depend on selected parameter $t \in$ called *temporal variable*, and the spatial variable $x \in \Omega \in \mathbb{R}^n$. If the fields do not depend on the parameter t, then they are called *stationary*.

The quantities characterizing liquids and gases are called *the field of velocity*, *the field of density* and *the field of pressure*. These fields are subjected to the laws of dynamics and thermodynamics. The universal law of conservation of mass (quantity of matter) is also applicable to them, as well as the laws of conservation of momentum and energy resulting from the dynamics. Another type of processes taking place in solids are the phenomena associated with transport of heat energy or neutrons, inter-diffusion, spread of pollution, etc., which may not be accompanied by flows of matter constituting the bulk of the substance under consideration. These processes can be observed in liquid and gaseous media as well, where they are accompanied, as a rule, by matter flows caused by gradients of field values. Equations defining the field quantities in transport phenomena are also derived from the aforementioned laws of dynamics, and the conservation laws associated with them. Almost everywhere for a complete and unambiguous description of the processes occurring in material substances, one also has to add the so-called *equations of state* (or determining equations), establishing relationships between individual fields (most often between the pressure field and the field of density). In contrast to the equations derived from the laws of mass, energy and momentum conservation, the determining equations do not follow merely from the basic principles of mechanics, and thermodynamics (although, undoubtedly, they must be compatible with them), but contain information about such subtle things as the structure of matter on the mesoscale, and the aggregate state which in turn depends on temperature and pressure. Factors such as the speed and intensity of external influences also play here an important role.

Experts studying the behavior of media (especially under the extreme conditions) know that each equation of state adequately describes the matter only within a limited range of parameters' changes. The inability to derive a universal equation of state describing the system in arbitrary range of field parameters' changes is undoubtedly a consequence of adopting the *continuum* hypothesis, which causes that the model cannot fully describe the real structure of the material medium consisting of individual atoms and molecules which form, depending on the state of aggregation, clusters, crystal lattices and other structures. On the other hand, the hypothesis of *continuum* allows to describe the medium in terms of field quantities satisfying partial differential equations, which is undoubtedly very favorable from the practical point of view.

5.2 (Gas-) Hydro-dynamic equations

As it was mentioned above, the states of both gas and liquid in n-dimensional case can be described using $n+2$ macroscopic parameters: velocity field components $u^i(t,\, x)$, $i = 1, \ldots, n$, density field $\rho(t,\, x)$ and field of pressure $p(t,\, x)$. In the description presented in this section, we ignore the tensor properties of liquids and gases, which are manifested rather in fast processes (such as impact, explosion, detonation, etc.), and processes in which the exchange of thermal energy plays an important role (as in the case of *thermoconvection* [Schuster (2005)]).

For a full description of the medium characterized by $n+2$ parameters, where n is a number of spatial variables, we need a system consisting of $n+2$ partial differential equations. Such a system can be found using the balance equations for mass and momentum and the equation of state relating the pressure field to the density field.

5.2.1 *Balance equation for mass*

The law of conservation of mass claims that matter does not disappear and does not form out of nothing (a person even superficially familiar with the theory of elementary particles may, of course, undermine a given thesis, but the said law works very accurately in the case of low speeds and moderate energies).

Mass contained in the set $\mathbb{R}^n \supset \Omega = \text{const}$ is expressed by the following equation:

$$M(t) = \int_{\Omega} \rho(t, x)\, d\,x, \tag{5.1}$$

where $\rho(t, x)$ is the density of media. Due to the aforementioned conservation law, changes of $M(t)$ can only be caused by a mass flow through the fixed boundary $\partial\Omega$. So the balance of mass equation will take the form

$$\frac{d\,M}{d\,t} = \int_{\Omega} \frac{\partial\,\rho(t, x)}{\partial\,t}\, d\,x = -\int_{\partial\Omega} \rho\,\vec{u}(t, S)\, d\,\vec{\sigma}(S), \tag{5.2}$$

where $S \in \partial\Omega$, $d\,\vec{\sigma}(S) = d\sigma\,\vec{n}(S)$, $d\sigma$ is the element of the surface $\partial\Omega$, $\vec{n}(S)$ is the field of unit vectors perpendicular to $\partial\Omega$ and directed outward Ω. Using the Green's theorem ([Maurin (1973)] Part 1, Ch. XIII, # 7), we get:

$$\int_{\Omega} \left\{ \frac{\partial\,\rho}{\partial\,t} + \operatorname{div}\,[\rho\,\vec{u}] \right\} d\,x = 0. \tag{5.3}$$

Arbitrariness of the set Ω implies that the above integral will be zero if and only if the expression under the integral nullifies at arbitrary point $x \in \Omega$, at any moment of time t, in other words, when the equations

$$\frac{\partial\,\rho(t, x)}{\partial\,t} + \sum_{k=1}^{n} \frac{\partial}{\partial\,x^k}\left[\rho(t, x)\,u^k(t, x)\right] = 0, \qquad n = 1, 2, 3 \tag{5.4}$$

hold.

5.2.2 *Balance of momentum equation. Complete systems of hydrodynamic-type equations*

Balance of momentum equation appears from the Newton's second law. For the point mass m, this law can be written as

$$\frac{d\,\left[m\,u^i\right]}{d\,t} = \sum_{k} F_k^i, \qquad i = 1, \ldots, n, \quad n = 1, 2, 3, \tag{5.5}$$

where $u^i = (\vec{u})^i$ is the i-th component of the velocity vector, $\sum_k F_k^i$ is the ith component of the sum of the forces acting on the point mass. Note that the expression $m\vec{u}$ describes the momentum of the point mass.

Due to the fact that the momentum of the material medium contained in the volume Ω is expressed by the formula

$$\vec{P}(t) = \int_{\Omega} \rho(t, x)\, \vec{u}(t, x)\, dx,$$

continual analog to the formula (5.5) takes the form

$$\frac{d}{dt} P(t)^i = \frac{d}{dt} \int_{\Omega} \rho(t, x)\, u(t, x)^i\, dx = \int_{\Omega} \sum_k f_k^i\, dx$$

$$- \int_{\partial\Omega} \rho\, u^i(t, x)\, \vec{u}(t, x)\, d\vec{\sigma} - \int_{\partial\Omega} p(t, x)\, \delta_{im}\, n^k\, d\sigma, \quad i = 1, \ldots, n, \quad n = 1, 2, 3,$$

where f_k^i is equal to the ith component of the external force per unit mass $\vec{f_k}$ (the most common external force is the gravitational force $\rho\,\vec{g}$), while the term $-p(t, x)\, \delta_{im}$ in the surface integral describes the projection of the pressure force perpendicular to the surface $\partial\Omega$ on the ith coordinate (according to the Pascal's law, the pressure is the same in all directions).

Using the Green-Gauss-Ostrogradsky's theorem, we obtain the equation

$$\int_{\Omega} \left\{ \frac{\partial}{\partial t} [\rho u^i] + \sum_{j=1}^{n} \frac{\partial}{\partial x_j} [\rho u^i u^j + p\, \delta_{ij}] - \sum_k f_k^i \right\} dx$$

$$= \int_{\Omega} \left\{ \frac{\partial}{\partial t} [\rho u^i] + \sum_{j=1}^{n} \frac{\partial}{\partial x_j} [\rho u^i u^j] + \frac{\partial p}{\partial x^i} - \sum_k f_k^i \right\} dx = 0.$$

It follows from the arbitrariness of the volume Ω that this integral will be equal to zero if and only if the integral expression is equal to zero at any point x, in arbitrary time t, in other words, when the equation holds

$$\frac{\partial}{\partial t} [\rho u^i] + \sum_{j=1}^{n} \frac{\partial}{\partial x_j} [\rho u^i u^j] + \frac{\partial p}{\partial x^i} = \sum_k f_k^i, \qquad i = 1, 2, 3.$$

The above equation can also be written as

$$\rho \left\{ \frac{\partial u^i}{\partial t} + \sum_{j=1}^{n} u^j \frac{\partial u^i}{\partial x^j} \right\} + \frac{\partial p}{\partial x^i} - F^i + u^i \left[\frac{\partial \rho}{\partial t} + \sum_{j=1}^{n} \frac{\partial}{\partial x^j} (\rho u^j) \right] = 0,$$

where $F^i = \sum_k f_k^i$. Note that the expression in the square brackets coincides with the l.h.s. of the equation (5.4). So ultimately, the momentum balance equation can be presented as follows:

$$\rho\left[\frac{\partial}{\partial t}u^i + \sum_{j=1}^{n} u^j \frac{\partial}{\partial x_j}u^i\right] + \frac{\partial p}{\partial x^i} = F^i, \qquad i = 1, \ldots, n, \quad n = 1, 2, 3. \tag{5.6}$$

The system (5.4), (5.6) consists of $n+1$ equations containing $n+2$ functions, so it is underdetermined. To complete it, one can use the equation of state $F(\rho, p) = 0$, which establishes the relationship between density and pressure. In the case of the description of adiabatic processes (i.e. "fast" processes in which heat exchange does not play a significant role), the equation of state takes the form

$$p = \frac{\beta}{\gamma+2}\rho^{\gamma+2}. \tag{5.7}$$

Using the equality (5.7), the function p can be eliminated from the momentum balance equation. The differential consequences of this equality can also be used. In the latter case, we obtain the classical (gas-) hydro-dynamic system, which in the absence of mass forces takes the form

$$\rho\left[\frac{\partial}{\partial t}u^i + \sum_{j=1}^{n} u^j \frac{\partial}{\partial x_j}u^i\right] + \frac{\partial p}{\partial x^i} = 0, \tag{5.8}$$

$$\frac{\partial \rho}{\partial t} + \sum_{k=1}^{n} \frac{\partial}{\partial x^k}\left(\rho u^k\right) = 0, \tag{5.9}$$

$$\frac{\partial p}{\partial t} + \sum_{k=1}^{n} u^k \frac{\partial p}{\partial x^k} + \beta\rho^{\gamma+1}\sum_{k=1}^{n} \frac{\partial}{\partial x^k}\left(\rho u^k\right) = 0, \tag{5.10}$$

where $i = 1, \ldots, n$, $n = 1, 2, 3$.

Note that if we express p in the equation (5.8) using the formula (5.7), then the system (5.8)–(5.9) will be complete. In the case of one spatial variable, this system will look as follows:

$$\frac{\partial}{\partial t}u + u\frac{\partial}{\partial x}u + \beta\rho^{\gamma}\frac{\partial \rho}{\partial x} = 0, \tag{5.11}$$

$$\frac{\partial \rho(t, x)}{\partial t} + \frac{\partial}{\partial x}\left[\rho(t, x)\, u(t, x)\right] = 0. \tag{5.12}$$

The system (5.11)–(5.12) is, apparently, the simplest nonlinear system of hydrodynamic type that can be obtained from the first principles.

For the incompressible medium ($\rho = \rho_0 = \text{const}$) in which the viscous forces act, the system analogous to (5.8)–(5.10) takes the form

$$\begin{cases} \rho_0 \left[\frac{\partial}{\partial t} u^i + \sum_k u^k \frac{\partial u^i}{\partial x_k}\right] + \frac{\partial p}{\partial x^i} = \nu \sum_{k=1}^n \frac{\partial^2}{\partial x_k^2} u^i, \\ \sum_{k=1}^n \frac{\partial}{\partial x^k} u^k(t, x) = 0, \qquad n = 1, 2, 3, \end{cases} \tag{5.13}$$

where ν is the viscosity coefficient. The system (5.13) is referred to as the Navier-Stokes system.

The state equation (5.7) used in (5.11) is not the only possible formula relating p and ρ. A much more general equation of state describing a wide class of heterogeneous substances (such as rocks, soils, multiphase mixtures) can be presented as [Vladimirov *et al.* (2012)]

$$p(t, x_l) = f[\rho(t, x_l)] + \int_{-\infty}^{t} K(t - t')\, g[\rho(t', x_l)]\, d t', \tag{5.14}$$

where x_l is the mass (Lagrange) coordinate, connected with the cartesian coordinate x by means of the formula

$$x_l = \int_c^x \rho(t, \xi)\, d\xi.$$

A simple calculations show that under the change of variables

$$t_l = t, \qquad x_l = \int_c^x \rho(t, \xi)\, d\xi$$

the partial derivatives are transformed as follows:

$$\frac{\partial}{\partial x} = \rho \frac{\partial}{\partial x_l}, \qquad \frac{\partial}{\partial t} = \frac{\partial}{\partial t_l} - \rho u \frac{\partial}{\partial x_l}.$$

Mass and momentum equations in the presence of external force ρF in new variables take the form:

$$\frac{\partial u}{\partial t_l} + \frac{\partial p}{\partial x_l} = F, \qquad \frac{\partial \rho}{\partial t_l} + \rho^2 \frac{\partial u}{\partial x_l} = 0 \tag{5.15}$$

(for the sake of simplicity we omit the lower indices further on). The system (5.15) is not closed, so we derive the closing equation from (5.14) assuming that

$$f(\rho) = \frac{\chi}{\tau(\nu + 2)} \rho^{\nu+2}, \quad g(\rho) = -\sigma \rho^{\nu+2}, \quad K(t - t') = \exp\left[-\frac{t - t'}{\tau}\right].$$

Differentiating the equation

$$p = \frac{\chi}{\tau(\nu + 2)} \rho^{\nu+2} - \sigma \int_{-\infty}^{t} \exp\left[-\frac{t - t'}{\tau}\right] \rho^{\nu+2}\, d t'$$

with respect to the variable t, we get the equation

$$\tau \frac{\partial}{\partial t}\left(p - \frac{\chi}{\tau(\nu+2)}\rho^{\nu+2}\right) = \kappa\,\rho^{\nu+2} - p,$$

where $\kappa = \chi/[\tau(\nu+2)] - \tau\,\sigma$. Of special interest is the system obtained by switching to the variable $V = \rho^{-1}$, under the assumption that $F = \gamma = \text{const}$ and $\nu = -1$. Then the closed system takes the following form [Vladimirov (2008)]:

$$\begin{cases} u_t + p_x = \gamma, \\ V_t - u_x = 0, \\ \tau\, p_t + \frac{\chi}{V^2}\, u_x = \frac{\kappa}{V} - p. \end{cases} \tag{5.16}$$

5.2.3 *Sub-models and their self-similar solutions*

As noted above, the system (5.11)–(5.12) is the simplest realistic system of the hydrodynamic type, and nevertheless in the general case solutions to this system can be obtained only numerically. However, some of the characteristics of these solutions can be explored by examining the equation

$$u_t + u\, u_x = 0, \tag{5.17}$$

serving as a sub-model to said system. Both the system (5.11)–(5.12) and the equation (5.17), called *the Hopf's equation*, possess the *shock-wave* solutions, demonstrating the discontinuities of the first kind. Such solutions may arise spontaneously from smooth initial conditions in the course of time evolution.

To obtain particular solutions of nonlinear differential equations, we will repeatedly use considerations based on the theory of similarity and dimensions [Barenblatt (2012)], the foundations of which are briefly outlined in this section.

As is known from elementary physics, physical quantities are divided into basic and derived quantities — the latter being expressed by basic ones. In mechanics, for example, time $[T]$, length $[L]$ and mass $[M]$ can be chosen as basic physical quantities ($[T]$, $[L]$, $[M]$ and other symbols in square brackets denote the corresponding physical units). All other mechanical quantities, for example velocity $[V] = \frac{L}{T}$, acceleration $[a] = \frac{L}{T^2}$, and force $[F] = \frac{M\,L}{T^2}$ can be expressed as a combination of the three basic quantities.

Remark. Note that the division into the basic and derived quantities is highly ambiguous. However, choosing basic quantities, or passing from one set of basic quantities to another one, we must ensure, that each basic quantity has independent dimension, and any other quantity can be expressed in the form of the algebraic combination of basic ones and dimensionless parameters.

Another important observation is that physical laws do not depend on what units they are expressed in: grams, seconds, centimeters, or, say, meters, hours, and kilograms, if to mention mechanics as an example. It is also possible to present the laws in the dimensionless form, and this leads, generally speaking, to simplification.

Theorem 5.1 (Buckingham's Π theorem). *Let us consider the formula*

$$u = f(W_1, W_2, ..., W_n), \tag{5.18}$$

expressing some relationship between the quantity u having some dimension and the quantities $W_1, \ldots, W_n$, whose dimensions are expressed by the basic quantities $L_1, L_2, ..., L_m$ $(m \le n)$:

$$[u] = L_1^{p_1} \cdot L_2^{p_2} \ldots \cdot L_m^{p_m}$$

$$W_j = L_1^{p_1^{j_1}} \cdot L_2^{p_2^{j_2}} \ldots \cdot L_m^{p_m^{j_m}}.$$

Then

- *this relationship can be presented in a dimensionless form*

$$\Pi = \Phi(\Pi_1, \Pi_2, ..., \Pi_k); \tag{5.19}$$

- *the number of parameters in the right part of the formula (5.19) does not exceed $n - m$.*

The main content of the Π theorem can be formulated as follows: the dependence (5.18) simplifies as a rule when being written in dimensionless variables.

Proof. Let us do the scaling

$$L_1 \to L_1^* = e^{\mu} L_1.$$

This scaling will change the parameters as follows:

$$u^* = e^{\mu\, p_1} u, \quad W_j^* = e^{\mu\, p_{1\,j}} W_j, \quad j = 1, \ldots, n.$$

As a result, the dependence (5.18) will turn into the dependence

$$e^{\mu\, p_1} u = f(e^{\mu\, p_{1\,1}} W_1, \ldots, e^{\mu\, p_{1\,n}} W_n).$$

And here the following cases can have place:

Case 1. $p_1 = p_{1\,1} = p_{1\,2} = ... = p_{1\,m} = 0$. Then the formula (5.18) does not depend on the quantity L_1, that is, the set of basic quantities for this formula consists of the parameters $L_2, L_3, ..., L_m$.

Case 2. One of the parameters $p_{1\,1}, p_{1\,2}, ..., p_{1\,n}$ is nonzero. Without the loss of generality we can assume that $p_{1\,1} \neq 0$. Let us introduce the following change of variables:

$$\begin{aligned} &X_{i-1} = W_i\, W_1^{-p_{1\,i}/p_{1\,1}}, \quad i = 2, 3, ..., n, \\ &X_n = W_1; \\ &v = u\, W_1^{-p_1/p_{1\,1}}. \end{aligned}$$

In new variables the formula (5.18) will take the form

$$v = F(X_1, X_2, \ldots, X_n), \tag{5.20}$$

where F is some function. The scaling $L_1^* = e^{\mu} L_1$ implies following change of variables:

$$\begin{aligned} &v^* = e^{p_1\,\mu}\, u\, (W_1\, e^{p_{1\,1}\mu})^{-p_1/p_{1\,1}} = v, \\ &X_i = W_i e^{\mu\, p_{1\,i}}\, (W_1\, e^{p_{1\,1}\mu})^{-p_{1\,i}/p_{1\,1}} = X_i, \quad i = 1, \ldots, n-1, \\ &X_n^* = W_1 e^{\mu p_{11}} = e^{\mu p_{11}} X_n. \end{aligned} \tag{5.21}$$

Thus,

$$v = F(X_1, X_2, \ldots, X_{n-1}, e^{\mu p_{11}} X_n). \tag{5.22}$$

Comparing (5.20) with (5.22), we conclude that F does not depend on $X_n = W_1$ and the formula (5.22) actually takes the form

$$v = G(X_1, X_2, \ldots, X_{n-1}). \tag{5.23}$$

Next, on virtue of (5.21) the variables $(v, X_1, \ldots, X_{n-1})$ are invariant with respect to the scaling $L_1^* = e^{\mu} L_1$, in other words, they do not depend on the basic quantity L_1.

Repeating the above procedure $m - 1$ times, we finally get the formula (5.19), in which the parameters $\Pi, \Pi_1, \Pi_2, \ldots, \Pi_k$ do not depend on the basic quantities $L_1, L_2, \ldots, L_m$, that is, they are dimensionless and this ends the proof.

Remark. Theorem 5.1 is associated with the name of Buckingham who formulated it for the first time in explicit form.

Example 5.1. Let us find the function expressing the drop of the pressure $\frac{dp}{dx}$ in a pipe with a diameter of $D[L]$, at which a liquid with a viscosity of $(\mu[\frac{M}{L\cdot T}])$ and density $\rho[\frac{M}{L^2}]$ moves with the uniform speed $v[\frac{L}{T}]$.

In other words, it is proposed to find the simplest form of the dependence

$$\frac{d\,p}{d\,x} = f(\rho,\, \mu,\, v,\, D) \tag{5.24}$$

using the Π theorem.

Let us choose as the base quantities $\mu\left[\frac{M}{LT}\right]$, $v\left[\frac{L}{T}\right]$ and $D[L]$. The quantities $\rho\left[\frac{M}{L^3}\right]$ and $\frac{d\,p}{d\,x}\left[\frac{M}{L^2\,T^2}\right]$ can be expressed as follows:

$$\rho = \Pi_1\,\frac{\mu}{v\,D}, \qquad \frac{d\,p}{d\,x} = \Pi\,\frac{\mu}{v\,D^2},$$

where Π and Π_1 denote the dimensionless parameters. Thus, we can rewrite the dependence (5.24) in the following equivalent form:

$$\Pi = \frac{\frac{d\,p}{d\,x}}{\mu\,v\,D^{-2}} = \frac{1}{\mu\,v\,D^{-2}}\,f(\rho,\, \mu,\, v,\, D)$$

and, since the parameter Π in the left-hand side is dimensionless, the right-hand side of the above formula can only depend on the parameter Π_1:

$$\Pi = \Phi(\Pi_1).$$

So the initial formula can be expressed in the form

$$\frac{d\,p}{d\,x} = \frac{\mu\,v}{D^2}\Phi(Re), \qquad Re = \frac{\rho\,v\,D}{\mu}.$$

Now we see that the drop of the pressure necessary for the liquid in the pipe to flow at a constant speed is expressed by the algebraic ratio of the basic quantities and some function depending on the dimensionless parameter $Re = \rho\,v\,D/\mu$ called the Reynolds number.

Example 5.2. And now we wish to demonstrate the effectiveness of the theory of similarity and dimensions for solving partial differential equations, using as an example *the problem of point explosion* for the Hopf equation: at $t = 0$ in the point $x = 0$ a perturbation $Q > 0$ is applied. We are interested in the evolution of the initial perturbation. Mathematically, the problem is formulated as follows:

$$u_t + u\,u_x = 0, \quad t > 0, \quad x \in R, \tag{5.25}$$

$$\partial_x^k\,u(0,\,x) = 0 \quad \text{as} \quad x \neq 0,\ k = 0, 1, \tag{5.26}$$

$$\int_{-\infty}^{+\infty} u(t, x)\,d\,x = Q. \tag{5.27}$$

As the basic units we choose

$$Q[L^2/T] \text{ and } t[T]$$

(let us remind that here and henceforth we use the following notation: $[L]$ is the unit of length, $[T]$ is the unit of time, $[M]$ is the unit of mass). The remaining values (that is $x[L]$ and $u[L/T]$) are expressed as follows:

$$u = \sqrt{\frac{Q}{t}}\varphi(\xi), \qquad x = \xi\sqrt{Q\,t}.$$

After substituting the above ansatz into the Hopf equation, we obtain the reduced equation

$$\frac{d}{d\,\xi}\,\varphi\cdot(\xi-\varphi)=0.$$

Integrating this equation, we obtain:

$$\varphi\cdot(\xi-\varphi)=C.$$

If we assume that $C=0$, then the only non-trivial solution will take the form $\varphi=\xi$. Next, in order to fulfill the condition (5.27), we should assume that the solution has compact support:

$$\varphi=\begin{cases}\xi & \text{as} \quad 0<\xi<\xi_f,\\ 0 & \quad \text{elsewhere},\end{cases}$$

where ξ_f is a non-negative parameter. Thus, the solution of the initial equation will have the form

$$u=\begin{cases}\frac{x}{t} & \text{as} \quad 0<x<x_f=\xi_f\sqrt{Q\,t},\\ 0 & \quad \text{elsewhere}.\end{cases}$$

The parameter ξ_f can be calculated from the condition (5.27):

$$Q=\int_0^{x_f}\frac{x}{t}\,d\,x=Q\int_0^{\xi_f}\xi\,d\,\xi,$$

hence $\xi_f=\sqrt{2}$.

Finally, we get the solution

$$u=\begin{cases}\frac{x}{t}, & \text{as} \quad 0<x<x_f=\sqrt{2\,Q\,t},\\ 0, & \quad \text{elsewhere}.\end{cases} \tag{5.28}$$

Next we consider the Burgers equation

$$\frac{\partial\,u}{\partial\,t}+u\frac{\partial\,u}{\partial\,x}=\nu\,\frac{\partial^2\,u}{\partial\,x^2}, \tag{5.29}$$

which is a model equation for the Navier-Stokes system (5.13) in the case of one spatial variable. Equation (5.29) has many interesting properties. It describes, inter alia, *diffuse shock waves.* Besides, the equation (5.29) belongs to the family of *completely integrable equations,* since the Cole-Hopf transformation $u(t,\, x) = -2\,\nu\partial_x \log \Phi(t,\, x)$ [Dodd *et al.* (1984)] reduces the problem to finding the function $\Phi(t,\, x)$ which satisfies the linear heat equation.

Example 5.3. Like (5.17), the equation (5.29) allows for the formulation of a problem of heat explosion that can be solved using the theory of dimensions and similarity: at the moment $t = 0$ at the point $x = 0$ the energy $A > 0$ is released.

Mathematically, the problem is formulated as follows:

$$u_t + u\, u_x = \nu\, u_{xx}, \quad t > 0, \quad x \in R, \tag{5.30}$$

$$\partial_x^k\, u(0,\, x) = 0 \quad \text{as} \quad x \neq 0,\; k = 0, 1, \tag{5.31}$$

$$\int_{-\infty}^{+\infty} u(t, x)\, d\, x = A. \tag{5.32}$$

Let us choose as basic the following units:

$$\nu[L^2/T] \text{ and } t[T],$$

where $[L]$ is the unit of length, $[T]$ is the unit of time. The remaining units, i.e. $x[L]$ and $u[L/T]$, can be expressed as follows:

$$u = \sqrt{\frac{\nu}{t}}\varphi(\xi), \qquad x = \xi\sqrt{\nu\, t}, \tag{5.33}$$

where $\varphi(\xi)$ is a function depending on the dimensionless parameter ξ. After substituting (5.33) into the Burgers equation, we get the equation

$$2\,\frac{d^2}{d\,\xi^2}\varphi = \frac{d}{d\,\xi}\left[\varphi(\varphi - \xi)\right].$$

Integrating this equation and equating to zero the constant of integration so that the initial conditions are met, we obtain the equation

$$\frac{1}{\varphi}\frac{d\,\varphi}{d\,\xi} = \frac{1}{2}(\varphi - \xi).$$

Introducing the new variable $p(\xi) = \log[\varphi]$, we obtain:

$$\frac{d\,p}{d\,\xi} - \frac{1}{2}e^p = -\frac{1}{2}\xi.$$

This equation can be solved using *the method of variation of constant.* Within this method we are looking first for the solution of the corresponding homogeneous equation

$$\frac{d\tilde{p}}{d\xi} - \frac{1}{2}e^{\tilde{p}} = 0.$$

Its solution is expressed as follows:

$$\tilde{p} = -\log[C - \xi/2].$$

Next we look for the solution of the heterogeneous equation in the form

$$\varphi = e^{p} = \frac{1}{C(\xi) - \xi/2}.$$

It is easily seen, that the function $C(\xi)$ satisfied the equation

$$C' - \frac{1}{2}\xi\, C = -\frac{1}{4}\,\xi^2.$$

It is the non-homogeneous linear equation. To solve it, we again use the method of variation of constant. As a result, we get the solution

$$C(\xi) = C_2\, e^{\xi^2/4} - \frac{1}{4}\int_0^{\xi} \tau^2 \,\exp\left[\frac{\xi^2 - \tau^2}{4}\right] d\,\tau.$$

To calculate the integral in the formula, we use the *Mathematica* package:

$$\int_0^{\xi} \tau^2 \,\exp\left[\frac{-\tau^2}{4}\right] d\,\tau = -2\,\xi\, e^{-\xi^2/4} + 2\sqrt{\pi}\,\mathrm{erf}[\xi/2]$$

(let us remind that $\mathrm{erf}[z] = \frac{2}{\sqrt{\pi}}\int_0^z e^{-\tau^2}\,d\,\tau$). Using this formula, we finally get the following solution:

$$u(t,\,x) = \sqrt{\frac{\nu}{t}}\frac{e^{-\xi^2/4}}{C_2 - \frac{\sqrt{\pi}}{2}\,\mathrm{erf}[\xi/2]}, \qquad \xi = \frac{x}{\sqrt{\nu\, t}}.$$

The constant C_2 is obtained from the condition (5.32). Inserting to this formula the above solution and performing the integration (again using the *Mathematica* package) we get:

$$A = \nu\int_{-\infty}^{+\infty}\frac{e^{-\tau^2/4}}{C_2 - \frac{\sqrt{\pi}}{2}\mathrm{erf}[\tau/2]}\,d\,\tau = 4\nu\,\mathrm{arccoth}\left[\frac{2}{\sqrt{\pi}\,C_2}\right].$$

Hence

$$C_2 = \frac{\sqrt{\pi}}{2}\cdot\frac{e^{\frac{A}{4\,\nu}} + e^{-\frac{A}{4\,\nu}}}{e^{\frac{A}{4\,\nu}} - e^{-\frac{A}{4\,\nu}}}.$$

Introducing the parameter $R = \frac{A}{2\nu}$ (being the analog of Reynolds number), we can write the solution in the form

$$u(t,\, x) = \sqrt{\frac{\nu}{t}}\, e^{-\xi^2/4}\, \frac{e^R - 1}{\frac{\sqrt{\pi}}{2}\left[1 + e^R + (1 - e^R)\,\mathrm{erf}(\xi/2)\right]}, \qquad \xi = \frac{x}{\sqrt{\nu\, t}}. \tag{5.34}$$

Asymptotics. Observing that $R \to 0$, as $\nu \to \infty$, we can use the approximation $e^R \approx 1 + R$ valid as $\nu \gg 1$. Under this assumption we get:

$$\mathrm{erf}\left[\frac{x}{\sqrt{4\,\nu\, t}}\right] = \frac{2}{\sqrt{\pi}} \int_0^{\frac{x}{\sqrt{4\,\nu\,t}}} \left(1 - \frac{\xi^2}{1} + \frac{\xi^4}{2} + \ldots\right) d\xi \approx \frac{2}{\sqrt{\pi}}\, \frac{x}{\sqrt{4\,\nu\, t}}.$$

So, $\mathrm{erf}\,[\frac{x}{\sqrt{4\,\nu\,t}}]$ tends to zero as ν tends to infinity. Assuming that ν is sufficiently large, we get:

$$u(t,\, x) \approx \sqrt{\frac{\nu}{t}} e^{-\frac{x^2}{4\,\nu\, t}}\, \frac{\frac{A}{2\,\nu}}{\sqrt{\pi}} = \frac{A}{\sqrt{4\,\pi\, t\, \nu}}\, e^{-\frac{x^2}{4\,\nu\,t}}. \tag{5.35}$$

Asymptotic solution obtained for large ν coincides with the solution of the heat explosion for the linear transport equation (5.48). It is also possible to show that at large R equation (5.34) tends to the solution

$$u(t,\, x) = \begin{cases} \frac{x}{t} & \text{as} \quad 0 < x < x_\Phi = \sqrt{2\, t\, A}, \\ \\ 0 & \text{as} \quad x > x_\Phi, \end{cases} \tag{5.36}$$

which the Burgers equation "shares" with the Hopf equation. We omit the proof of this statement for it is rather tedious. Yet one can check the validity of this assertion building the graphic image of (5.34) in the *Mathematica* package. For comparison, the solutions of the Burgers equation (5.34) corresponding to the values $\nu = 0.25$ and $R = 25$ are presented in Fig. 5.1 together with the solutions of the equation (5.36) corresponding to $A = 12.5$.

Another sub-model can be obtained by linearizing the system (5.11)–(5.12). For this purpose, let us consider small disturbances of the stationary solution $u = 0,\ \rho = \rho_0 > 0$. We are looking for a solution of the form

$$u(t,\, x) = \epsilon \hat{u}(t,\, x), \qquad \rho(t,\, x) = \rho_0 + \epsilon \hat{\rho}(t,\, x), \quad 0 < \epsilon \ll 1.$$

Inserting these ansatze into the source system we get, drooping out the terms of the order $O(|\epsilon|^2)$, the following system:

$$\frac{\partial}{\partial t}\, \hat{u} + \beta\, \rho_0^\gamma\, \frac{\partial\, \hat{\rho}}{\partial\, x} = 0,$$

$$\frac{\partial\, \hat{\rho}(t,\, x)}{\partial\, t} + \frac{\partial}{\partial\, x}\, (\rho_0\, \hat{u}(t,\, x)) = 0.$$

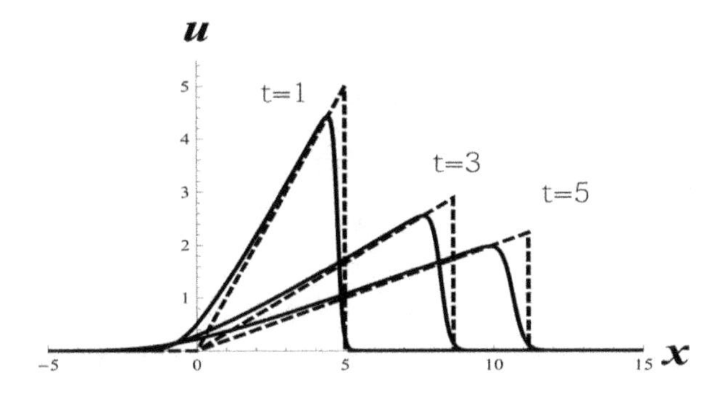

Fig. 5.1 Graphs of the function (5.34) obtained for $\nu = 0.25$ and $R = 25$ (solid lines) and the graphs of the function (5.36) (dashed lines) obtained for the corresponding values of the parameters.

Differentiating the first equation with respect to x and the second equation with respect to t, dividing the second equation by ρ_0 and then subtracting the equations side by side, we get a one-dimensional *wave equation*

$$\frac{\partial^2 \hat{\rho}(t,\, x)}{\partial\, t^2} - c^2\, \frac{\partial^2 \hat{\rho}(t,\, x)}{\partial\, x^2} = 0, \qquad c^2 = \beta\, \rho_0^{\gamma+1}. \tag{5.37}$$

This equation describes the propagation of small longitudinal disturbances of the density.

A natural question arises: what additional conditions need to be met for the solution to the problem (5.37) be unique? We will get the answer to this question when solving the wave equation by means of the *method of characteristics*. Introducing the new variables

$$\xi = x - c\,t, \qquad \eta = x + c\,t,$$

expressing the partial derivatives in terms of ξ, η and then performing the simple algebraic manipulations, we get the equation

$$\frac{\partial^2 \hat{\rho}}{\partial\, \xi \partial\, \eta} = 0.$$

Integrating this equation with respect to the variable η, we get:

$$\frac{\partial\, \hat{\rho}}{\partial\, \xi} = F(\xi).$$

Integrating the above equation with respect to ξ, we finally get the solution of the following form:

$$\hat{\rho} = f(x - c\,t) + g(x + c\,t), \tag{5.38}$$

where the functions $f(\cdot)$, $g(\cdot)$ are arbitrary. Thus, for the solution to be unique, it is enough, for example, at time $t = 0$ to define two functions. Usually, the initial value problem takes the form of

$$\hat{\rho}(0,\, x) = \varphi(x), \qquad \frac{\partial}{\partial t}\hat{\rho}(0,\, x) = \psi(x), \tag{5.39}$$

where $\varphi(\cdot)$, $\psi(\cdot)$ are known functions. On virtue of the above conditions, $f(\cdot)$, $g(\cdot)$ will satisfy the system

$$f(x) + g(x) = \varphi(x), \qquad c\dot{f}(x) - c\dot{g}(x) = \psi(x).$$

Integrating the second equation within the limits $(A,\, x)$, we get the system of algebraic equations:

$$f(x) + g(x) = \varphi(x), \qquad g(x) - f(x) = \frac{1}{c}\int_A^x \psi(y)\,dy.$$

Solving them, we obtain:

$$g(x) = \tfrac{1}{2}\left[\varphi(x) + \tfrac{1}{c}\int_A^x \psi(y)\,dy\right],$$

$$f(x) = \tfrac{1}{2}\left[\varphi(x) - \tfrac{1}{c}\int_A^x \psi(y)\,dy\right].$$

Substituting it into the formula (5.38), after simple transformations we get the unique solution to the initial value problem (called *the Cauchy* problem) in the form:

$$\hat{\rho}(t,\, x) = \varphi(x) = \frac{1}{2}\left[\varphi(x - ct) + \varphi(x + ct)\right] + \frac{1}{2c}\int_{x-ct}^{x+ct} \psi(y)\,dy. \tag{5.40}$$

Example 5.4. Consider the following initial conditions:

$$\hat{\rho}(0,\, x) = \begin{cases} 1 + \cos x, & \text{as } \ |x| < \pi, \\ 0, & \text{elsewhere}, \end{cases} \tag{5.41}$$

$$\partial_t\,\hat{\rho}(0,\, x) = \psi(x) = 0.$$

A non-zero initial condition can be written in the *Mathematica* package using the command φ [x _] = (1+ Cos [x]) ThetaHeaviside[x + π] ThetaHeaviside[π - x]. Substituting this function into the formula (5.40) and then plotting graphs corresponding to different values of the time variable, we get the result shown in Fig. 5.2. One can see that the initial disturbance splits into two impulses moving in opposite directions. These impulses in the course of evolution retain the shape of the initial disturbance, but their maximum amplitudes are two times smaller than the maximum amplitude of the initial one.

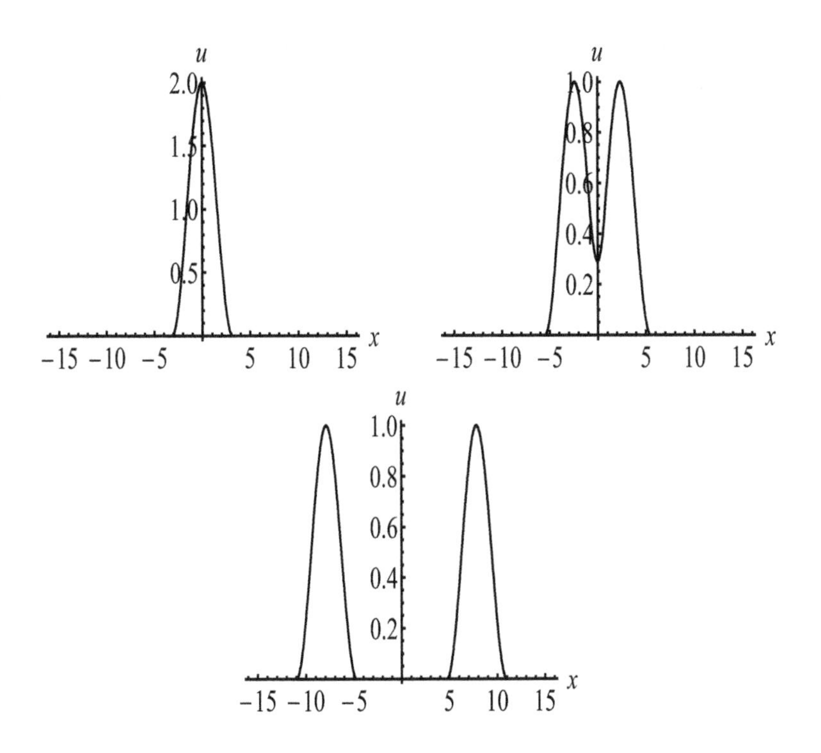

Fig. 5.2 Solution of the wave equation (5.37) satisfying the initial conditions (5.41) in the case of $c = 1$. Left panel: $t = 0$; middle panel: $t = 0.75$; right panel: $t = 2.5$.

Problem.

(1) A liquid with density ρ and a viscosity μ flows out of the pipe in which is kept a constant pressure drop. The liquid fills the vessel of volume Q at time τ. Using the theory of dimensions and similarities, express the dependence

$$\frac{dp}{dx} = f(\mu, \rho, Q, \tau)$$

in the form of an algebraic combination of parameters characterizing the process and certain function dependent on a dimensionless parameter.

(2) A body with the characteristic cross-section S moves with a constant velocity v in a liquid with density ρ and viscosity μ. Express the dependence of the resistance force exerted by the body during the movement as an algebraic combination of parameters characterizing the process and a function depending on the Reynolds number $Re = \frac{\rho v \sqrt{S}}{\mu}$.

(3) Give a proof of the Pythagorean theorem based on the theory of similarity and dimensions.

5.3 Transport equations

5.3.1 *Derivation of the heat transport equation*

Thermal energy accumulated in a set $\Omega \in \mathbb{R}^n$, $n = 1, 2, 3$, is expressed by the integral

$$E = \int_{\Omega} c\,T(t,\,x)\,d\,x, \tag{5.42}$$

where c is the thermal capacity, $T(t,\,x)$ is the temperature field. We assume that the changes in thermal energy accumulated in the set Ω is caused by the following factors:

- heat flow $\vec{Q}$ through the surface $\partial\,\Omega$;
- heat source $\tilde{f}(T;\,t,x)$ inside Ω.

With such assumptions, the energy balance equation takes the form

$$\frac{d\,E}{d\,t} = -\int_{\partial\,\Omega} \vec{Q}\,d\,\vec{\sigma} + \int_{\Omega} \tilde{f}(T;t,\,x)\,d\,x, \tag{5.43}$$

where $d\,\vec{\sigma} = d\sigma\,\vec{n}$, $d\sigma$ is area of the infinitesimal surface element, $\vec{n}$ is the unit vector normal to $\partial\,\Omega$ and directed outwards.

In accordance with the Fick's law,

$$\vec{Q} = -\tilde{\kappa}\,(T;\,t,\,x)\vec{\nabla}\,T, \tag{5.44}$$

where $\vec{\nabla} = \left(\frac{\partial}{\partial\,x_1}, \ldots, \frac{\partial}{\partial\,x_n}\right)^{tr}$, $0 \le \tilde{\kappa}\,(T;\,t,\,x)$ is the transport coefficient. Using the above equality, and applying Green's theorem to the surface integral on the right-hand side of the equality (5.43), we get:

$$\int_{\Omega} \left\{ \frac{\partial\,T}{\partial\,t} - \vec{\nabla}\,\kappa\,(T;\,t,\,x)\,\vec{\nabla}\,T - f(T;t,\,x) \right\} d\,x = 0, \tag{5.45}$$

where $\kappa\,(T;\,t,\,x) = \tilde{\kappa}\,(T;\,t,\,x)/c$, $f(T;t,\,x) = \tilde{f}(T;t,\,x)/c$. The arbitrariness of the set Ω implies that this integral will be equal to zero if and only if the expression under the integral nullifies at each x. So, we get the equation

$$\frac{\partial\,T(t,\,x)}{\partial\,t} = \vec{\nabla}\,\kappa\,(T;\,t,\,x)\vec{\nabla}\,T(t,\,x) + f(T;t,\,x). \tag{5.46}$$

If the process under consideration is time independent and $\kappa = \text{const}$, then (5.46) turns into the *Poisson's equation*

$$\Delta T(x) = \rho(T,\, x), \quad \Delta = \sum_{i=1}^{n} \frac{\partial^2}{\partial x_i^2}, \quad \rho(T,\, x) = -\frac{1}{\kappa}\, f(T;\, x). \tag{5.47}$$

In addition to stationary heat flows, this equation describes many physical processes such as stationary flow of viscous fluid, electrostatic field, etc.

5.3.2 *Thermal explosion problem for linear transport equation*

Let us consider the simplest equation with one spatial variable belonging to the family (5.46):

$$\frac{\partial u}{\partial t} = \kappa \frac{\partial^2 u}{\partial x^2}. \tag{5.48}$$

An analogue of the point explosion problems for the Burgers and Hopf equations in the case of the equation (5.48) is called the *thermal explosion.* The problem is formulated as follows: at the moment $t = 0$ at the point $0 = x \in \mathbb{R}$ the energy Q is released. We are looking for the solution describing the propagation of the heat energy under the assumption that $\kappa = \text{const}$. So we are looking for the solution to the equation (5.48), satisfying the conditions

$$u(0,\, x) = 0, \qquad x \neq 0, \tag{5.49}$$

and the energy conservation law

$$\int_R u(t,\, x)\, d\, x = Q, \qquad Q = \text{const} \tag{5.50}$$

(we have denoted the temperature as u since T is reserved in dimensional theory for the unit of time).

When solving the problem formulated above, we will again use the methods of the theory of dimensions and similarity. Let us note, that in addition to independent variables, the desired solution depends also on the constant parameters appearing in the equations (5.48) and (5.50):

$$u = f\left(t,\, x,\, \kappa,\, Q\right), \tag{5.51}$$

where $t[T]$ is the temporal variable, $x[L]$ is the spatial coordinate, $u[E]$ is the temperature (which is defined as thermal energy per unit mass), κ has the dimension $\left[\frac{L^2}{T}\right]$, Q has the dimension $[E \cdot L]$. Our goal is to reduce the problem to a dimensionless form. Let's choose as the basic units

$$t[T], \quad \kappa\left[\frac{L^2}{T}\right], \quad \text{and} \quad Q\,[E \cdot L]\,.$$

The other units can be expressed as follows:

$$u = \Pi\,\frac{Q}{\sqrt{\kappa\,t}}, \qquad x = \xi\,\sqrt{\kappa\,t},$$

where Π, ξ are dimensionless parameters. In accordance with the Buckingham Π theorem, the dependence (5.51) will take the form

$$u = \frac{Q}{\sqrt{\kappa\,t}}\,\varphi(\xi), \qquad \xi = \frac{x}{\sqrt{\kappa\,t}}, \tag{5.52}$$

where $\varphi(\xi)$ is the dimensionless function to be determined.

Calculating the partial derivatives, we obtain:

$$u_t = -\frac{Q}{2\,t\sqrt{\kappa\,t}}\varphi - \frac{Q}{\sqrt{\kappa\,t}}\left(\frac{\xi}{2\,t}\right)\dot{\varphi}(\xi),$$

$$u_x = \frac{Q}{\sqrt{\kappa\,t}}\dot{\varphi}(\xi)\,\frac{1}{\sqrt{\kappa\,t}}, \qquad u_{xx} = \frac{Q}{\sqrt{\kappa\,t}}\ddot{\varphi}(\xi)\,\frac{1}{\kappa\,t}.$$

Inserting these formulas into (5.48), we get after some algebraic manipulations the following ODE:

$$2\ddot{\varphi} + \varphi + \xi\,\dot{\varphi} = \frac{d}{d\,\xi}\left\{2\dot{\varphi} + \xi\,\varphi\right\} = 0.$$

Integrating this equation, we obtain the first order ODE

$$\dot{\varphi} + \xi\,\varphi = C.$$

We must take the constant of integral C equal to zero. This is due to the symmetry of the initial problem with respect to the transformation $x \to -x$. This symmetry is inherited by the solution itself, so $\varphi(\xi)$, being an even function, has a local extremum at zero. Therefore,

$$0 = \lim_{\xi \to 0}\left[\dot{\varphi}(\xi) + \xi\,\varphi(\xi)\right] = C.$$

Integrating the equation

$$\dot{\varphi}(\xi) + \xi\,\varphi = 0,$$

we obtain:

$$\varphi = C_1\,e^{-\xi^2/4}. \tag{5.53}$$

The integration constant C_1 can be determined from the condition (5.50). One can easily see that in dimensionless variables this condition takes the form

$$\int_R \varphi(\xi)\, d\xi = 1.$$

Inserting the formula (5.53) into this integral, we get:

$$1 = C_1 \int e^{-\xi^2/4}\, d\xi = 2\, C_1 \sqrt{\pi}.$$

So the solution takes the form

$$u(t,\, x) = \frac{Q}{\sqrt{4\,\pi\,\kappa\, t}} \exp\left[\frac{-x^2}{4\,\kappa\, t}\right]. \tag{5.54}$$

5.3.3 *Thermal explosion problem for nonlinear transport equation*

Let us consider the following problem:

$$\frac{\partial\, u}{\partial\, t} = \kappa \frac{\partial}{\partial\, x}\left[u\,\frac{\partial\, u}{\partial\, x}\right], \qquad t > 0, \qquad x \in R, \tag{5.55}$$

$$u(0,\, x) = 0, \qquad x \neq 0, \tag{5.56}$$

$$\int_R u(t,\, x)\, d\,x = Q, \qquad Q = \text{const}. \tag{5.57}$$

As in the linear case, we are looking for the solution of the form

$$u = f(t,\, x,\, \kappa,\, Q), \tag{5.58}$$

where $t[T]$ is time, $x[L]$ is the spatial variable, $u[E]$ is the temperature (that is the internal energy per unit mass), κ has the dimension $\left[\frac{L^2}{T\,E}\right]$, Q has the dimension $[E \cdot L]$. As the basic units we choose

$$t[T], \quad \kappa\left[\frac{L^2}{T\,E}\right], \quad \text{and} \quad Q\,[E\,L]\,.$$

The parameters u and x can be expressed as

$$u = \Pi\,\frac{Q^{2/3}}{(\kappa\, t)^{1/3}}, \qquad x = \xi\,(\kappa\, t\, Q)^{1/3},$$

where Π and ξ are dimensionless parameters. In accordance with the Π theorem, (5.58) takes the following form:

$$u = \frac{\lambda}{(t)^{1/3}}\,\varphi(\xi), \qquad \xi = \frac{x}{(\kappa\, t\, Q)^{1/3}}, \qquad \lambda = \frac{Q^{2/3}}{\kappa^{1/3}}. \tag{5.59}$$

We calculate the derivatives of the function (5.59):

$$u_t = -\frac{\lambda}{3\,t^{4/3}}\,(\varphi + \xi\,\dot{\varphi}),$$
$$u_x = \frac{\lambda}{t^{1/3}}\dot{\varphi}(\xi)\,\frac{1}{(Q\,t\,\kappa)^{1/3}},$$
$$u_{xx} = \frac{\lambda}{t^{1/3}}\dot{\varphi}(\xi)\,\frac{1}{(Q\,t\,\kappa)^{2/3}}.$$

Inserting these formulas into (5.58), we obtain:

$$-\frac{\lambda}{3\,t^{4/3}}\,(\varphi + \xi\,\dot{\varphi}) = \kappa\frac{\partial}{\partial x}\left[\frac{\lambda^2}{(t)^{2/3}}\,\varphi(\xi)\,\dot{\varphi}(\xi)\,\frac{1}{(Q\,t\,\kappa)^{1/3}}\right]$$

$$= \kappa\,\frac{\lambda^2}{(t)^{2/3}}\,\frac{1}{(Q\,t\,\kappa)^{2/3}}\,\left[\varphi(\xi)\,\ddot{\varphi}(\xi) + \dot{\varphi}^2\right].$$

Performing the elementary transformations, we get the equation

$$\frac{d}{d\,\xi}\,(3\varphi\dot{\varphi} + \xi\,\varphi) = 0.$$

Integrating it, we obtain:

$$3\varphi\dot{\varphi} + \xi\,\varphi = C.$$

From the same considerations as in the linear case (symmetry of the initial conditions) it follows that φ is an even function, and this, in turn, implies that

$$\lim_{\xi\to 0}\,(3\varphi(\xi)\dot{\varphi}(\xi) + \xi\,\varphi(\xi)) = 0 = C.$$

So we get the equation

$$\varphi\,(3\,\dot{\varphi} + \xi) = 0,$$

whose nontrivial solution takes the form

$$\varphi = C_0 - \frac{\xi^2}{6}.$$

Assuming that the function u is non-negative, we finally obtain the following generalized solution:

$$\varphi(\xi) = \begin{cases} \frac{1}{6}\left(\xi_f^2 - \xi^2\right), & \text{if } |\xi| < \xi_f = \text{const}, \\ 0, & \text{elsewhere.} \end{cases}$$

We can determine the constant $\xi_f > 0$ using the condition (5.66), which in dimensionless variables takes the form

$$\int_R \varphi(\xi)\, d\xi = 1.$$

Thus we get the equality

$$1 = 2\int_0^{\xi_f} \frac{1}{6}\left(\xi_f^2 - \xi^2\right) d\xi = \frac{2}{9}\,\xi_f^3,$$

from which follows that $\xi_f = (9/2)^{1/3}$. So the solution of the problem (5.55) takes the form

$$u = \frac{1}{6\,\kappa\, t}\begin{cases} x_f^2 - x^2, & \text{as } |x| < x_f = \left(\frac{9\,\kappa\, Q\, t}{2}\right)^{1/3}, \\ 0, & \text{elsewhere.} \end{cases} \tag{5.60}$$

Corollary 5.1. *In case when the coefficient of thermal conductivity is equivalent to $\kappa\, u$, the solution of the thermal explosion problem has a compact support confined in the segment $|x| < \left(\frac{9\,\kappa\, Q\, t}{2}\right)^{1/3}$. The fronts of the compact heat wave spread to the left and to the right with the velocity $d\,x_f/d\,t = (\kappa\, Q/6)^{1/3}\, t^{-2/3}$, see Fig. 5.3.*

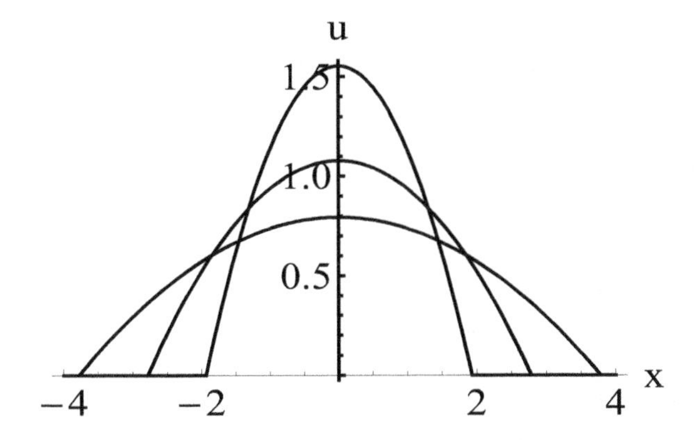

Fig. 5.3 Function (5.60) corresponding to $\kappa = 1$, $Q = 4$. The maximal amplitude of the perturbation decreases with time while the support grows. Graphs correspond to $t = 0.4$, 1.2 and 3.

Problem.

(1) Solve the thermal explosion problem in the following setting:

$$\frac{\partial u}{\partial t} = \kappa \frac{\partial}{\partial x}\left[u^n \frac{\partial u}{\partial x}\right], \quad n \in \mathbb{N} \quad t > 0, \qquad x \in R, \tag{5.61}$$

$$u(0, x) = 0, \qquad x \neq 0, \tag{5.62}$$

$$\int_R u(t, x)\, d\, x = q, \qquad q = \text{const}\,. \tag{5.63}$$

(2) Solve the thermal explosion problem

$$\frac{\partial u}{\partial t} = \kappa \frac{\partial}{\partial x}\left[u^{-1} \frac{\partial u}{\partial x}\right], \quad t > 0, \qquad x \in R, \tag{5.64}$$

$$u(0, x) = 0, \qquad x \neq 0, \tag{5.65}$$

$$\int_R u(t, x)\, d\, x = q, \qquad q = \text{const}\,. \tag{5.66}$$

5.3.4 *Generalized transport models*

The application of the equation (5.46) is not restricted to descriptions of heat transfer processes. Analogous arguments, based on the integral equations describing the conservation of matter, can be applied to the processes of neutron diffusion in solids, the spread of pollutants in liquids and soils, the filtration of gases and liquids in porous media, and the like. Besides, the equation (5.46) can be modified such as to reflect the peculiarities of the processes described and/or properties of the medium. For example, if the medium has an internal structure on mesoscale, then the hypothesis about the arbitrariness of the volume Ω will be incorrect. A consequence of the presence of the structure on mesoscale is that, when deriving the equations, it is impossible to pull the volume Ω down to the point. In other words, if we monitor a change of some magnitude at a point inside a structure element, then information about the flows at the element's boundary will arrive at that point with the delay τ, which is call the *relaxation time.* Formally, the presence of a structure can be taken into account by changing the Fick's law (5.44) by its generalization [Joseph and Preziozi (1989); Cattaneo (1948)]:

$$\tau \frac{\partial}{\partial t}\vec{Q} + \vec{Q} = -\tilde{\kappa}\,(T;\; x)\vec{\nabla}\, T. \tag{5.67}$$

We assume here that κ does not depend on t, because in general case the equation becomes too complex.

Inserting (5.67) instead of (5.44) into the balance equation (5.43), we get the equation that contains second-order partial derivatives with respect to the variable t. In case of one spatial variable, the appropriate equation will be as follows:

$$\tau \frac{\partial^2 u}{\partial t^2} + \frac{\partial u}{\partial t} = \frac{\partial}{\partial x}\left[\kappa(u;\, x)\, \frac{\partial u}{\partial x}\right] + f(u;\, t,\, x). \tag{5.68}$$

Apart from the transport of thermal energy, the equation (5.68) can also describe other diffusion processes in the *structured media*. Therefore, we have replaced the symbol T, commonly denoting temperature, with the function u, describing diffusion process of arbitrary nature.

Keeping in mind, that in liquid and gaseous media, along with the transport processes, convective flows are possible, caused, for example, by temperature gradients, we introduce a different kind of generalization. The most general form of the equation in this case will be as follows:

$$\frac{\partial u}{\partial t} + g(u)\frac{\partial u}{\partial x} = \frac{\partial}{\partial x}\left[\kappa\,(u;\, x)\frac{\partial u}{\partial x}\right] + f(u;\, t,\, x). \tag{5.69}$$

Equation (5.69) is often referred to as *convection-reaction-diffusion* equation.

Another class of models closely related to the transport phenomena is described by the system

$$\tau \frac{\partial^2 u}{\partial t^2} + \frac{\partial u}{\partial t} = \kappa \frac{\partial^2 u}{\partial x^2} + f(u, w), \tag{5.70}$$

$$\frac{\partial w}{\partial t} = g(u,\, w). \tag{5.71}$$

For $\tau = 0$, the model (5.70) turns into the celebrated FitzHugh-Nagumo system [Rocsoreanu *et al.* (2012)]. This system describes the propagation of an electrical signal along a nerve fiber of living organism in the presence of an electro-chemical reaction that maintains this process. Note that the presence of a term proportional to τ is substantiated in the works of Engelbrecht [Engelbrecht (1992); Maugin and Engelbrecht (1994)] and his co-workers for electrically conductive lines with specific characteristics.

5.4 Classification of the second order partial differential equations

Among the equations describing continual media, the scalar second order PDEs [Tikhonov and Samarskii (1990); Perestyuk and Marynets (2001)], as we have repeatedly seen, appear in a natural way. In particular, the wave equation (5.37), the heat transport equation (5.46) and the Poisson's equation (5.47) belong to, but do not exhaust, this family. These equations are quite simple and have analytical solutions in some cases. However, such a relatively simple description of the processes in physical media can easily be lost due to the change of variables, which is often used in order to simplify the solution of the boundary value problems. The problem of description of all second-order scalar partial differential equations that can be reduced to the simplest (canonical) form by means of non-singular changes of variables is therefore of prime importance.

Let us consider the following equation

$$a_{11}u_{xx} + 2a_{12}u_{xy} + a_{22}u_{yy} = f(x, y, u, u_x, u_y), \tag{5.72}$$

where $a_{ij} = a_{ij}(x, y)$, lower indices x, y at the function u mean the partial derivatives with respect to the corresponding variables. Our aim is to show that the left hand side of the equation (5.72) can be simplified, using the change of variables. Let us introduce the new variables

$$\xi = \varphi(x, y), \qquad \eta = \psi(x, y), \tag{5.73}$$

such that

$$J = \det \begin{vmatrix} \varphi_x & \varphi_y \\ \psi_x & \psi_y \end{vmatrix} \neq 0,$$

where x, y belong to some open set $U \subset \mathbb{R}^2$.

In order to present (5.72) in new variables ξ, η, we calculate the partial derivatives of the first and the second order:

$$\begin{aligned} u_x = u_\xi \varphi_x + u_\eta \psi_x, \qquad u_y &= u_\xi \varphi_y + u_\eta \psi_y, \\ u_{xx} = \varphi_x^2 u_{\xi\xi} + 2\varphi_x \psi_x u_{\xi\eta} &+ \psi_x^2 u_{\eta\eta} + \varphi_{xx} u_\xi + \psi_{xx} u_\eta, \\ u_{yy} = \varphi_y^2 u_{\xi\xi} + 2\varphi_y \psi_y u_{\xi\eta} &+ \psi_y^2 u_{\eta\eta} + \varphi_{yy} u_\xi + \psi_{yy} u_\eta, \\ u_{xy} = \varphi_x \varphi_y u_{\xi\xi} + (\varphi_x \psi_y + \varphi_y \psi_x)\, u_{\xi\eta} &+ \psi_x \psi_y u_{\eta\eta} + \varphi_{xy} u_\xi + \psi_{xy} u_\eta. \end{aligned}$$

Using these formulas and grouping the coefficients at the same partial derivatives, we get the following equation:

$$\alpha_{11}u_{\xi\xi} + 2\alpha_{12}u_{\xi\eta} + \alpha_{22}u_{\eta\eta} = F\left(\xi, \eta, u, u_\xi, u_\eta\right), \tag{5.74}$$

where

$$\begin{cases} \alpha_{11} = a_{11}\varphi_x^2 + 2a_{12}\varphi_x\varphi_y + a_{22}\varphi_y^2, \\ \alpha_{12} = a_{11}\varphi_x\psi_x + a_{12}(\varphi_x\psi_y + \varphi_y\psi_x) + a_{22}\varphi_y\psi_y, \\ \alpha_{22} = a_{11}\psi_x^2 + 2a_{12}\psi_x\psi_y + a_{22}\psi_y^2. \end{cases} \tag{5.75}$$

Aiming at the elimination of the terms proportional to $u_{\xi\xi}$ and $u_{\eta\eta}$, we choose the functions φ and ψ satisfying the equation

$$a_{11}z_x^2 + 2a_{12}z_x z_y + a_{22}z_y^2 = 0. \tag{5.76}$$

We can simplify this equation by representing it in the factorized form

$$\left(a_{11}z_x + (a_{12} + \sqrt{\Delta})z_y\right)\left(a_{11}z_x + (a_{12} - \sqrt{\Delta})z_y\right) = 0, \tag{5.77}$$

where $\Delta = a_{12}^2 - a_{11}a_{22}$ (we assume that $a_{11} \neq 0$). As a result, the equation (5.77) splits into two first-order equations, which in turn can be represented in a characteristic form [Elsgolts (1977)]

$$a_{11}z_x + (a_{12} + \sqrt{\Delta})z_y = 0 \Longleftrightarrow \frac{dx}{a_{11}} = \frac{dy}{a_{12} + \sqrt{\Delta}},$$
$$a_{11}z_x + (a_{12} - \sqrt{\Delta})z_y = 0 \Longleftrightarrow \frac{dx}{a_{11}} = \frac{dy}{a_{12} - \sqrt{\Delta}}.$$

Assuming that $a_{11} \neq 0$, we can rewrite the last equation in the following equivalent form

$$a_{11}(dy)^2 - 2a_{12}dxdy + a_{22}(dx)^2 = 0, \tag{5.78}$$

which is called *the characteristic form* of the equation (5.72). To facilitate the process of remembering this equation, the following "mnemonic" can be used. Note that the order of the differentials in the expression (5.78) is inverse to the order of the second derivatives in the equation (5.72). Of course, the sign for mixed differentials should be changed to the opposite.

Definition 5.1. Solutions of the characteristic equation (5.78) are called the characteristics of the equation (5.72).

The above definition allows for the following classification of quasilinear equations of the form (5.72):

- if $\Delta > 0$, then the equation is called hyperbolic;
- if $\Delta = 0$, then the equation is called parabolic;
- if $\Delta < 0$, then the equation is called elliptic.

The equation (5.75) implies the following identity:

$$\alpha_{12}^2 - \alpha_{11}\alpha_{22} = \Delta J^2,$$

from which appears that the type of the equation is invariant with respect to the change of the variables (5.73).

Example 5.5. Let us determine the type of the equation

$$u_{xx} + x\, u_{yy} - 3y\, u = 0.$$

Calculating we get that $\Delta = -x$ and conclude from this equality that the above equation is hyperbolic when $x < 0$, elliptic when $x > 0$ and parabolic when $x = 0$.

After examining the type of the equation (5.72), we can reduce it to the canonical form. Assume first that the equation is hyperbolic, that is, $\Delta > 0$. Integrating the characteristic equation (5.78), we get in this case two families of characteristics defining the change of variables (5.73), which leads to the equality $\alpha_{11} = \alpha_{22} = 0$. So the equation (5.72) comes down to the *first canonical form*

$$U_{\xi\eta} = -F(\xi, \eta, U, U_\xi, U_\eta)/(2\alpha_{12}). \tag{5.79}$$

Using the change of variables

$$\alpha = \xi + \eta, \quad \beta = \xi - \eta, \qquad J = -2 \neq 0,$$

we can get the *second canonical form*, which is as follows:

$$U_{\alpha\alpha} - U_{\beta\beta} = \Phi(\alpha, \beta, U, U_\alpha, U_\beta). \tag{5.80}$$

Now let us assume that $\Delta = a_{12}^2 - a_{11}a_{22} = 0$ (i.e. the equation is parabolic). From this appears that $a_{12} = \sqrt{a_{11}a_{22}}$ and the equation (5.76) possesses only one family of characteristics $C = \varphi(x, y)$. In this case the change of variables $\xi = \varphi$, η is any function chosen such that the condition $J \neq 0$ is fulfilled (most often the substitutions $\eta = x$ or $\eta = y$ are used), leads to the equality $\alpha_{11} = \alpha_{12} = 0$. Thus in this case Eq. (5.72) can be reduced to the *third canonical form*

$$U_{\eta\eta} = F(\xi, \eta, U, U_\xi, U_\eta)/\alpha_{22}.$$

Example 5.6. Let us find the type of equation

$$u_{xx} - 2\sin x\, u_{xy} + \sin^2 x\, u_{yy} - \mathrm{ctg} x\, u_x = 0.$$

In this case the characteristic equation takes the form

$$(dy)^2 + 2\sin x dx dy + \sin^2 x (dx)^2 = 0.$$

Since $\Delta = \sin^2 x - 1 \cdot \sin^2 x \equiv 0$, it is parabolic. The only family of characteristics of this equation that satisfies the equation $d\,y + \sin x\, d\,x = 0$ has the form $y - \cos x = \text{const}$. Using this family of characteristics, we define the change of the variables

$$\xi = y - \cos x, \qquad \eta = y,$$

such that

$$J = \det \begin{vmatrix} \sin x & 1 \\ 0 & 1 \end{vmatrix} = \sin x \neq 0 \text{ as } x \neq \pi n, n \in \mathcal{Z}.$$

Passing to the new variables, we get the equation $u_{\eta\eta} = 0$, whose general solution takes the form $u = \mu_1(\xi)\eta + \mu_2(\xi)$.

Now let us consider the elliptic case $\Delta < 0$. Under such a condition, the equation (5.78) possesses two families of complex characteristics $v(x,y) \pm i\omega(x,y) = C_{1,2} = \text{const}$. Using the change of variables $\xi = v(x,y)+i\omega(x,y)$, $\eta = v(x,y) - i\omega(x,y)$ we get the following equation:

$$\alpha_{11} = a_{11}(v_x + i\omega_x)^2 + 2a_{12}(v_x + i\omega_x)(v_y + i\omega_y) + a_{22}(v_y + i\omega_y)^2 = 0.$$

Equating the real and complex parts of this equation to zero, we get the system

$$\begin{aligned} a_{11}(v_x)^2 + 2a_{12}v_x v_y + a_{22}(v_y)^2 &= a_{11}(\omega_x)^2 + 2a_{12}\omega_x\omega_y + a_{22}(\omega_y)^2, \\ a_{11}v_x\omega_x + a_{12}(v_x\omega_y + v_y\omega_x) + a_{22}v_y\omega_y &= 0. \end{aligned} \tag{5.81}$$

And now, instead of complex variables, we introduce a pair of real variables

$$\alpha = \frac{\xi + \eta}{2} = v(x,y), \qquad \beta = \frac{\xi - \eta}{2i} = w(x,y).$$

Equation (5.81) implies the equalities $\alpha_{11} = \alpha_{22}$, $\alpha_{12} = 0$, from which appear that the equation (5.78) can be reduced to the following canonical form:

$$U_{\alpha\alpha} + U_{\beta\beta} = F(\alpha, \beta, U, U_\alpha, U_\beta)/\alpha_{11}.$$

Note that the coefficients at the second derivatives in the canonical representation are equal.

Example 5.7. Let us determine the type of the following equation:

$$e^y u_{xx} + e^x u_{yy} - (e^y u_x + e^x u_y)/2 = 5xy$$

and then reduce it to the canonical form. For this equation $\Delta = -e^{x+y} < 0$, so it is elliptic. Solutions to the characteristic equations take the form $C_{1,2} = e^{y/2} \pm e^{x/2} i$. The ansatz $\xi = e^{y/2}$, $\eta = e^{x/2}$ leads us to the following canonical form:

$$U_{\eta\eta} + U_{\xi\xi} = \frac{80}{\xi^2\eta^2} \ln\eta \ln\xi.$$

Note that the above classification can be naturally generalized onto the multidimensional case [Vladimirov (1984); Tikhonov and Samarskii (1990); Perestyuk and Marynets (2001)]. This classification is the basis for the method of reducing the equation to the canonical form. For example, the quasilinear equation

$$\sum_{i=1}^{m}\sum_{j=1}^{m} a_{ij}u_{x_i x_j} = f(x, u, u_x), \quad x = x_1, x_2, \ldots, x_m \tag{5.82}$$

can be reduced to the following from:

$$\sum_{k=1}^{m} \lambda_k U_{\xi_k \xi_k} = f(\xi, U, U_\xi). \tag{5.83}$$

The above equation allows for the following classification:

- if all the coefficients λ_k are nonzero and have the same signs, then the equation is elliptic;
- if at least one among the coefficients λ_k is zero, then the equation is parabolic;
- if all the coefficients λ_k are nonzero and the sign of one of the coefficients is opposite to the signs of the remaining $m-1$ coefficients, that the equation is hyperbolic;
- if r coefficients $(2 \leq r < m-1)$ are positive, while the other coefficients are negative, then the equation is ultrahyperbolc.

Implementing the procedure of reduction of the equation (5.82) to the canonical form, one can use the Lagrange method enabling to reduce the quadratic form to the diagonal representation. This method uses the algorithmic procedure enabling to remove the non-diagonal elements of the matrix defining the quadratic form [Gelfand (1971)].

Problem.

Consider the following equations;

a)

$$x^2 u_{xx} - y^2 u_{yy} = 0, \quad xy \neq 0,$$

b)

$$u_{xx} - 2u_{xy} + u_{yy} + 3u_x = 0,$$

c)

$$u_{xx} - yu_{yy} + u = 0, \quad y < 0,$$

d)

$$x^2 u_{xx} - 2xyu_{xy} - 3y^2 u_{yy} = 0, \quad xy \neq 0.$$

Determine the type of each of them. Reduce the equations to canonical form. Plot the families of the characteristic curves whenever it is possible. In cases (a) and (d) solve the Cauchy problem, assuming that $u(x, y = 1) = \varphi(x)$, $u_t(x, y = 1) = \psi(x)$.

Hints.

Ad (a): $u = \sqrt{x\,y}\,(f(xy) + g(y/x))$;

Ad (b): $W_{\eta\eta} + 3W_\xi = 0$, $W = \exp(-1.5\eta - 0.25\xi)u$, $\xi = y + x$, $\eta = x$;

Ad (c): $U_{\xi\xi} + U_{\eta\eta} - \eta^{-1}U_\eta + U = 0$;

Ad (d): $u = \sqrt[4]{x^3 y}\,\big(f(yx^3) + g(y/x)\big)$.

Chapter 6

Methods of solving linear partial differential equations

There is a whole range of methods enabling to solve initial- and boundary-value problems for linear partial differential equations. Among these methods, the most popular and frequently used are those based on the Fourier and Laplace integral transformations, because their employment in many cases allows to reduce the complexity of the problem. Effective tools for solving the initial- and boundary-value problems possessing certain *symmetry* is the *method of separating variables.* This chapter covers all of the above mentioned methods.

6.1 Method of solving Cauchy problem based on the Fourier transform

For any function $\phi(x): \mathbb{R}^n \to \mathbb{R}^1$, whose modulus is integrable the following map can be defined:

$$F[\phi](\xi) = \int_{\mathbb{R}^n} \phi(x)\, e^{i(\xi,\, x)}\, d\, x, \tag{6.1}$$

where $(\xi,\, x) = \sum_{k=1}^n \xi_k\, x_k$. The mapping F (called the *Fourier transform*) is defined in a natural way on the Schwarz space $\mathcal{S}(\mathbb{R}^n)$ consisting of C^∞ functions tending to zero as $|x| \to \infty$ together with their partial derivatives quicker than $(1+|x|)^{-m}$, where m is arbitrary natural number. The set $\mathcal{S}(\mathbb{R}^n)$ is a topological vector space in which convergency is defined with the help of countable set of seminorms: a sequence of functions $\phi_1, \ldots, \phi_k, \ldots$ belonging to $\mathcal{S}(\mathbb{R}^n)$ tends to the function $\phi \in \mathcal{S}(\mathbb{R}^n)$, if for any pair of multiindices

$$\alpha = (\alpha_1, \ldots, \alpha_n), \quad \beta = (\beta_1, \ldots, \beta_n), \quad \alpha_i,\ \beta_i \in \mathbb{N} \cup \{0\},\ i = 1, 2, \ldots, n$$

$$x^\beta\, D^\alpha \phi_k(x) \underset{k\to\infty}{\to} x^\beta\, D^\alpha \phi(x) \quad \forall\, x \in \mathbb{R}^n,$$

where $x^\beta = x^{\beta_1} \cdot x^{\beta_2} \dots \cdot x^{\beta_n}$, $D^\alpha = \partial^{\alpha_1}_{x_1} \dots \partial^{\alpha_n}_{x_n}$. This topology is constructed in such a way that the operation of taking derivative D^α and the change of variables $\phi(x) \to \phi(A\,x + b)$, where A is a square matrix with a non-zero determinant, are continuous in $\mathcal{S}(\mathbb{R}^n)$.

The Fourier transformation (6.1) maps the space $\mathcal{S}(\mathbb{R}^n)$ into itself, therefore function $F[\phi](\xi)$ can be differentiated and multiplied by the polynomials of any finite order and the function $\xi^\alpha\, D^\beta\, F[\phi](\xi)$ remains an element of the space $\mathcal{S}(\mathbb{R}^n)$.

Basic properties of the Fourier transform

(1) Linearity:

$$F[\alpha_1\,\phi_1 + \alpha_2\,\phi_2](\xi) = \alpha_1\,F[\phi_1](\xi) + \alpha_2\,F[\phi_2](\xi), \qquad \forall\,\alpha_1,\,\alpha_2 \in \mathbb{C}, \quad \forall\,\phi_1,\,\phi_2 \in \mathcal{S}. \tag{6.2}$$

(2) The derivative of the Fourier transform is expressed by the formula:

$$D^\alpha_\xi F[\phi](\xi) = \int\limits_{\mathbb{R}^n} (i\,x)^\alpha \phi(x) e^{i(\xi,\,x)}\,d\,x = F\left[(i\,x)^\alpha\,\phi\right](\xi). \tag{6.3}$$

(3) The Fourier transform of the derivative is given by the formula

$$F[D^\alpha_x\,\phi](\xi) = \int\limits_{\mathbb{R}^n} e^{i(\xi,\,x)} D^\alpha_x\,\phi(x) = (-i\,\xi)^\alpha\,F[\phi](\xi). \tag{6.4}$$

Using the formulas (6.3) and (6.4), we get:

$$\xi^\beta\,D^\alpha_\xi\,F[\phi](\xi) = \xi^\beta F\left[(i\,x)^\alpha\,\phi\right](\xi) = i^{|\alpha|+|\beta|} F\left[D^\beta_x\,(x^\alpha\,\phi)\right](\xi). \tag{6.5}$$

It follows from the equality (6.5) that for any α, β functions $\xi^\beta\,D^\alpha\,F[\phi](\xi)$ are uniformly bounded with respect to the ξ, because

$$\left|\xi^\beta\,D^\alpha\,F[\phi](\xi)\right| \le \int\limits_{\mathbb{R}^n} \left|D^\beta\,(x^\alpha\,\phi)\right|\,d\,x < \infty.$$

From this follows that $F[\phi] \in \mathcal{S}$, and since $F[\phi]$ is absolutely integrable, then $\phi(x)$ is connected with $F[\phi]$ by means of the inverse Fourier transform F^{-1}:

$$\phi(x) = F^{-1}\left[F[\phi]\right] = F\left[F^{-1}[\phi]\right], \tag{6.6}$$

where

$$\begin{aligned} F^{-1}[\psi](x) &= \frac{1}{(2\,\pi)^n}\int\limits_{\mathbb{R}^n} \psi(\xi)\,e^{-i(\xi,\,x)}\,d\,\xi = \frac{1}{(2\,\pi)^n} F[\psi](-x) \\ &= \frac{1}{(2\,\pi)^n}\int\limits_{\mathbb{R}^n} \psi(-\xi) e^{i(\xi,\,x)}\,d\,\xi = \frac{1}{(2\,\pi)^n}\,F[\psi(-\xi)]. \end{aligned} \tag{6.7}$$

When attempting to use the Fourier transform to solve Cauchy problems for linear partial differential equations, one should realize that in general there is no reason to expect the solution to be an element of the Schwarz space $\mathcal{S}$. This circumstance does not discredit in any significant way the Fourier method, since, as it is shown in Appendix B, the above-mentioned construction can be extended to much more general functions, which justifies the use of the Fourier transform not only in the search for classical solutions, but even in the case when they belong to the set of generalized function.

In the *Mathematica* package the Fourier transform is implemented by means of the commands

```
FourierTransform[f[t], t, w, FourierParameters → {1, 1}],
InverseFourierTransform[f[w], w, t, FourierParameters → {1, 1}].
```

Example 6.1. Let us show how it is possible to solve the initial-value problem

$$u_{tt} = c^2 u_{xx} \qquad u(0,x) = e^{-x^2/4}, \qquad u_t(0,x) = 0 \tag{6.8}$$

using the Fourier transform.

We perform the Fourier transform of the problem (6.8) in the *Mathematica* package:

Cell 6.1.

In[1]:
```
f[x_]=Exp[ - x^2/a^2];
```
In[2]:
```
FourierTransform[f[x], x, k, FourierParameters → {1, 1}]
```
Out[2]:

$$\frac{e^{-\frac{1}{4}a^2k^2}\sqrt{\pi}}{\sqrt{\frac{1}{a^2}}}$$

In[3]:
```
%/. a → 2
```
Out[3]:

$$2e^{-k^2}\sqrt{\pi}$$

Thus, we get the initial value problem

$$\hat{u}_{tt} + \omega^2\hat{u} = 0, \qquad \hat{u}(0,k) = 2\sqrt{\pi}e^{-k^2}, \quad \hat{u}_t(0,k) = 0, \tag{6.9}$$

where $\hat{u}(t,\,k) = F[u](t,\,k)$. The expression $\omega^2 = c^2k^2$ is called *dispersion relationship*. There are two parameters closely related to the dispersion relationship, namely the phase velocity v_{ph} and the group velocity v_{gr}, which are defined as follows:

$$v_{ph} = \frac{\omega}{k}, \quad v_{gr} = \frac{d\omega}{dk}. \tag{6.10}$$

Let us note that in the case of the wave equation (6.8) $v_{ph} = v_{gr} = \pm c$, where $c = \text{const}$.

Solutions of the equation (6.9) can be presented in the form

$$\hat{u}(k,\, t) = C(k)e^{-i\omega t} + D(k)e^{i\omega t}.$$

Using the initial condition, we get $C = D = \hat{u}(0,k)/2$. Hence $\hat{u}(k,\, t) = \hat{u}(0,k)\cosh(i\omega t)$. Finally, the solution to the problem (6.8) takes the form

$$u(t,x) = \frac{1}{2\pi}\int_{-\infty}^{\infty} \hat{u}(k,\, t)e^{ikx}dk.$$

Graphs of the solutions at $t = 0, 2, 4$ are presented in Fig. 6.1.

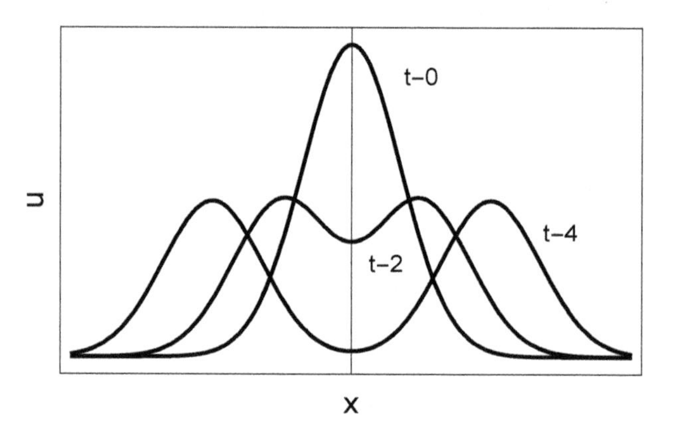

Fig. 6.1 Graphical representation of the solution to the initial value problem (6.8).

Example 6.2. We consider the multidimensional wave equation

$$u_{tt} = c^2\Delta u$$

together with the initial condition

$$u(\vec{r},0) = u_0(\vec{r}), \quad \text{and} \quad u_t(\vec{r},0) = v_0(\vec{r}).$$

In order to solve this problem, we use Fourier transform method:[1]

$$\hat{u} = \int_{-\infty}^{\infty} e^{-ik_x x}dx \int_{-\infty}^{\infty} e^{-ik_y y}dy \int_{-\infty}^{\infty} e^{-ik_z z}\, u(x,y,z,t)\, dz$$
$$= \int_{-\infty}^{\infty} u(\vec{r},t)e^{-i\vec{k}\cdot\vec{r}}d^3\vec{r},$$

[1]For some reasons it is convenient to use the argument $-i\vec{k}\,\vec{r}$.

where $\vec{k} = (k_x, k_y, k_z)$ denotes the wave vector (we also use the notation $k = \sqrt{k_x^2 + k_y^2 + k_z^2}$). The equation transformed in this way takes the form

$$\hat{u}_{tt} + \omega^2 \hat{u} = 0, \qquad \omega^2 = c^2\, k^2. \tag{6.11}$$

Equation (6.11) has the periodic solution with the period $2\,\pi/\omega$. Applying the inverse Fourier transformation to this solution, we get:

$$u(r,t) = \frac{1}{(2\pi)^3}\left(\int C(k) e^{i(\vec{k}\cdot\vec{r}-\omega t)}\, d^3k + \int D(k) e^{i(\vec{k}\cdot\vec{r}+\omega t)}\, d^3k\right).$$

Solution describing the wave moving in the direction of $\vec{k}$ can be presented in more simple form:

$$u = 2\,\mathrm{Re} \int \frac{d^3k}{(2\pi)^3} \tilde{C}(k) e^{i(\vec{k}\cdot\vec{r}-\omega t)}, \tag{6.12}$$

where

$$\tilde{C}(k) = \frac{1}{2}\left(\bar{u}_0(k) + i\frac{\bar{v}_0(k)}{\omega}\right).$$

Let us consider the evolution of spherically symmetric perturbations $u(r,0) = U_0\, e^{-r^2/a^2}$, $v_0(r,0) = 0$, $r^2 = x^2 + y^2 + z^2$. Applying the Fourier transform to them, we get $\hat{u}(k,0) = a^3\pi^{3/2} U_0 e^{-a^2k^2/4}$. In accordance with (6.12), the solution takes the following form

$$u(r,t) = a^3\pi^{3/2} U_0\ \mathrm{Re} \int \frac{d^3\vec{k}}{(2\pi)^3} e^{i(\vec{k}\cdot\vec{r}-\omega t)-a^2k^2/4}, \quad \vec{k} = (k_x,\, k_y,\, k_z), \tag{6.13}$$

where $k_x = k\cos\phi\sin\theta$, $k_y = k\sin\phi\sin\theta$, $k_z = k\cos\theta$. Note that the Jacobian $J = k^2\sin\theta$. The coordinate system is chosen such that the direction of the k_z coordinate coincides with the direction of the vector $\vec{r}$. And then $\vec{k}\cdot\vec{r} = k\,r\,\cos\theta$, where $r = |\vec{r}|$. Eventually we will obtain the formula

$$u(r,t) = \frac{a^3\pi^{3/2} U_0}{(2\pi)^3}\ \mathrm{Re} \int_0^\infty k^2 dk \int_0^{2\pi} d\phi \int_0^\pi \sin\theta d\theta e^{ik(r\cos\theta - ct) - a^2k^2/4}.$$

Integrating now over the variable θ, we obtain the expression

$$u(r,t) = \frac{a^3\pi^{3/2} U_0}{(2\pi)^2}\ \mathrm{Re} \int_0^\infty \frac{i\,k}{r}\left(e^{-ik(r+ct) - e^{-ik(r-ct)}}\right) e^{-a^2k^2/4} d\,k.$$

We perform further calculations in the *Mathematica*, using the command Integrate[...,k,0,Infinity]. Finally we obtain the solution of the form

$$u(r,t) = \frac{U_0}{2r}\left(e^{-(r+ct)^2/a^2}(r+ct) + e^{-(r-ct)^2/a^2}(r-ct)\right). \tag{6.14}$$

Graphical illustration of the solution obtained is presented in Fig. 6.2. In the left panel the initial perturbation is shown together with the solution corresponding to $t = 0.05$. In the right panel a two-dimensional projection of this solution obtained in the *Mathematica* package with the help of the command ContourPlot is presented (the construction of the graphs can be found in the file PM2_1.nb).

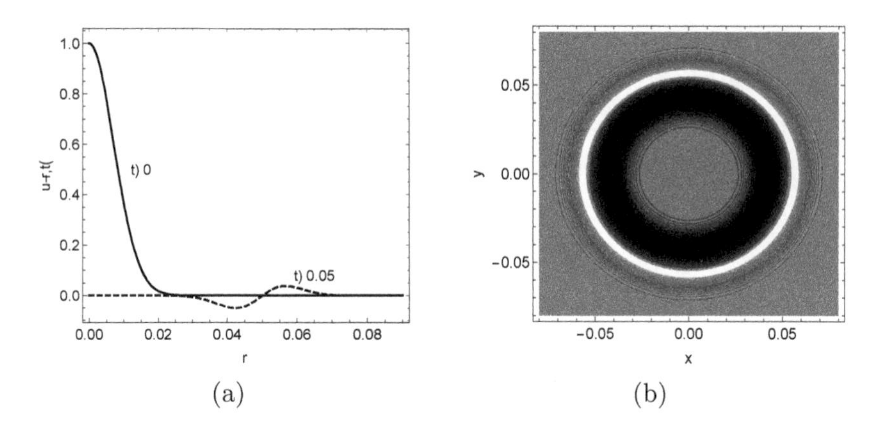

Fig. 6.2 Solution of the equation (6.14) satisfying the initial conditions $u(0, r) = U_0 \exp(-r^2/a^2)$, $u_t(0, r) = 0$, $U_0 = 1$, $a = 0.01$. (a): graph of the function $u(r,t)$ for times $t = 0$; 0.05; (b): two-dimensional projection of the function $u(x, y, 0, t)$ taken at time $t = 0.05$.

Example 6.3. Let's apply the Fourier transform for solving the heat transport equation

$$T_t = a(T_{xx} + T_{yy} + T_{zz}) + S(x, y, z, t), \quad t \geq 0, \quad (x,\, y,\, z) \in M, \quad (6.15)$$

where $M = \{(x,\, y,\, z) :\; x^2 + y^2 + z^2 \leq R^2\}$, S is a known function that describes a heat source.

We are looking for the solution satisfying the initial condition $T(\vec{r}, t = 0) = T_0(\vec{r})$. Using the Fourier transform $\hat{T}(k,t) = \int T(\vec{r},t) e^{-i\vec{k}\vec{r}} d\vec{r}$, we get the ordinary differential equation

$$\hat{T}_t = -k^2 a \hat{T} + \hat{S}(k,t).$$

After solving this equation, we apply the inverse transformation to $\hat{T}$ and this way obtain the solution in the initial representation.

Let us find the solution to (6.15) assuming that $S(\vec{r},t) = \delta(r-r_0)\delta(t-t_0)$, and posing the initial condition $T_0(\vec{r}) = 0$. Applying the Fourier transform, we get:

$$\hat{S}(k,t) = \int S(r,t)e^{-ikr}dr = \int \delta(r-r_0)\delta(t-t_0)e^{-ikr}dr = e^{-ikr_0}\delta(t-t_0).$$

Hence

$$\hat{T}(k,t) = \int_0^t \int e^{-i\,k\,r_0}\delta(t-t_0)e^{ak^2(\tau-t)}d\tau = e^{-i\,k\,r_0+a\,k^2(t_0-t)}H(t),$$

where $H(t)$ is the Heaviside function. Using the inverse transformation, we obtain:

$$T(r,t) = \frac{1}{(2\pi)^3}\int e^{-ikr_0+ak^2(t_0-t)}e^{ikr}dk = \frac{1}{(2\pi)^3}\int e^{ik(r-r_0)-ak^2(t-t_0)}dk.$$

Integration in the above formula can be performed in spherical coordinates $k_x = k\cos\phi\sin\eta$, $k_y = k\sin\phi\sin\eta$, $k_z = k\cos\eta$. Jacobian in the spherical coordinates is expressed by the formula $J = k^2\sin\eta$. We assume that the vectors $\vec{r}-\vec{r}_0$ and k_z are parallel, and $k(r-r_0) = k\Delta r\cos\eta$. After the integration over the interval $\eta \in [0;\pi]$, we get the expression

$$T(r,t) = \frac{1}{(2\pi)^2}\int \frac{ki}{\Delta r}\left(e^{-i\Delta rk} - e^{i\Delta rk}\right)e^{-ak^2(t-t_0)}dk.$$

Using the command Integrate[...,k,0,Infinity] in the *Mathematica* package, we finally obtain

$$T(r,t) = \frac{1}{8\pi^{3/2}[a(t-t_0)]^{3/2}}\exp\left(-\frac{(r-r_0)^2}{4a(t-t_0)}\right).$$

The profiles of the function $T(r,t)$ presented in Fig. 6.3 indicate that the evolution of the initial perturbation in the medium described by the equation (6.15) differs significantly from the propagation of an exponentially localized initial disturbance in the case of the wave equation.

Problem.

(1) Using the Fourier transform, find the solution to the one-dimensional wave equation $y_{tt} - y_{xx} = 0$ assuming that:

$$y(x,0) = 0, \qquad \frac{\partial y}{\partial t}(x,0) = xe^{-|x|}.$$

Illustrate the result obtained with an animation over the period of time $0 < t < 2$.

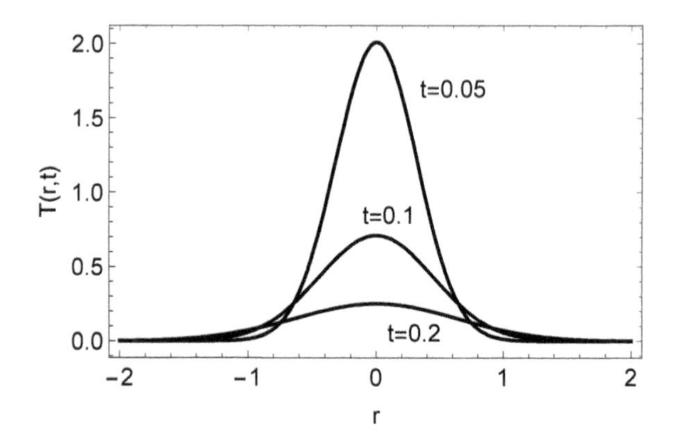

Fig. 6.3 Solutions of the heat transport equation (6.15) with the source term in the form of δ-function. At the initial moment of time the temperature in the entire space is assumed to be zero. The parameters take the following values: $t_0 = 0$, $r_0 = 0$, $a = 1$. The profiles of the function T correspond to the moments of time $t = 0.05$; $t = 0.1$; $t = 0.2$.

(2) Using the Fourier transform, find the solution to the equation $u_t = u_{xx}$, satisfying the initial condition $T(x, 0) = e^{-|x|}$. Illustrate the result obtained with an animation on the segment $-5 < x < 5$ over the period of time $0 < t < 2$.

(3) Surface waves in shallow water satisfy the dispersion relations $\omega(k) = \sqrt{gk}$, where g is the gravitational acceleration. Find the evolution of surface waves in the case of cylindrical symmetry, assuming that the Fourier transform of the initial perturbation is given by the formula $A(k) = e^{-k^2/36}$. Illustrate the result obtained by constructing the graph of the solution in the initial variables.

6.2 Laplace transform

Definition 6.1. Function $f : [0, +\infty) \to \mathbb{R}$ is said to be the exponential type function of the finite order if there exist a real numbers $M > 0,\ C > 0$ such that

$$\forall\, t \geq 0 \quad |f(t)| < M\, e^{C\, t}.$$

The smallest number $C_0 \geq 0$ for which the above property is satisfied is called the order of growth of the function f.

Let us note that for any bounded function $C_0 = 0$.

For a function $f(t)$ of the finite order C_0 the transformation called the *Laplace transform* (see Appendix B) is defined by the following formula:

$$F(p) = \int_0^\infty f(t)e^{-p\,t}dt, \tag{6.16}$$

where p is a complex number satisfying the condition $\mathrm{Re}[p] > C_0$. From the above definition appears that the integral in the formula (6.16) exists and is absolutely convergent. In the following we'll use the notation $F(p) = \mathcal{L}[f](p)$.

Example 6.4. For the function $f(t) = \sin \omega t$ following formula holds:

$$\mathcal{L}[f][p] = \int_0^\infty \sin \omega t \cdot e^{-pt} dt = \frac{\omega}{p^2 + \omega^2}.$$

Calculation of the Laplace transform in this case can be reduced to the integration of exponential functions by taking advantage of the Euler representation

$$\sin \omega\, t = \frac{1}{2\,i}\left(e^{i\,\omega\,t} - e^{-i\,\omega\,t}\right).$$

In the *Mathematica* package obtaining the Laplace transform for the above function is performed by means of the command

```
LaplaceTransform[Sin[ω t], t, p]
```

Basic properties of the Laplace transform

Theorem 6.1. *Let Laplace transforms exist for all the functions considered*

below. Then the following formulas hold true:

$$\mathcal{L}(\alpha f(t) + \beta g(t))[p] = \alpha \mathcal{L}(f(t))[p] + \beta \mathcal{L}(g(t))[p]; \tag{6.17}$$

$$\mathcal{L}(f(\alpha t))[p] = \frac{1}{\alpha} \mathcal{L}(f(t)) \left[\frac{p}{\alpha}\right]; \tag{6.18}$$

$$\mathcal{L}(f'(t))[p] = p\,\mathcal{L}(f(t))[p] - f(0); \tag{6.19}$$

$$\mathcal{L}(f''(t))[p] = p^2 F(p) - pf(0) - f'(0); \tag{6.20}$$

$$\mathcal{L}(f^{(n)}(t))[p] = p^n\, \mathcal{L}(f(t))[p] - \sum_{i=0}^{n-1} p^{n-1-i} f^{(i)}(0); \tag{6.21}$$

$$\mathcal{L}(f(t)e^{-\alpha t})[p] = \mathcal{L}(f(t))[p + \alpha]; \tag{6.22}$$

$$\mathcal{L}\left[\int_0^t f_1(\tau) f_2(t-\tau) d\tau\right] = F_1(p) \cdot F_2(p). \tag{6.23}$$

In the last equation the symbols $F_{1,2}$ denote the Laplace transforms of the functions $f_{1,2}$, correspondingly.

Proof.

The property (6.17) results from the linearity of the integral. To obtain the formula (6.18), it is sufficient to use the change of variable $z = \alpha t$. To obtain the formula (6.19), one uses the method of integration by parts. In the proof of the formula (6.20), the property (6.19) is used twice:

$$\begin{aligned}\mathcal{L}(f''(t)) &= p\,\mathcal{L}(f'(t)) - f'(0) = p\,[p\,F(p) - f(0)] - f'(0)\\ &= p^2 F(p) - pf(0) - f'(0).\end{aligned}$$

Validity of the formula (6.21) can be shown using the method of induction. The formula (6.22) follows from the properties of the exponential function.

Now let us prove the formula (6.23). From the definition of the Laplace transform, we get the identity

$$\mathcal{L}\left[\int_0^t f_1(\tau) f_2(t-\tau) d\tau\right] = \int_0^\infty \int_0^t f_1(\tau)\, f_2(t-\tau) e^{-pt}\, d\tau\, dt.$$

Then we notice that the integral at the r.h.s. coincides with the following double integral

$$I = \iint_\Omega f_1(\tau) f_2(t-\tau) e^{-pt} dt d\tau,$$

where $\Omega = \{(\tau, t) \in \mathbb{R}^2 : 0 \le t < \infty; 0 \le \tau \le t\}$ is a set bounded by the half-lines $\tau = 0$ and $t = \tau$, see Fig. 6.4.

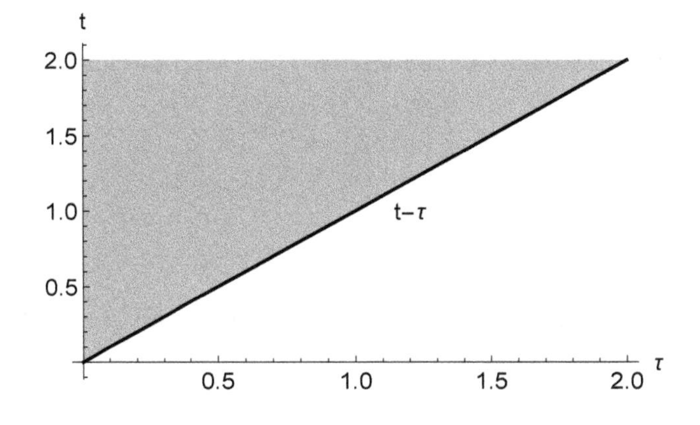

Fig. 6.4 The set Ω (see proof of the validity of equation (6.23)).

By changing the order of integration, we get:

$$I = \int_0^\infty d\tau \int_\tau^\infty dt f_1(\tau) f_2(t-\tau) e^{-pt} = \int_0^\infty d\tau f_1(\tau) \int_\tau^\infty f_2(t-\tau) e^{-pt} dt$$

$$= \left| \begin{array}{l} \text{integration by} \\ \text{substitution} \\ t-\tau = z,\ dt = dz \\ t = \tau \to z = 0, \\ t = \infty \to z = \infty \end{array} \right| = \int_0^\infty d\tau f_1(\tau) d\tau \int_0^\infty f_2(z) e^{-p(\tau+z)} dz$$

$$= \int_0^\infty f_1(\tau) e^{-p\tau} d\tau \int_0^\infty f_2(z) e^{-pz} dz = F_1(p) \cdot F_2(p).$$

The inverse Laplace transform is given by the formula

$$f(t) = \frac{1}{2\pi i} \int_{\gamma - i\infty}^{\gamma + i\infty} F(p) e^{+pt} dp, \tag{6.24}$$

where γ is an arbitrary positive number greater than C_0. Integration is performed along the line $Re(p) = \gamma > C_0$. The above expression, called the *Mellin's formula*, is not very advantageous. Therefore, one often uses in practice the procedures based on presenting the function $F(p)$ in the form of a series and the following formula

$$f(t) = \sum_{k \in \mathcal{M}} \operatorname{Res} F(p_k) e^{p_k t},$$

where $\operatorname{Res} F(p_k)$ denotes the *residuum* of the function $F(\cdot)$ in the singular point p_k (which is assume to be pole) [Maurin (1973); Aramanovich *et al.* (2004)]. The preceding formula is well defined if the function $F(p)$ satisfies certain conditions, which are formulated in the Appendix B. Note that if c is the n-th order pole of the function $f(z)$, then the following formula holds:

$$\operatorname{Res} f(c) = \frac{1}{(n-1)!} \lim_{z \to c} \frac{d^{n-1}}{dz^{n-1}} (z-c)^n f(z). \tag{6.25}$$

For $n = 1$ we get the formula

$$\operatorname{Res} f(c) = \lim_{z \to c} (z-c) f(z). \tag{6.26}$$

If $f(z) = g(z)/h(z)$, then, using the l'Hospital rules, we get:

$$\operatorname{Res} f(c) = \lim_{z \to c} \frac{g(z)}{h'(z)} = \frac{g(c)}{h'(c)}. \tag{6.27}$$

Let us consider the implementation of this formula for the function $f[z] = \exp[z]/(z^2(z^2+9))$ in the *Mathematica* package.

Cell 6.2.

In[1]:
```
f[z_] = Exp[z]/(z^2 (z^2 + 9));
```
In[2]:
```
Residue[f[z], z, 0]
```
Out[2]:
$\frac{1}{9}$

In[3]:
```
Limit[D[z^n f[z], z], z → 0]/Factorial[n - 1] /. n → 2
```
Out[3]:
$\frac{1}{9}$

In[4]:
```
Residue[f[z], z, 3 I]
```
Out[4]:
$\frac{1}{54} i e^{3i}$

In[5]:
```
Numerator[f[z]]/D[Denominator[f[z]], z] /. z → 3 I
```
Out[5]:
$\frac{1}{54} i e^{3i}$

Note that on virtue of the formula (6.21), the Laplace transform maps time dependent linear PDEs into the ODEs.

Example 6.5. Let us apply the Laplace transform for finding the temperature field in a heat conductor. When solving this problem, it is necessary to analyze the fulfillment of the conditions enabling the expression of the inverse Laplace transform through residua, see Appendix B. The issue is formulated as follows. A heat conductor of half-infinite length and negligible cross-section, placed at the half-line $[0, +\infty)$, has zero initial temperature. At the left end of the heat conductor, the following temperature regime is maintained: $T(t, 0) = \cos \omega t$. The thermal conductivity coefficient is assumed to be equal to 1. We are going to find the temperature field over the entire length of the conductor.

The initial-value problem is formulated as follows:

$$\begin{aligned} &T_t = T_{xx}, \qquad x \in [0, \infty), \\ &T(0, x) = 0, \qquad T(t, 0) = \cos \omega t. \end{aligned} \tag{6.28}$$

Performing the Laplace transform $U(p, x) = \mathcal{L}[T](p, x)$ with respect to the temporal variable, we get:

$$U_{xx} = p\,U, \qquad x \in [0, \infty), \quad p > 0, \tag{6.29}$$

$$U(0, x) = 0, \qquad U(p, x) = \mathcal{L}[\cos \omega t](p, x) = \frac{p}{p^2 + \omega^2}. \tag{6.30}$$

Solution of the equation (6.29) is given by the formula

$$U(p, x) = c_1(p)\, e^{x\sqrt{p}} + c_2(p)\, e^{-x\sqrt{p}}.$$

It follows from the physical considerations, that the function we are looking for should be equal to zero at infinity, and therefore we assume that $c_1(p) = 0$. With this assumption, using the initial condition, we get the following solution:

$$U(p, x) = \frac{p}{p^2 + \omega^2} e^{-x\sqrt{p}}. \tag{6.31}$$

Inserting (6.31) into the formula (6.24), we get:

$$T(t, x) = \lim_{R \to \infty} \frac{1}{2\pi i} \int_{\gamma - i\,R}^{\gamma + i\,R} e^{+pt} \frac{p}{p^2 + \omega^2} e^{-x\sqrt{p}}\, d\,p. \tag{6.32}$$

Let us consider the closed loop $\mathcal{G}$ presented in Fig. 6.5. The function (6.31) has the following singular points inside the set bounded by $\mathcal{G}$: the simple poles in the points $p_{\pm} = \pm i\,\omega$; the branching point at the origin. So, in accordance with theorem B.1 (see Appendix B),

$$\frac{1}{2\pi i} \int_{\mathcal{G}} e^{+pt} \frac{p}{p^2 + \omega^2} e^{-x\sqrt{p}}\, d\,p = \text{sum of residua of the integrand.} \tag{6.33}$$

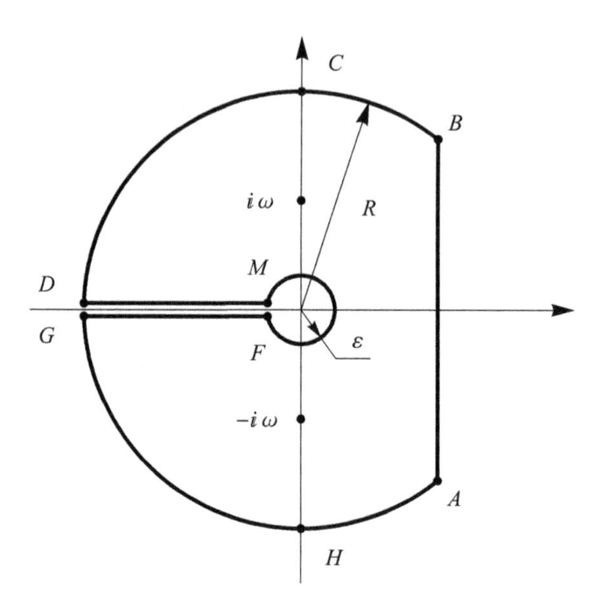

Fig. 6.5 The outlook of the contour $\mathcal{G}$ used to obtain the inverse Laplace transform in the problem of the heat conduction in a semi-infinite rod with negligible cross-section and periodically changing temperature source applied to its left end.

Calculating the residua with the help of the formula (6.27), we obtain:

$$\frac{1}{2\pi i}\int_{\mathcal{G}} e^{+pt}\frac{p}{p^2+\omega^2}e^{-x\sqrt{p}}\,dp = \frac{1}{2}\left[e^{i\omega t - x\sqrt{i\omega}} + e^{-i\omega t - x\sqrt{-i\omega}}\right]$$

$$= \frac{1}{2}\left[e^{i\omega t - x(1+i)\sqrt{\omega/2}} + e^{-i\omega t - x(1-i)\sqrt{\omega/2}}\right] = e^{-x\sqrt{\omega/2}}\cos\left(\omega t - x\sqrt{\omega/2}\right). \tag{6.34}$$

Let's estimate the limits that the integrals strive to on the individual contour segments of $\mathcal{G}$, as $\epsilon \to 0$ i $R \to \infty$. First we consider the integral along the arc CD. If $p = R e^{i\varphi} \in CD$, then $\sqrt{p} = \sqrt{R}e^{i\varphi/2}$ will lie in the right half-plane, hence $\mathrm{Re}\sqrt{p} \geq 0$ and

$$\left|e^{-x\sqrt{p}}\right| = e^{-x\mathrm{Re}\sqrt{p}} < 1.$$

For $R \gg 1$,

$$\frac{p}{p^2+\omega^2} \sim \frac{1}{p},$$

therefore the function $U(p, x)$ tends to zero as $|p| \to \infty$ uniformly over the variable φ and the integral along the arc CD tends to zero. The reasoning for the integral calculated along the arc GH is analogous, except that it must be taken into account that the continuous function $\arg\sqrt{p}$ belonging to the segment $(3\,\pi/2,\,7\,\pi/4)$ attains at the point G the value close to $3\pi/2$, which ensures that $\mathrm{Re}\sqrt{p} \geq 0$.

The functions $e^{t\,p}$ and $e^{-x\sqrt{p}}$ are bounded on the arcs BC and HA so the integrals along these segments tend to zero. Let us show this for the segment $B\,C$. The argument φ on this segment changes within the interval $(\pi/2 - \delta_0,\,\pi/2)$, where $\delta_0 = \arcsin\gamma/R$. For $e^{t\,p}$ the following estimation holds:

$$\left|e^{t\,p}\right| = \left|e^{t\,R\,\cos(\pi/2-\delta)}\right| = e^{t\,R\,\sin\delta} \leq e^{t\,R\,\sin\delta_0} = e^{t\,\gamma}$$

(in the above formula we use the notation $\delta = \pi/2 - \varphi$). Hence the function $|e^{t\,p}|$ is bounded on $B\,C$. For the function $e^{-x\sqrt{p}}$ we have the estimation

$$\left|e^{-x\sqrt{p}}\right| = e^{-x\sqrt{R}}\cos\frac{\pi/2-\delta}{2} = e^{-\frac{x\sqrt{R}}{2}}\,(\cos\,\delta/2 + \sin\,\delta/2) < 1$$

as long as $0 < \gamma/R$ is sufficiently small. Estimation for HA are performed in the analogous way. Thus the presence of the factor $p/(p^2+\omega^2)$ ensures that the integrals along the arcs BC i HA tend to zero uniformly with respect to φ. Let us note that the lengths of the arcs BC and HA tend to γ as $R \to 0$, which ensures convergence of both of the integral to zero. On the arc MF all the integrands are bounded, while the length of the arc tends to zero as $\epsilon \to 0$. Therefore this integral also tends to zero.

So after taking the limit in the l.h.s. of the formula (6.33), there remains the integral along the line $\mathrm{Re}\,p = \gamma$, which is equal to the function $T(t,\,x)$ we are looking for, and the integrals along the segments DM and FG. Therefore, in the limiting case we will get the following

$$\begin{aligned} T(t,\,x) &= e^{-x\sqrt{\omega/2}}\cos\left(\omega\,t - x\sqrt{\omega/2}\right) \\ &\quad - \frac{1}{2\,\pi}\left[\int_{-\infty}^{0} e^{t\,p}U(p,\,x) + \int_{0}^{-\infty} e^{t\,p}U(p,\,x)\right]. \end{aligned} \tag{6.35}$$

When processing the integrals in square brackets, it should be taken into account that as a result of circling the origin, the argument of the function $\sqrt{p}$ in the second integral changes by π. So performing the substitution $p = -r$ in the first integral we get:

$$\begin{aligned} \int_{-\infty}^{0} e^{t\,p}U(p,\,x) &= \int_{\infty}^{0} e^{-t\,r-i\,x\sqrt{r}}\frac{r}{r^2+\omega^2}\,d\,r \\ &= -\int_{0}^{\infty} e^{-t\,r-i\,x\sqrt{r}}\frac{r}{r^2+\omega^2}\,d\,r. \end{aligned}$$

Using the analogous substitution in the second integral and taking into account the change of the argument $\sqrt{p}$, we obtain the equality

$$\int_0^{-\infty} e^{t\,p} U(p,\,x) = \int_0^{\infty} e^{-t\,r + i\,x\sqrt{r}} \frac{r}{r^2+\omega^2}\,d\,r.$$

Finally, we get the solution of the form

$$T(t,\,x) = e^{-x\sqrt{\omega/2}} \cos\left(\omega\,t - x\sqrt{\omega/2}\right) - \frac{1}{\pi}\int_0^{-\infty} e^{-t\,r}\frac{r}{r^2+\omega^2}\sin(x\sqrt{r})\,d\,r. \tag{6.36}$$

Note that the integral on the right side of the formula (6.36) tends to zero when $t \to \infty$, so after the transition period the first term on the right side will dominate in this formula, which means that at sufficiently large $t > 0$ the temperature at each point $x > 0$ will change periodically with the frequency ω, demonstrating the phase shift by the factor $x\sqrt{\omega/2}$. The amplitude of oscillations at the point $x > 0$ is equal to $e^{-x\sqrt{\omega/2}}$ as one moves away from a heat source. In addition, the amplitude depends on the value of ω, so for higher frequency the decay is more intensive.

Example 6.6. Let us consider a model describing the heat flow in a rod of length L, assuming (without loss of generality) that the transport coefficient is 1. The temperature of $T_l = 0°\mathrm{C}$ is kept on the left border of the rod, while on the other end of the rod the temperature $T_p = 1°\mathrm{C}$ is maintained. It is assumed that $T(x,\,0) = 0°\mathrm{C}$ for $x\ \ in\,(0,\,L)$.

Thus, the initial-boundary problem can be presented as

$$T_t = T_{xx}, \quad T(0,\,x) = 0, \quad T(t,\,0) = 0, \quad T(t,\,L) = 1. \tag{6.37}$$

Using the Laplace transform with respect to t variable, we get the following boundary value problem:

$$U_{xx} - p\,U = 0, \qquad U(p,\,0) = 0, \quad U(p,\,L) = \frac{1}{p}, \tag{6.38}$$

where $U(p,x) = \mathcal{L}(T(t,x))$. Solution of (6.38) is given by the following formula:

$$U(p,x) = \frac{1}{p}\cdot\frac{\sinh x\,\sqrt{p}}{\sinh L\,\sqrt{p}}, \qquad 0 < x < L. \tag{6.39}$$

To justify the formula for inverse Laplace transform, Theorem B.1 formulated in Appendix B should be used. We are going to show that the following formula holds:

$$u(t,\,x) = \frac{1}{2\,\pi\,i}\int_{\gamma - i\,\infty}^{\gamma + i\,\infty} \frac{1}{p}\,\frac{\sinh x\,\sqrt{p}}{\sinh L\,\sqrt{p}} e^{p\,t}\,d\,p = \sum_{k\,\in\,\mathcal{M}} \operatorname{res}\left[\frac{1}{p_k}\,\frac{\sinh x\,\sqrt{p_k}}{\sinh L\,\sqrt{p_k}} e^{p_k\,t}\right]. \tag{6.40}$$

In order to show the correctness of the formula (6.40), we check if the conditions formulated in Theorem B.1 are fulfilled.

(1) It is evident that the function $U(p, x)$ from the formula (6.39) is analytic in $\mathbb{C}$ everywhere except the origin and the set of points satisfying the equation

$$e^{-2L\sqrt{p}} - 1 = 0.$$

This set consists of isolated points

$$p_k = -\left(\frac{\pi k}{L}\right)^2, \qquad k = 0, 1, ... \tag{6.41}$$

belonging to the left half-line $\mathbb{R}_-$.

(2) We select the parabola-shaped contours

$$r_k = \frac{2\,a_k^2}{1 - \cos\varphi} = \frac{2\,a_k^2}{\sin^2\frac{\varphi}{2}}, \quad a_k = \frac{\pi\left(k - \frac{1}{2}\right)}{L}, \quad k = 1, 2, ...,$$

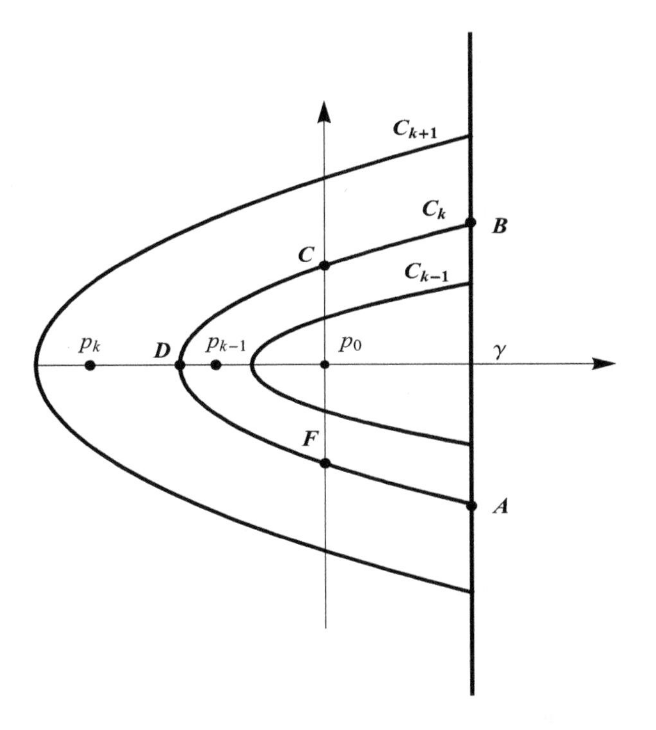

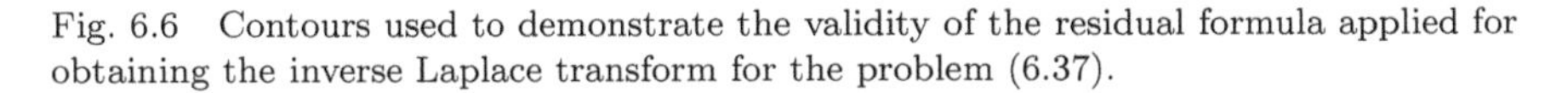
Fig. 6.6 Contours used to demonstrate the validity of the residual formula applied for obtaining the inverse Laplace transform for the problem (6.37).

based on a straight line $\mathrm{Re}\, p = \gamma > 0$ (see Fig. 6.6). The parameters of the parabola are selected in such a way that the following inequalities hold:

$$-a_{k+1}^2 < p_k = -\left(\frac{\pi\, k}{L}\right) < -a_k^2.$$

Thus, each pole p_k lies between two adjacent parabolas.

(3) The points belonging to the parabola C_k satisfy the equation $p = r_k\, e^{i\,\varphi}$. Using the identity

$$\left|\sinh(x + i\, y)\right|^2 = \cosh^2\, x - \cos^2\, y,$$

we get the following chain of relations for the integrand appearing in the formula (6.40):

$$\left|\frac{\sinh x\sqrt{p_n}}{\sinh L\sqrt{p_n}}\right|^2 = \left|\frac{\sinh x\sqrt{r_n}\left(\cos\frac{\varphi}{2} + i\,\sin\frac{\varphi}{2}\right)}{\sinh L\sqrt{r_n}\left(\cos\frac{\varphi}{2} + i\,\sin\frac{\varphi}{2}\right)}\right|^2 = \frac{\cosh^2\left(x\sqrt{r_n}\cos\frac{\varphi}{2}\right) - \cos^2\left(x\sqrt{r_n}\sin\frac{\varphi}{2}\right)}{\cosh^2\left(L\sqrt{r_n}\cos\frac{\varphi}{2}\right) - \cos^2\left(L\sqrt{r_n}\sin\frac{\varphi}{2}\right)}$$

$$= \frac{\cosh^2\left(\frac{x\pi(n-\frac{1}{2})}{L}\cot\frac{\varphi}{2}\right) - \cos^2\left(\frac{x\pi(n-\frac{1}{2})}{L}\right)}{\cosh^2\left(\pi(n-\frac{1}{2})\cot\frac{\varphi}{2}\right) - \cos^2\left(\pi(n-\frac{1}{2})\right)} \le \frac{\cosh^2\left(\pi(n-\frac{1}{2})\cot\frac{\varphi}{2}\right) - \cos^2\left(\frac{x\pi(n-\frac{1}{2})}{L}\right)}{\cosh^2\left(\pi(n-\frac{1}{2})\cot\frac{\varphi}{2}\right)}$$

$$\le 1 + \frac{\cos^2\left(\frac{x\pi(n-\frac{1}{2})}{L}\right)}{\cosh^2\left(\pi(n-\frac{1}{2})\cot\frac{\varphi}{2}\right)} \le 2.$$

For the integral along CD we get the estimation

$$\left|\int_{CD} \frac{1}{p_n}\frac{\sinh x\sqrt{p_n}}{\sinh L\sqrt{p_n}}\, e^{p_n\, t} p_n\, e^{i\,\varphi}\, d\,\varphi\right| \le \int_{\pi/2}^{3\,\pi/2}\left|\frac{1}{p_n}\frac{\sinh x\sqrt{p_n}}{\sinh L\sqrt{p_n}}\, e^{p_n\, t} p_n\, e^{i\,\varphi}\right| d\,\varphi$$

$$\le \sqrt{2}\int_{\pi/2}^{3\,\pi/2}\frac{1}{r_n} e^{t\, r_n\,\cos\varphi} r_n\, d\,\varphi \le \sqrt{2} \underset{\pi/2\le\varphi\le 3\,\pi/2}{\left|\frac{1}{r_n}\right|} \int_0^{\pi/2} e^{-t\, a_n^2\,\sin\psi} a_n^2\, d\,\psi$$

$$\le \frac{\sqrt{2}}{a_n^2}\int_0^{\pi/2} e^{-t\, a_n^2\,\sin\psi} a_n^2\, d\,\psi \le \frac{\sqrt{2}}{\pi(n-\frac{1}{2})\, t} \underset{n\to\infty}{\to} 0\,.$$

In the last expression we used the inequality

$$\int_0^{\pi} r\, e^{-a\, r\,\sin\varphi}\, d\,\varphi \le \frac{\pi}{a},$$

which is shown during the proof of Jordan's lemma (see Theorem A.4).

(4) Let us now estimate the integral on the segment BC, for which, analogically to the segment CD, the following inequality holds:

$$\left|\frac{\sinh x\sqrt{p_n}}{\sinh L\sqrt{p_n}}\right|^2 \le 2.$$

Taking into account that the angle varies along the line CD within the interval $\pi/2 - \delta < \varphi < \pi/2$, $\delta = \arctan\left(\frac{\gamma L^2}{2\pi^2\left(n-\frac{1}{2}\right)^2}\right)$, we get the estimation

$$\left|\int_{BC} \frac{1}{p_n} \frac{\sinh x\sqrt{p_n}}{\sinh L\sqrt{p_n}} e^{p\,t} p_n\, e^{i\,\varphi}\, d\varphi\right| \le \frac{\sqrt{2}}{2}\int_{\pi/2-\delta}^{\pi/2} e^{a_n^2\, t\,\cos\varphi}\, d\varphi$$

$$\le \frac{\sqrt{2}}{2} e^{t\,a_n\,\sin\delta} \arctan\left(\frac{\gamma L^2}{2\pi^2\left(n-\frac{1}{2}\right)^2}\right) \le \frac{\sqrt{2}}{2} e^{t\,\gamma} \arctan\left(\frac{\gamma L^2}{2\pi^2\left(n-\frac{1}{2}\right)^2}\right) \underset{n\to\infty}{\to} 0\,.$$

Similarly, it can be shown that the integral on the segment DF also tends to zero when $n \to \infty$.

We have thus shown that the inverse Laplace transform is expressed by the formula (6.40). Carrying out the appropriate calculations, we finally get the solution in the following form:

$$u(t,x) = \frac{x}{L} + \frac{2}{\pi}\sum_{k=1}^{\infty} \frac{(-1)^k}{k} \sin\left(\frac{k\pi x}{L}\right) e^{\frac{-k^2\pi^2\,t}{L^2}}.$$

Below we present the solution of the same problem obtained in the *Mathematica* package.

Cell 6.3.

In[1]:

```
u[x] /. DSolve[D[u[x], x, 2] - p u[x] == 0, u[0] == 0, u[L] == 1/p, u[x], x] //
Simplify
```

Out[1]:

$$\left\{e^{\sqrt{p}(L-x)}\left(-1+e^{2\sqrt{p}x}\right)\left(-1+e^{2\sqrt{p}L}\right)p\right\}$$

In[2]:

```
First[%] - Sinh[Sqrt[p] x]/(p Sinh[Sqrt[p] L]) // Simplify
```

Out[2]:

```
0
```

The obtained solution can be presented as an infinite sum according to the following formula [Maurin (1973); Aramanovich *et al.* (2004)]:

$$f(z) = f(0) + \sum_{n=1}^{\infty} A_n\left(\frac{1}{z-a_n} + \frac{1}{a_n}\right), \tag{6.42}$$

to which the inverse Laplace transform can be applied. First, we find the representation of the function

$$g(p) = \frac{\sinh\sqrt{p}\,x}{\sinh\sqrt{p}\,L}$$

in the form of a power series. To obtain $g(0)$, we employ the l'Hospitala rule:

$$g(0) = \lim_{p \to 0} g(p) = \frac{x}{L}.$$

The poles of the function will coincide with zeroes of the denominator:

$$\left(-1 + e^{2\sqrt{p}L}\right) p = 0.$$

Solving the equation $e^{2\sqrt{p}L} = e^{2\,\pi\,k\,i}$, $k \in \mathbb{N}$, we get:

$$2\sqrt{p}L = 2\,\pi\,k\,i, \quad k = 1, 2, \ldots.$$

Finally, we obtain the solution of the form $p_k = -\left(\frac{\pi k}{L}\right)^2$. Residuals of the function $g(p)$ at the poles are as follows:

$$A_k = \operatorname{Res} g(p_k) = (-1)^{k+1} \frac{2\pi k}{L^2} \sin \frac{\pi k}{L} x.$$

Thus,

$$g(p) = g(0) + \sum_{k=1}^{\infty} \frac{A_k p}{p_k(p - p_k)} = \frac{x}{L} + \frac{2}{\pi} \sum_{k=1}^{\infty} \frac{(-1)^{k+1}}{k} \frac{p \sin \pi k x/L}{p + \pi^2 k^2/L^2}.$$

We will continue the calculations with the use of the *Mathematica* package.

Cell 6.4.

In[1]:
```
g[p_]=Sinh[Sqrt[p] x]/Sinh[Sqrt[p] L];
```
In[2]:
```
Limit[g[p],p → 0]
```
Out[2]:

$\frac{x}{L}$

In[3]:
```
Ak=Sinh[Sqrt[p] x]/D[Sinh[Sqrt[p] L], p] /. p → - Pi^2 a k^2/L^2 // FullSimplify
```
Out[3]:

$$-\frac{2k\pi Sec[k\pi]Sin[k\pi x/L]}{L^2}$$

So we have obtained the following result:

$$u(p, x) = \frac{x}{pL} + \frac{2}{\pi} \sum_{k=1}^{\infty} \frac{(-1)^k}{k} \frac{\sin \pi k x/L}{p + \pi^2 k^2/L^2}.$$

Applying the inverse Laplace transforms to the individual terms, we finally get the solution in the following form

$$u(t, x) = \frac{x}{L} + \frac{2}{\pi} \sum_{k=1}^{\infty} \frac{(-1)^k}{k} \sin\left(\frac{k\pi x}{L}\right) e^{\frac{-k^2 \pi^2 t}{L^2}}.$$

Example 6.7. We consider an infinite cylinder of diameter R. A constant temperature $u_0 \neq 0$ is maintained on the lateral surface of the cylinder. We are going to find the temperature field, under the assumption that at the initial moment of time $t = 0$ the temperature inside the cylinder is zero.

We choose the coordinate system in such a way that the axis OZ coincides with the axis of symmetry of the cylinder. We denote by $u(M, t)$ the temperature at the point $M = (x, y, z)$.

The problem of heat conductivity in an infinite cylinder boils down to solving the following initial-boundary value problem:

$$u_t = a\left[u_{xx} + u_{yy} + u_{zz}\right], \quad u(M, 0) = 0, \quad u(M_R, t) = u_0, \tag{6.43}$$

where $M = \{(x,\, y,\, z) \in R^3 : x^2 + y^2 \leq R\}$, $M_R = \{(x,\, y,\, z) \in R^3 : x^2 + y^2 = R\}$, $t \geq 0$.

It is convenient to solve the problem with cylindrical symmetry in cylindrical coordinates $(r,\, \varphi,\, z)$, related to the Cartesian variables as follows:

$$r = \sqrt{x^2 + y^2}, \qquad \varphi = \operatorname{arctg}\frac{y}{x}, \qquad z = z.$$

The Laplace operator $\Delta = \frac{\partial^2}{\partial\, x^2} + \frac{\partial^2}{\partial\, y^2} + \frac{\partial^2}{\partial\, z^2}$ in new variables takes the form [Korn and Korn (1968)]:

$$\Delta = \frac{1}{r}\frac{\partial}{\partial\, r}\, r\, \frac{\partial}{\partial\, r} + \frac{1}{r^2}\frac{\partial^2}{\partial\, \varphi^2} + \frac{\partial^2}{\partial\, z^2}.$$

Due to the symmetry of the problem, the temperature field depends neither on the variable φ, nor on the variable z, therefore the problem ultimately comes down to the following one:

$$u_t = a\left[u_{rr} + \frac{1}{r}\, u_r\right], \quad u(r, 0) = 0, \quad 0 \leq r \leq R, \quad u(R, t) = u_0. \tag{6.44}$$

Applying to (6.44) the Laplace transform, we get the modified Bessel equation

$$U_{rr} + r^{-1} U_r - (p/a) U = 0. \tag{6.45}$$

The initial condition attains the form $U(R, 0) = u_0/p$. The general solution to (6.44) takes the form

$$U = c_1 I_0(qr) + c_2 K_0(qr),$$

where $c_{1,2}$ are constants, $q = \sqrt{p/a}$, I_0, K_0 are the modified Bessel functions, which can be invoked in the *Mathematica* package using the commands

```
BesselI[0,x], BesselK[0,x]
```

Since $K_0(z) \underset{z\to+0}{\to} \infty$, and we are looking for the bounded solution, it should be assumed that $c_2 = 0$. Hence $U = (u_0 I_0(qr))/(pI_0(qR))$. In order to obtain the solution in physical variables, we should apply the inverse Laplace transform to the function found. In this situation, instead of using the Mellin formula, it is advisable to use the residues theorem, which leads to the following result:

$$u(r,t) = \sum \operatorname{Res} \frac{u_0 I_0(qr)}{pI_0(qR)} e^{pt},$$

where $p = \{0\} \cup \{\mu_{0k}\}_{k\in I}$, μ_{0k} are points in which $I_0(\mu_{0k} R) = 0$. The values μ_{0k} can be obtained from the equality $I_0(qR) = J_0(iqR) = 0$. Using the tabulated zeros of the Bessel function μ_{0k} (they are invoked by the command BesselJZero[n, k], giving the value of the k-th zero of the function J_n), we can get the equality $iqR = \mu_{0k}$, or $p_k = -\mu_{0k}^2 a/R^2$, $k = 1, 2, \ldots$. So,

$$u(r,t) = u_0 \sum \frac{e^{p_k t} I_0(q_k r)}{p_k (\frac{d}{dp}) I_0(qR)_{p=p_k}}.$$

Employing the equation

$$p\frac{d}{dp} I_0(qR)_{p=p_k} = -J_1(\mu_{0k})\mu_{0k}/2,$$

being the consequence of the identities $d\,J_0/dx = J_1$ and $I_0(q_k r) = J_0(\mu_{0k} r/R)$, we get the final version of the solution to the problem:

$$u(r,t) = u_0 \left(1 - 2\sum_{k=1}^{\infty} \frac{J_0(\mu_{0k} r/R) e^{-\mu_{0k}^2 a R^{-2} t}}{\mu_{0k} J_1(\mu_{0k})}\right). \tag{6.46}$$

Interpretation of the formula (6.46) in the *Mathematica* package as well as the construction of the corresponding graphs are performed in the file PM2_2.nb. Plots of the function $u(r,t)$ for fixed values of r are presented in Fig. 6.7. Note that for $r = 1$ the stationary solution corresponds to the boundary condition $u_0 = 1$.

Remark. Solving the above problem, we have not provided a proof of the applicability of the formula for residua for computing the inverse Laplace transform. Such a proof can, of course, be made in full analogy to how it was done in the previous examples, but since we are primarily concerned with solving a specific boundary value problem, we can just directly check that the obtained formula satisfies the problem (6.44) and use the theorem on existence and uniqueness of the solution to this problem.

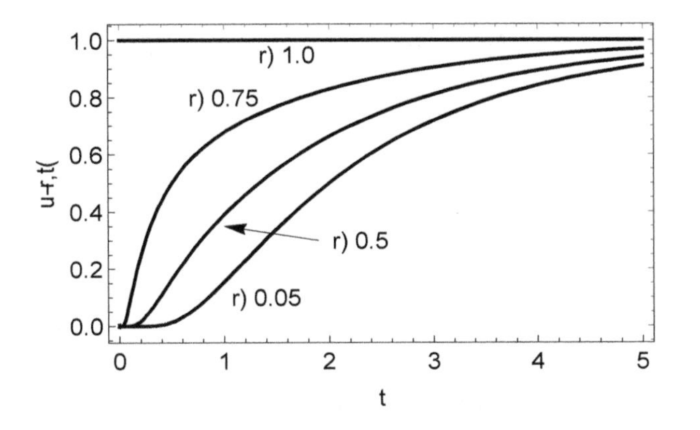

Fig. 6.7 Profiles of the solution to the equation (6.44) for $a = 0.1$, $R = 1$, $u_0 = 1$ and $r = 0.05$; 0.5; 0.75; 1.0. The total number of terms used in series decomposition is equal to 250 ($k \in (1,\ 250)$).

Problem.

(1) Solve the equation describing the heat flow in a thin wire of length l heated by an electric current of constant intensity. The model equation takes the form

$$u_t = \kappa u_{xx} - bu + a, \quad 0 < x < l, \quad t > 0,$$

where the κ, b, a are positive. At the ends of the wire zero temperature is maintained. The initial temperature is also zero. Assume that $a = b = \kappa = 1$

(the answer: $u = 1 - \frac{\operatorname{ch}(\ell/2-x)\sqrt{1/k}}{\operatorname{ch}\frac{\ell}{2}\sqrt{\frac{1}{k}}} - \frac{4}{\pi}\sum_{n=0}^{\infty} \frac{\exp(-(1+k\pi^2\ell^{-2}(2n+1)^2))t}{(2n+1)1+k\pi^2\ell^{-2}(2n+1)^2}$).

(2) The temperature at the left end of the rod is zero. At the right end the temperature changes according to the formula $u(\ell, t) = \sin\omega t$. Find the temperature change in the rod assuming that the initial temperature is zero and the lateral surface is insulated

(the answer: $u(t,x) = \frac{x}{\ell}\sin\omega t + \frac{2}{\pi}\sum_{k=1}^{\infty}\frac{(-1)^k}{k}\sin\left(\frac{k\pi x}{\ell}\right)e^{\frac{-k^2\pi^2 at}{\ell^2}}$).

(3) Find the longitudinal oscillations of the rod, assuming that one of its ends is immobilized and the other end is subjected to the action of the force $F = \sin\omega t$ directed along the symmetry axis. In the initial moment of time, the rod is in a state of equilibrium. Hint: the boundary condition takes the form $E\,u_x(\ell,t) = F = \sin\omega t$, where E is the Young's modulus

(the answer: $u(x,t) = \frac{a^2}{E^2}\frac{\sin(\omega x/a)}{\cos(\omega \ell/a)}\sin\omega t + \frac{2\omega a}{E\ell}\sum_{n=1}^{\infty}\frac{(-1)^k}{k_n}\frac{\sin k_n x}{\omega^2 - k_n^2 a^2}$ $\sin k_n at$, $k_n = \pi\ell^{-1}(n-0.5)$).

(4) Create the animation exposing the evolution of the solution to the problem

$$u_t = a u_{xx}, \quad x \in [0,\, L], \quad u(x,0) = 0, \quad u(0,t) = 0, \quad u(L,t) = 1.$$

(5) Create the animation showing the evolution of the solution to the problem posed in Example 6.7.

(6) Solve the problems 2 and 3 using the *Mathematica* package.

6.3 The method of separating variables

6.3.1 *One dimensional case*

The method of separating variables works effectively for the following class of the problems:

$$a(t)u_{tt} + b(t)u_t + c(x)u_{xx} + d(x)u_x + [e(x) + f(t)]\, u = 0, \tag{6.47}$$

$$u(0,x) = \varphi(x), \quad u_t(0,x) = \psi(x), \quad x \in [0, L], \tag{6.48}$$

$$\alpha u(t,0) + \beta u_x(t,0) = 0, \quad \gamma u(t,L) + \delta u_x(t,L) = 0. \tag{6.49}$$

We search for the solution to (6.47), in the form

$$u(t,x) = X(x)T(t). \tag{6.50}$$

Employment of the ansatz (6.50) occurs to lead to the separation of variables in the equation (6.47). Indeed, inserting (6.50) into the equation (6.47), we get:

$$a(t)X\ddot{T} + b(t)X\dot{T} + c(x)X^{''}T + d(x)X'T + [e(x) + f(t)]\, XT = 0.$$

Dividing this equation by XT and grouping the result appropriately, we obtain the equation

$$\left(a(t)\frac{\ddot{T}}{T} + b(t)\frac{\dot{T}}{T} + f(t)\right) = -\left\{c(x)\frac{X^{''}}{X} + d(x)\frac{X'}{X} + e(x)\right\}.$$

Due to the fact that the left side depends only on the t variable, while the right side depends only on the x variable, this equation is equivalent to the following system of equations:

$$a(t)\frac{\ddot{T}}{T} + b(t)\frac{\dot{T}}{T} + f(t) = -\lambda; \qquad c(x)\frac{X^{''}}{X} + d(x)\frac{X'}{X} + e(x) = \lambda,$$

where $\lambda = \text{const}$. This system can also be presented as

$$a(t)\ddot{T} + b(t)\dot{T} + (\lambda + f(t))T = 0, \tag{6.51}$$

$$c(x)X'' + d(x)X' + (e(x) - \lambda)X = 0. \tag{6.52}$$

In the equations defining the boundary conditions, the variables also split:

$$\alpha X(0) + \beta X'(0) = 0, \quad \gamma X(L) + \delta X'(L) = 0. \tag{6.53}$$

This way the spectral problem (or the *Sturm-Liouville problem*) arises [Tikhonov and Samarskii (1990); Simon (2005)]:

find the values of the parameter λ for which a non-trivial solutions to the problems (6.51)–(6.53) exist.

Among the properties of the solutions of the spectral problem, the orthogonality of independent solutions is particularly important. To demonstrate this property, we will reduce the equation (6.52) to the standard form, multiplying the equation by the function $0 \neq \rho(x)$, which is non-negative on the segment $[0, L]$. The function $\rho(x)$ is chosen such that $\rho\, d = (\rho c)'$ is satisfied (if possible). Then the equation (6.52) can be rewritten as follows:

$$(p(x)X')' - g(x)X + \lambda\rho(x)X = 0, \tag{6.54}$$

where $p = -\rho c$, $g = e\rho$.

Let us consider two eigenfunctions X_1 and X_2 satisfying the equation (6.54) and corresponding to the eigenvalues λ_1 i λ_2:

$$(pX_1')' - gX_1 + \lambda_1\rho X_1 = 0, \tag{6.55}$$

$$(pX_2')' - gX_2 + \lambda_2\rho X_2 = 0. \tag{6.56}$$

Multiplying (6.55) by X_2, and (6.56) by X_1, and subtracting the results by sides, we get:

$$(\lambda_1 - \lambda_2)\rho\, X_1\, X_2 = (p\, X_1'\, X_2 - p\, X_2'\, X_1)'.$$

Integrating this expression on the segment $[0, L]$ we get, after taking into account the boundary conditions, the equality:

$$(\lambda_1 - \lambda_2)\int_0^L \rho\, X_1\, X_2\, dx = (p\, X_1'\, X_2 - p\, X_2'\, X_1)\Big|_0^L = 0.$$

So, if $\lambda_1 \neq \lambda_2$ then the appropriate eigenfunctions are orthogonal in the weighted norm. From this also follows that all eigenvalues are real and the conditions $\alpha\beta \leqslant 0$, $\gamma\delta \geqslant 0$, $e(x) \geqslant 0$ are fulfilled. Besides, it is possible to show that the set $\{X_n\}_{n\in I}$ forms a basis in C$[0; L]$ [Tikhonov and Samarskii (1990)].

Let us consider the use of this method for finding solutions to the wave equation

$$u_{tt} = a^2 u_{xx}, \tag{6.57}$$

satisfying the boundary conditions $u(t,0) = u(t,L) = 0$ ($\alpha = \gamma = 1$, $\beta = \delta = 0$). Employing the ansatz (6.50), we get after some manipulations the separated equation:

$$\frac{\ddot{T}}{a^2 T} = \frac{X''}{X}.$$

Since the left side depends only on the variable t, while the right side depends only on the variable x, equality is possible if and only if both sides are equal to some constant:

$$\frac{\ddot{T}}{a^2 T} = \frac{X''}{X} = -\lambda, \tag{6.58}$$

that is, when the following system is satisfied:

$$\ddot{T} + a^2\lambda T = 0, \qquad X'' + \lambda X = 0. \tag{6.59}$$

The solution of the second equation should satisfy in additional the boundary conditions $X(0) = X(L) = 0$ and there are several options.

- If $\lambda < 0$, then

  $$X(x) = C_1 e^{\sqrt{-\lambda}\,x} + C_2 e^{-\sqrt{-\lambda}x}, \qquad C_{1,2} = \text{const}.$$

  It appears from the boundary conditions that C_1 and C_2 should satisfy the system

  $$C_1 + C_2 = 0,$$
  $$C_1 e^{\sqrt{-\lambda}L} + C_2 e^{-\sqrt{-\lambda}L} = 0.$$

  The only solution to this system is the trivial solution $C_1 = C_2 = 0$. Therefore, $X(x) \equiv 0$, and this, in turn, implies that $\lambda < 0$ cannot be an eigenvalue.
- If $\lambda = 0$, then

  $$X(x) = C_1 + C_2 x, \qquad C_{1,2} = \text{const}.$$

  The constants C_1, C_2 in this case satisfy the system

  $$C_1 = 0,$$
  $$C_1 + C_2 L = 0$$

  having merely zero solutions. Hence $X(x) \equiv 0$, and the case $\lambda = 0$ should be dropped out.

- If $\lambda > 0$, then
$$X(x) = C_1 \sin\sqrt{\lambda}x + C_2 \cos\sqrt{\lambda}x.$$
The constants $C_{1,2}$ satisfy the system
$$C_2 = 0,$$
$$C_1 \sin\sqrt{\lambda}L = 0.$$
This system has a non-zero solution provided that $\sqrt{\lambda}\,L = \pi n$, $n = 1, 2, \ldots$. Hence $\lambda = \lambda_n = (\pi n/L)^2$, $n = 1, 2, \ldots$. These eigenvalues correspond to the eigenfunctions
$$X_n(x) = C_1 \sin(\pi n x/L).$$

Now we can easily find a solution to the first equation of the system (6.59) corresponding to the eigenvalue λ_n:
$$T_n(t) = C_2 \cos\left(\frac{\pi\, n\, a\, t}{L}\right) + C_3 \sin\left(\frac{\pi\, n\, a\, t}{L}\right).$$

Finally, using the superposition principle, we can present the general solution to the boundary problem in the form
$$u(x,t) = \sum_{n=1}^{\infty} X_n(x) T_n(t) = \sum_{n=1}^{\infty} \left[a_n \cos\left(\frac{\pi n a t}{L}\right) + b_n \sin\left(\frac{\pi n a t}{L}\right)\right] \sin\left(\frac{\pi n x}{L}\right). \tag{6.60}$$

To define the constants a_n, b_n, we use the initial conditions:
$$\sum_n a_n \sin(\pi n x/L) = u(x,0) = \varphi(x),$$
$$\sum_n b_n (\pi n a/L) \sin(\pi n x/L) = u_t(x,0) = \psi(x).$$

Multiplying these equations by $X_k(x) = \sin(\pi\, k\, x/L)$, and next integrating the expressions obtained over the interval $[0; L]$, we get:
$$a_k = \frac{2}{L}\int_0^L \varphi(x) \sin\frac{\pi k x}{L}dx, \qquad b_k = \frac{2}{\pi k a}\int_0^L \psi(x) \sin\frac{\pi k x}{L}dx.$$

Example 6.8. Let us find the solution to the equation
$$u_{tt} = c^2 u_{xx} \tag{6.61}$$
satisfying the initial conditions
$$u(0,\, x) = \varphi(x) = \begin{cases} \cos^2\left[\frac{(x-b)\pi}{2a}\right] & \text{as } |x-b| < a, \\ 0 \ \text{elsewhere}, \end{cases} \qquad u_t(0,\, x) = 0 \tag{6.62}$$
and the boundary conditions
$$u(t,\, 0) = u(t,\, L) = 0. \tag{6.63}$$

The condition $u_t(0,x) = 0$ implies that $b_n = 0$, and hence

$$u(t,x) = \sum_{n=1}^{\infty} a_n \cos(\lambda_n ct) \sin \lambda_n x, \qquad \lambda_n = \pi n/L.$$

Using nonzero initial condition, we get:

$$a_n = \frac{2}{L}\int_0^L \varphi(x) \sin(\lambda_n x)\, dx.$$

The solution to the above problem is presented in Fig. 6.8, which shows that the initial disturbance splits into two identical impulses moving in opposite directions. After bouncing off the edges of the area, the waves restore their original shape, but are mirrored about the horizontal axis. After the second bounce, the waves completely restore their shape.

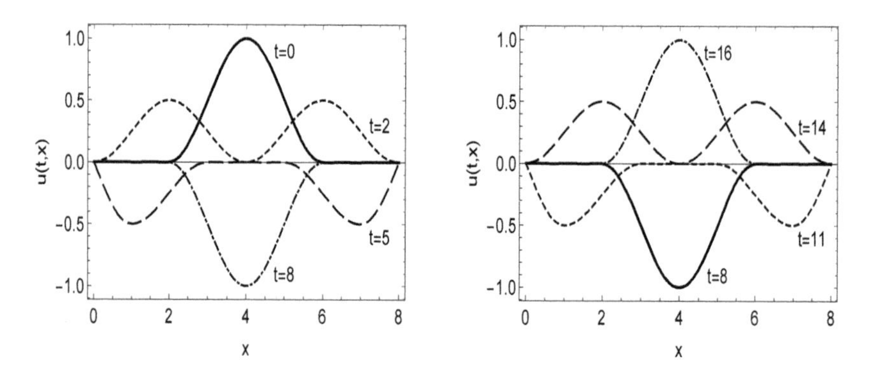

Fig. 6.8 Graphical illustration of the solution to the problem (6.61), (6.62), and (6.63). The solution was obtained with the following parameter values: $L = 8$, $b = 4$, $a = 2$, $c = 1$. The profiles correspond to the moments of time $t = 0;\ 2;\ 5;\ 8;\ 11;\ 14;\ 16$.

Example 6.9. Let us consider the initial value problem

$$u_{tt} - u_{xx} = 0, \tag{6.64}$$

$$u(0,\, x) = \phi(x) = \begin{cases} 1 + \cos[x], & \text{as} \quad |x| < \pi, \\ 0, & \text{elsewhere}, \end{cases} \tag{6.65}$$
$$u_t(0,\, x) = 0.$$

We are looking for the solutions satisfying boundary conditions

$$u(t,\, -L) = u(t,\, L) = 0, \qquad L \gg \pi. \tag{6.66}$$

The problems (6.64), (6.65), and (6.66) describe evolution of a small transverse disturbances in an unstretchable string, the attachment points of which are located in the same distance from the origin.

Remark. In case when $x \in \mathbb{R}$, solution of the problems (6.64), (6.65) are given by the d'Alembert formula

$$u(t,\, x) = \frac{1}{2}\left[\phi(x-t) + \phi(x+t)\right],$$

which was derived at the end of Section 5.2.3 (see Example 5.4).

After adding the boundary condition (6.66), we can use the spectral method. Due to the symmetry of the initial conditions, the solution is presented as a series

$$u(t,\, x) = \sum_{n=0}^{\infty} a_n \cos\left[\frac{\pi(n+\frac{1}{2})}{L}x\right] \cos\left[\frac{\pi(n+\frac{1}{2})}{L}t\right], \tag{6.67}$$

where

$$a_n = \frac{1}{L}\int_{-\pi}^{\pi} \phi(x)\cos\left[\frac{\pi(n+\frac{1}{2})}{L}x\right]d\,x = \frac{16\,L^3 \sin\left[\frac{(1+2\,n)\pi^2}{2\,L}\right]}{\pi(1+2\,n)\{[\pi(1+2\,n)]^2 - 4\,L^2\}}.$$

Animation of the solution (6.67) and the solution to the problems (6.61)–(6.63) are shown in the PM2_3.nb.

Example 6.10. Let us solve the problem of longitudinal vibration of an elastic bar, the deformations of which satisfy the equation $u_{tt} = c^2\, u_{xx}$ and the following conditions: $u(0,\, x) = 0$, $u_t(0,\, x) = \psi(x)$, $u_x(t,\, 0) = 0$ (left edge not fastened), $u_x(t,\, L) = -h\, u(t,\, L)$ (elastic force with the elasticity coefficient h acts on the right edge).

Using the ansatz $u = X(x)T(t)$, we get the spectral problem

$$X'' + \lambda^2 X = 0, \quad X'(0) = 0, \quad X'(L) = -h\cos\lambda L.$$

Nonzero solutions of this problem are as follows:

$$X_n = \cos\lambda_n x,$$

where the eingenvalues $\lambda_n > 0$, $n = 1, 2, \ldots$ satisfy the transcendental equation

$$\tan(\lambda_n L) = \frac{h}{\lambda_n} \to \tan(\theta_n) = \frac{h\,L}{\theta_n}, \qquad \theta_n = \lambda_n L, \qquad n = 1, 2, 3 \ldots.$$

Figure 6.9(a) shows the graphs of the functions $y_1(x) = \tan(x\,L)$ and $y_2(x) = h/x$. Successive points of intersection of $y_2(x)$ with the branches of the function $y_1(x)$ determine the eigenvalues λ_n. The appropriate calculations performed in the *Mathematica* package are presented in the cell 6.5.

Cell 6.5.

```
In[1]: L = 2; h = 1; n = 20;
In[2]:
  roots = Table[ x/L /. FindRoot[ Tan[x] == L h/x, {x, 0.1 + (j - 1) Pi}], {j, 1,
n}]
Out[2]:
  {0.538437, 1.8218, 3.28917, 4.81478, ..., 29.8619}
In[3]:
  pln = Table[{roots[[j]], Tan[L roots[[j]]] }, {j, 1, Length[roots] }];
In[4]:
  EigenValues = Plot[{Tan[x L], h/x }, {x, 0, 10}, Epilog → {PointSize[Large],
Point[pln]}, PlotRange → {All, {-0.5, 2.5}}, PlotStyle → Thick, BaseStyle →
{FontSize → 14}, AxesLabel → {"x", "φ(x)"}, AxesStyle → Directive[Black]
```

Ultimately, the solution takes the form

$$u = \sum_{n=0}^{\infty} B_n \sin(\lambda_n\, c t) \cos(\lambda_n\, x), \tag{6.68}$$

where

$$B_n = \frac{\int_0^L \psi(x) \cos \lambda_n x dx}{\lambda_n c \int_0^L \cos^2 \lambda_n x dx}.$$

For

$$\psi(x) = \begin{cases} -x^2 + 2x - 1 + a^2, & \text{as } |x-1| < a, \\ 0, & \text{elsewhere} \end{cases} \qquad a = 0.5,$$

solution is shown in Fig. 6.9(b). Figure 6.9(c) presents the solution in the point $x = 1$ as a function of time. The solution of the transcendental equation and the visualization of the temporal evolution of the initial disturbance are presented in the file PM2_4.nb.

Example 6.11. Let us find the solutions to the equation

$$u_t = a u_{xx},$$

describing the heat transport in a thin bar, satisfying the initial conditions

$$u(0, x) = \frac{3}{7}x^2 - x^3 + \frac{3}{4}x^4 - \frac{6}{35}x^5.$$

In addition, we require that both the side surface of the bar and its ends are insulated from the environment.

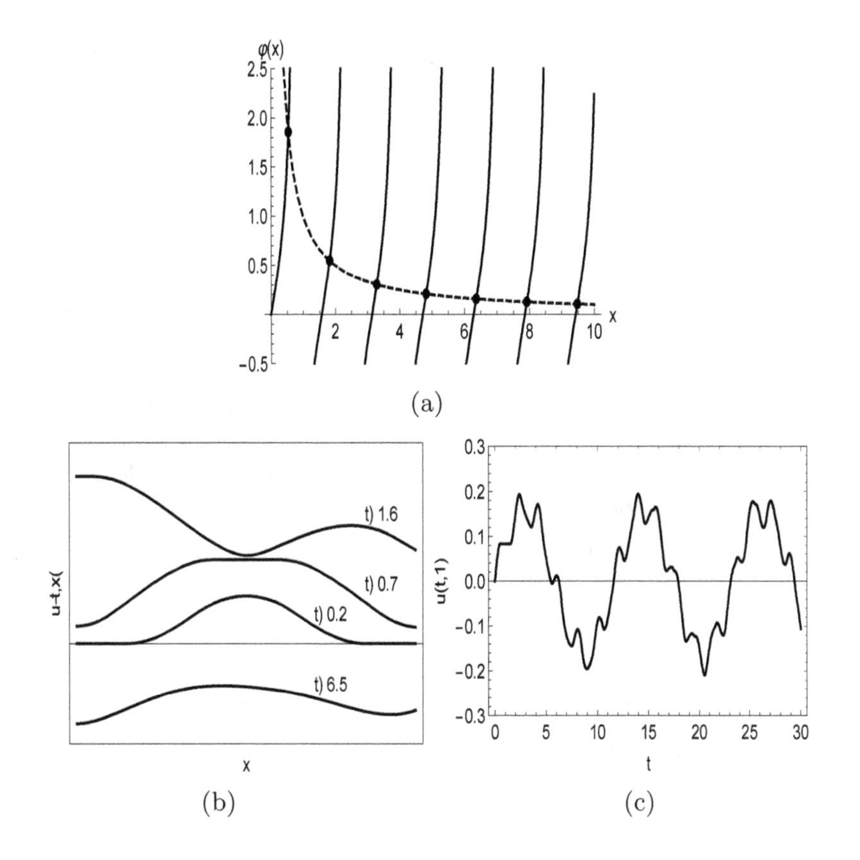

Fig. 6.9 Panel (a): computation of the eigenvalues λ_n; panel (b): graphs of the function $u(t, x)$ represented by a finite number of series decomposition (6.68) calculated at $t = 0\,; 0.2\,; 0.7\,; 1.6\,; 6.5$. The other parameters have the following values: $L = 2$, $c = 1$, $h = 1$; the number of the terms of the series is equal to 20; panel (c): graph presenting the approximation of the function $u(t, 1)$ performed with the same accuracy.

Solution of the problem comes down to finding eigenfunctions that satisfy the boundary conditions

$$u_x(t,\,0) = u_x(t,\,L) = 0. \tag{6.69}$$

As in the previous example, we are looking for a solution of the form $u(t,\,x) = T(t) \cdot X(x)$. Substituting this ansatz into the output equation, we get the following system:

$$\frac{\dot{T}}{a\,T} = -\lambda^2 = \frac{X''}{X}.$$

The solutions of this system take the form

$$X_n = C_{1\,n} \sin \lambda_n\, x + C_{2\,n} \cos \lambda_n\, x, \qquad T_n = C_{3\,n} e^{-a\lambda_n^2\, t}.$$

The condition (6.69) implies that $C_{1\,n} = 0,\ n = 1,\, 2, ...,$ while $\lambda_n = \pi\, n/L$, $n = 0,\, 1,\, 2, ...$

Thus, the solution to the initial-boundary problem is as follows:

$$u(t,\, x) = \sum_{n=0}^{\infty} b_n\, e^{-a\lambda_n^2\, t} \cos \lambda_n\, x,$$

where

$$b_0 = \frac{1}{L} \int_0^L \varphi(x)\, d\,x, \qquad b_n = \frac{2}{L} \int_0^L \varphi(x) \cos \lambda_n\, x\, d\,x,\ n = 1,\, 2, ...$$

The solution $u(t,\, x)$ calculated for $L = 1$, $a = 0.01$ is presented in Fig. 6.10, from which it is seen that with increasing time the temperature profile tends to a constant value. The procedure for solving this problem and the animation of the process are presented in the file PM2_5.nb.

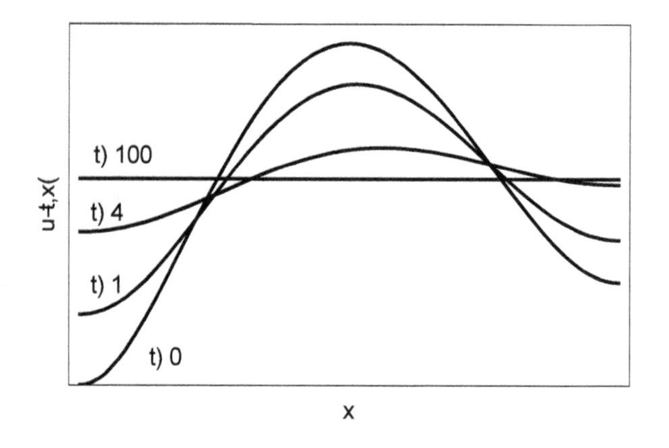

Fig. 6.10 Dynamics of the heat transport process in a thin insulated bar. The graphs show the temperature profiles at times $t = 0$; 1; 4; 100. As time increases, the temperature profile tends to a constant value.

Example 6.12. Vibrations of a string in a viscous medium are described by the equation

$$u_{tt} = a^2 u_{xx} - 2\nu u_t + \Phi(x) \sin \omega t. \tag{6.70}$$

We want to find the solution to this equation that satisfies the boundary conditions $u(0,t) = u(L,t) = 0$, assuming that at the initial moment of time $t = 0$ the string is in the rest, that is, $u(0,\, x) = u_t(0,\, x) = 0$.

Due to the linearity of the problem, we can look for the solution to the complex equation

$$U_{tt} = a^2 U_{xx} - 2\nu U_t + \Phi(x)e^{i\omega t} \tag{6.71}$$

(which is a simpler task), and then define the solution to the initial problem as $u(t,x) = \text{Re } (U(t,x))$.

We look for solutions to the equation (6.71) in the form $U = X(x)e^{i\omega t}$. It is easily see that $X(x)$ satisfies the equation

$$X'' - \Omega^2 X = -\frac{1}{a^2}\Phi(x), \qquad \Omega = \sqrt{-\frac{\omega^2 - 2\nu\omega i}{a^2}} = \alpha + \beta i.$$

We find the solution to this equation using the Laplace transform. Let $\mathcal{L}X(x) = Y(p)$, $\mathcal{L}X''(x) = p^2Y - pA - B$, $A = X(0)$, $B = X'(0)$, $\mathcal{L}\Phi(x) = \tilde{\Phi}(p)$. Then Y satisfying the equation

$$p^2Y - pA - B - \Omega^2 Y = -\frac{1}{a^2}\tilde{\Phi},$$

is as follows:

$$Y = \frac{pA + B - \tilde{\Phi}/a^2}{p^2 - \Omega^2} = \frac{(B + \Omega A)/2\Omega}{p - \Omega} + \frac{(\Omega A - B)/2\Omega}{p + \Omega} - \frac{\tilde{\Phi}/a^2}{p^2 - \Omega^2}.$$

Returning to the initial representation, we get the solution in the following form:

$$X = C_1\, e^{\Omega x} + C_2\, e^{-\Omega x} - \frac{1}{a^2\Omega}\int_0^x \Phi(s)\sinh(\Omega(x-s))ds.$$

Using the boundary conditions, we obtain that

$$C_1 = -C_2 = \frac{2}{a^2\Omega\, \sinh(\Omega L)}\int_0^L \Phi(s)\, \sinh(\Omega(L-s))ds.$$

So, finally we get

$$X(x) = \frac{1}{a^2\Omega}\left(\frac{\sinh(\Omega x)}{\sinh(\Omega L)}\int_0^L \Phi(s)\, \sinh(\Omega(L-s))ds \right. \\ \left. - \int_0^x \Phi(s)\, \sinh(\Omega(x-s))ds\right). \tag{6.72}$$

Assuming that $\Phi(x) = 1$, $\omega = 1.2$, $\nu = 0.1$, $a = 1$, $L = 2$, we build up the solution marked in Fig. 6.11 with the black circles.

The numerical solution of the problem can be obtained by direct integration using the code

```
sim=NDSolve[{D[u[t, x], {t, 2}] == a^2 D[u[t, x], {x, 2}] - 2 ν D[u[t, x], t] +
Sin[w t], u[0, x] == 0, (D[u[t, x], t] /. t -> 0) == 0, u[t, 0] == u[t, L] ==
0}, u, {x, 0, L}, {t, 0, 150}]
```

Figure 6.11(b) shows the time evolution of the numerical solution at $x = L/2$ for $t \in (130; 150)$. Figure 6.11(c) shows the profiles $\mathrm{Re}X(L/2)e^{i\omega t}$. The comparison of the curves obtained from the analytical solution with the curves $u(t, x)$ obtained as a result of numerical interpolation confirms that the numerical solution is close to the solution described by the function $\mathrm{Re}X(L/2)e^{i\omega t}$.

Using the command

```
FindMaximum[{First[Evaluate[u[t, L/2] /. sim]], 142 <= t <=150}, {t, 142}]
```

it is possible to determine the moment of time t_m where the closest maximum of the function $u(t,\ L/2)$ occurs on the segment $t \in$ 142, 150). This way, the maximum value of 1.19138 is obtained, attained at $t_m = 142,858$. Returning to Fig. 6.11(a), we draw on it $u(t_m,\ L/2)$ (continuous line), and convince that solution obtained using numerical methods is consistent with the stationary solution described by the formula (6.72).

An important issue from the point of view of applications is the determination of the dependence of the maximum amplitude of the function $u(t, x)$ on the vibration frequency. Let's find this relationship for a point with a spatial coordinate $x = L/2$. We can use the solution (6.72) for this purpose, analyzing the dependence of $|X(L/2)|$ on frequency ω in the range $\omega \in (1; 2)$. The result is shown in Fig. 6.11(c). It is seen that in the selected range of variation, the amplitude curve has the only maximum corresponding to the resonant frequency. Solution to the above problem in the *Mathematica* package can be found in the file PM2_6.nb.

Problem.

(1) Consider the transverse vibrations of the string attached at the points $x = 0$ and $x = L > 0$, assuming that the initial velocity of all points is zero, and the deviation from the equilibrium at the initial moment has the shape of a parabola with the vertex at $(x,\ u) = (L/2; h)$. Find the maximum deviation of the points from the equilibrium.

(2) Consider the string fixed at $x = 0$ and $x = L > 0$ and being in the equilibrium. At the moment $t = 0$ it is hit by a flat hammer of length $0 \le d = \beta - \alpha \le L$. Consider the vibrations of the string in the case when $\alpha = 0$ and $\beta = L$ and the second extreme case when $\beta - \alpha = 0$

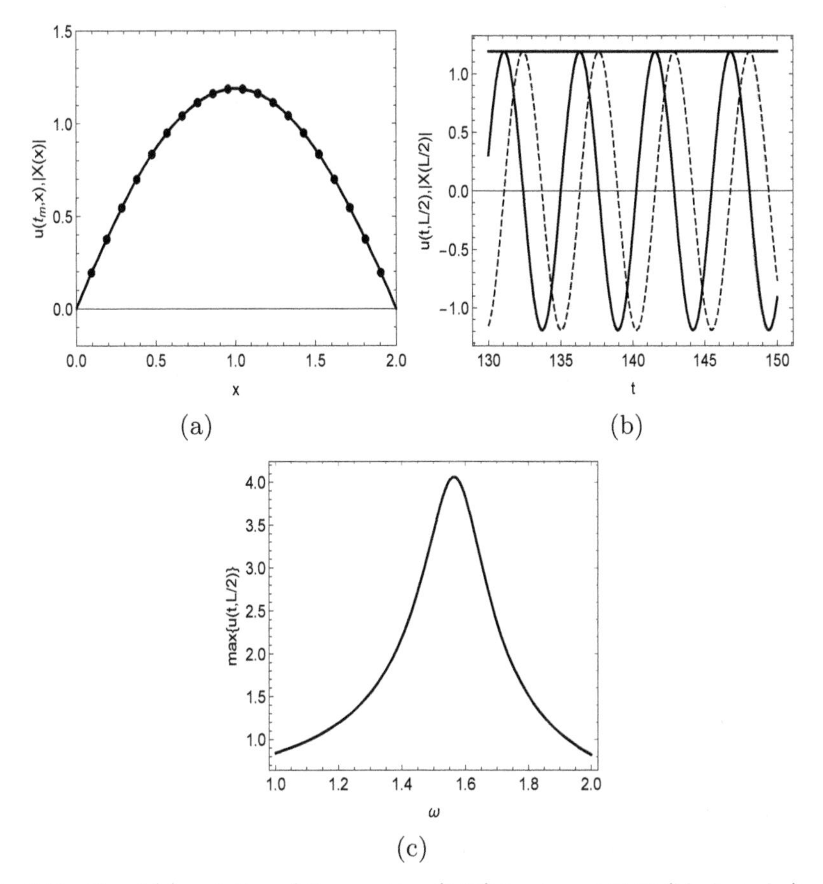

Fig. 6.11 Panel (a): solution of the equation (6.70) $t = t_m = 142.86$ (black circles) and the dependence of $|X(x)|$ on the parameter ω (continuous curve). Panel (b): Graphs of $u(t,\, L/2)$ and $Re\left[X(L/2)e^{i\omega\, t}\right]$ for sufficiently large values of t, when the solution reaches the maximum (solid horizontal line) equal to 1.18994 (in fact, both graphs perfectly match each other, and to distinguish them we slightly shifted one of them). Panel (c): amplitude of vibrations at point $x = L/2$ versus the excitation frequency ω.

(the hammer has a negligible area, so it hits a point with the coordinate c, $0 < c < L$). Find the maximum deviations of the points from the equilibrium state.

Hint. The initial velocity of each point being in contact with the hammer at $t = 0$ is v_0 (which is equal to the hammer speed at the time of impact).

(3) Consider longitudinal vibrations of the rod under the assumption that its left end is rigidly fixed, while the right end is subjected to the action

of the horizontal exciting force $F = A \sin \omega t$ (the force acts along the axis coinciding with the rod's symmetry).

(4) Consider a homogeneous bar of length L, cross-section S and density ρ, whose left end (placed at $x = 0$) is rigidly fixed, while the right end (having the coordinate $L > 0$ in the state of equilibrium) is free. The bar is set in motion by pushing the free end in the horizontal direction. Describe the vibration of the bar, assuming that at the moment $t = 0$ the bar attains the impulse P.
Hint. If at the moment $t = 0$ a segment of length $\epsilon \geq 0$ closing to the right edge is set into motion with velocity v_0, then $P = \epsilon S v_0$ (the direction of speed must be correctly determined). The problem corresponds to the limiting case $\epsilon \to 0$.

(5) Solve the following problem: $u_t = u_{xx}$, $t \geq 0$, $x \in [0, L]$, $u(0,x) = x^2 + ax^3 + bx^4 + cx^5$, $u_x(t,0) = u_x(t,L) = 0$.
Hint. One can choose the parameters a, b, i c arbitrarily, ensuring only that in the moment $t = 0$ the boundary conditions and the condition $u(0,x) \geq 0$, $x \in (0, L)$ are met.

(6) Derive the amplitude characteristics for the problem (6.70).

6.3.2 *Method of separating variables: evolutionary equations with two spatial variables*

Let us now consider the application of the method of separating the variables to the description of the membrane vibrations. The statement of the initial value problem is as follows:

$$u_{tt} = a^2(u_{xx} + u_{yy}), \tag{6.73}$$

$$u(0,x,y) = \varphi(x,y), \qquad u_t(0,x,y) = \psi(x,y), \quad (x,\, y) \in \Omega \subset \mathbb{R}^2. \tag{6.74}$$

We assume in addition that the membrane is rigidly fixed along the closed curve Γ, which is the boundary of the set Ω:

$$u(t;\, x,\, y) \Big|_{(x,\, y) \in \Gamma} = 0. \tag{6.75}$$

In case when Ω is a rectangle, (see Fig. 6.12), $\Gamma = \{x = 0 \cup x = b \cup y = 0 \cup y = c\}$.

We are looking for the solution to the problem in the form $u = T(t)V(x,y)$. Substitution of this ansatz into the equation (6.73) leads to the separation of variables:

$$\begin{aligned} \ddot{T} + a^2\lambda^2 T = 0, &\qquad V_{xx} + V_{yy} + \lambda^2 V = 0, \\ V(0,y) = V(b,y) = 0, &\qquad V(x,0) = V(x,c) = 0, \end{aligned} \tag{6.76}$$

where λ is the spectral parameter.

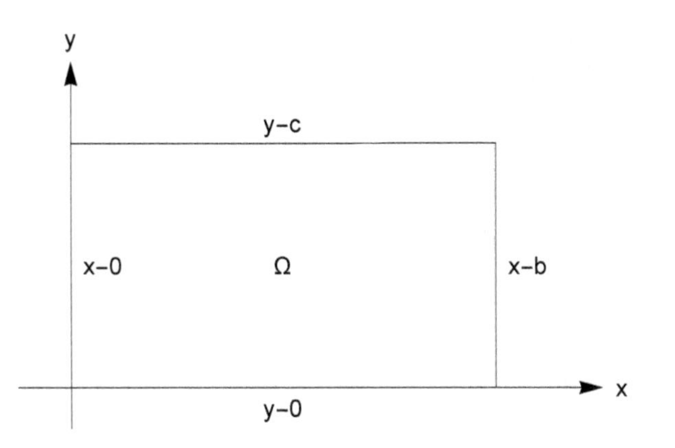

Fig. 6.12 Rectangular membrane.

The equation containing the variable V can be splitted by means of the ansatz $V(x,\, y) = X(x)Y(y)$, leading to the system

$$\begin{aligned} X'' + \mu^2 X = 0, \qquad & X(0) = X(b) = 0, \\ Y'' + (\lambda^2 - \mu^2)Y = 0, \qquad & Y(0) = Y(c) = 0. \end{aligned} \tag{6.77}$$

Implementing the construction described in the previous section, we get the solution

$$\begin{aligned} X_n = \sin\frac{\pi n x}{b}, \quad \mu = \frac{\pi n}{b}, \quad n = 1, 2, \ldots \\ Y_m = \sin\frac{\pi m y}{c}, \quad \lambda^2 - \mu^2 = \left(\frac{\pi m}{c}\right)^2, \quad m = 1, 2, \ldots. \end{aligned} \tag{6.78}$$

So, the boundary value problem (6.76) has the eigenvalues

$$\lambda^2_{n,m} = \left(\frac{\pi n}{b}\right)^2 + \left(\frac{\pi m}{c}\right)^2, \tag{6.79}$$

corresponding to the eigenvectors

$$V_{n,m}(x,\, y) = \sin\frac{\pi n x}{b}\sin\frac{\pi m y}{c}.$$

Note that one and the same eigenvalue can correspond to more than one eigenfunction. For example, if $b = c = \pi$ then $\lambda^2_{1,2} = \lambda^2_{2,1} = 5$; this number corresponds to the eigenfunctions $V_{1,2} = \sin x \sin 2y$ and $V_{2,1} = \sin 2x \sin y$. Such eigenvalues are called multiple (in this particular case the eigenvalue 5 has the multiplicity 2). It can be shown that the eigenvectors $V_{n,m}$ corresponding to different eigenvalues are mutually orthogonal.

General solution of the equation (6.73) satisfying the boundary condition (6.75) can be presented in the form

$$u(t,x,y) = \sum_{n=1}^{\infty}\sum_{m=1}^{\infty}[A_{n,m}\cos(a\lambda_{n,m}t) + B_{n,m}\sin(a\lambda_{n,m}t)]V_{n,m}(x,\,y). \tag{6.80}$$

The constants in the formula (6.74) are defined form the initial conditions:

$$A_{n,m} = \frac{\int_0^b\int_0^c \varphi(x,y)V_{n,m}dx\,dy}{\int_0^b\int_0^c V_{n,m}^2 dx\,dy}, \qquad B_{n,m} = \frac{\int_0^b\int_0^c \psi(x,y)V_{n,m}dx\,dy}{a\lambda_{n,m}\int_0^b\int_0^c V_{n,m}^2 dx\,dy}.$$

Another two-dimensional problem is the description of the vibration of a circular membrane. The geometry of the area clearly indicates that the problem should be represented in polar coordinates

$$\rho = \sqrt{x^2+y^2}, \qquad \theta = \mathrm{arctg}\frac{y}{x}.$$

In these coordinates, the wave equation takes the form

$$u_{tt} = c^2\Delta u, \qquad \Delta = (\cdot)_{\rho\rho} + \frac{1}{\rho}(\cdot)_\rho + \frac{1}{\rho^2}(\cdot)_{\theta\theta}. \tag{6.81}$$

We assume that at $t = 0$ the initial conditions $u(0,\rho,\theta) = \varphi(\rho,\theta)$, $u_t(0,\rho,\theta) = \psi(\rho,\theta)$ are fulfilled, where $\varphi(\rho,\theta)$ and $\psi(\rho,\theta)$ are known functions. The boundary conditions when the edge is rigidly fixed take the form $u(t,\rho=a,\theta)=0$. The problem consists in finding the smooth bounded solutions of (6.81) satisfying the additional conditions posed. We are looking it in the form of the product $u = T(t)V(\rho,\theta)$. The functions T, V satisfy the equations

$$\ddot{T} + \lambda^2c^2T = 0, \tag{6.82}$$

$$\Delta V + \lambda^2 V = 0, \tag{6.83}$$

where λ is the spectral parameter.

Inserting into the equation (6.83) the ansatz of the form $V = R(\rho)Y(\theta)$, we reach the complete separation of variables:

$$\ddot{Y} + \mu^2 Y = 0, \tag{6.84}$$

$$R'' + \frac{1}{\rho}R' + \left(\lambda^2 - \frac{\mu^2}{\rho^2}\right)R = 0. \tag{6.85}$$

In addition, we assume that $Y(\theta) = Y(\theta+2\pi)$, $R(\rho=a)=0$, and $R(\rho = 0) < \infty$.

The general solution of (6.84) takes the form $Y = C_1\cos\mu\theta + C_2\sin\mu\theta$, $\mu \in \mathbb{N}$. The fact that the parameter μ belongs to the set of natural numbers ensures the required periodicity of the function Y.

The Bessel-type equation (6.85) is described in Appendix C. Its general solution takes the form

$$R(\rho) = C_1 J_n(\lambda\rho) + C_2 Y_n(\lambda\rho),$$

where J_n, Y_n are the Bessel functions of the firs and the second kind, correspondingly. From the requirement that the solution be bounded at the origin, it immediately follows that $C_2 = 0$, since Y_n tends to infinity as the argument approaches zero.

Thus, the solution satisfying the boundedness condition will have the form $R_n(\rho) = J_n(\lambda\rho)$. Let us note that fulfillment of the condition $J_n(\mu_{k,s}) = 0$, ensures the orthogonality of the functions $J_n(\mu_k \cdot \rho)$ and $J_n(\mu_s \cdot \rho)$ when $k \neq s$ on the set $[0, 1]$ with respect to the scalar product defined by the integral with the weight ρ. Let us show this.

Let $J_k(\mu_k\rho)$ and $J_s(\mu_s\rho)$ be the solutions to the equations

$$\rho^2 J_k'' + \rho J_k' + \left(\rho^2\mu_k^2 - \mu^2\right) J_k = 0,$$
$$\rho^2 J_s'' + \rho J_s' + \left(\rho^2\mu_s^2 - \mu^2\right) J_s = 0.$$

Multiplying the first equation by J_s/ρ, and the second by J_k/ρ, and then subtracting the resulting equations, we get the equation

$$\rho(J_k'' J_s - J_k J_s'') + J_k' J_s - J_k J_s' + J_k J_s \rho(\mu_k^2 - \mu_s^2) = 0,$$

which can be also presented in the form

$$\left[\rho\left(J_k' J_s - J_k J_s'\right)\right]' = J_k J_s \rho(\mu_s^2 - \mu_k^2).$$

From this we get the equality

$$(\mu_s^2 - \mu_k^2)\int_0^1 J_k J_s \rho d\rho = \rho(J_k' J_s - J_k J_s')\Big|_{\rho=0}^{\rho=1} = 0.$$

The boundary conditions imply that $J_n(\lambda a) = 0$. This, in turn, implies that $\lambda_{n,m} = \mu_m^{(n)}/a$ coincides with m-th root of the Bessel function $J_n(x)$. So we ultimately have two eigenfunctions

$$\bar{V}_{n,m} = R_{n,m}\cos n\theta \quad \text{and} \quad \bar{\bar{V}}_{n,m} = R_{n,m}\sin n\theta.$$

Using the superposition principle, we can present an arbitrary solution to the boundary value problem for the wave equation (6.73) in the form

$$\begin{aligned} u = \sum_{n=0}^{\infty}\sum_{m=0}^{\infty} &\bar{V}_{n,m}(A_{n,m}\cos\lambda_{n,m}ct + B_{n,m}\sin\lambda_{n,m}ct) \\ &+ \bar{\bar{V}}_{n,m}(C_{n,m}\cos\lambda_{n,m}ct + D_{n,m}\sin\lambda_{n,m}ct) \end{aligned} \tag{6.86}$$

where the coefficients $A_{n,m}$, $B_{n,m}$, $C_{n,m}$, $D_{n,m}$ are calculated as follows: we equate the r.h.s. of the expression (6.86) taken at $t = 0$ to the corresponding initial condition, then multiply both sides of the equality by the eigenfunction with fixed indices n, m, integrate the equality obtained over the set $\rho \times \theta \in [0; a] \times [0; 2\pi]$, and taking advantage of the orthogonality conditions for the eigenfunctions.

As an example, let us consider the concentric vibrations of a circular membrane described by the function $u(t, \rho)$, satisfying the wave equation inside the circle with the radius a, initial conditions

$$u(0, \rho) = J_0(\mu_1 \rho) + J_0(\mu_2 \rho), \qquad u_t(0, \rho) = 0, \qquad \rho < a$$

and the boundary condition $u(t, \rho = a) = 0$. Solution to this problem is given by the formula (6.86), which, due to the symmetry, should be independent of the variable θ. So we put $n = 0$ in the above formula, and obtain the equality $\bar{\bar{V}}_{n,m} = 0$. From the second initial condition appears that $B_{n,m} = 0$. Thus, the solution can be presented as follows:

$$u = \sum_{m=1}^{\infty} A_m R_m \cos \lambda_m ct = \sum_{m=1}^{\infty} A_m J_0(\lambda_m \rho) \cos \lambda_m ct, \quad \lambda_m = \frac{\mu_m^0}{a}.$$

The parameters A_n, $n = 1, 2, \ldots$ can be obtained from the equation

$$\sum_{m=1}^{\infty} A_m J_0(\lambda_m \rho) = u(0, \rho). \tag{6.87}$$

We multiply the above solution by $\rho J_0(\lambda_n \rho)$ and then integrate both sides of the equation over the interval $(0; a)$. Using the orthogonality conditions

$$\int_0^a \rho J_0(\mu_n^0 \rho/a) J_0(\mu_k^0 \rho/a) d\rho = 0, \text{ for } n \neq k,$$

we get, after the simple algebraic manipulation, the following expression:

$$A_n = \frac{\int_0^a \rho J_0(\mu_n^0 \rho/a) u(0, \rho) d\rho}{\int_0^a \rho [J_0(\mu_n^0 \rho/a)]^2 d\rho}, \qquad n = 1, 2, 3 \ldots.$$

The initial condition is the sum of the eigenvectors, so only two of the coefficients A_n in the equation (6.87), namely, $A_{1,2}$, are nonzero. Ultimately, we obtain the solution

$$u = J_0(\mu_1^0 \rho/a) \cos \lambda_1 ct + J_0(\mu_2^0 \rho/a) \cos \lambda_2 ct.$$

Using the *Mathematica* command Table[N[BesselJZero[0, m]/a], {m, 1, 2}], we get, under the assumption that $a = c = 1$, the following parameters' values: $\mu_1^0 = 2.40483$, $\mu_2^0 = 5.52008$. Geometric configurations of a circular membrane in moments of time $t = 0$ and $t = 0.6$ is presented in Fig. 6.13. Solution of the problem assisted by the *Mathematica* package, as well as the animation of the solution obtained can be found in the file PM2_7.nb.

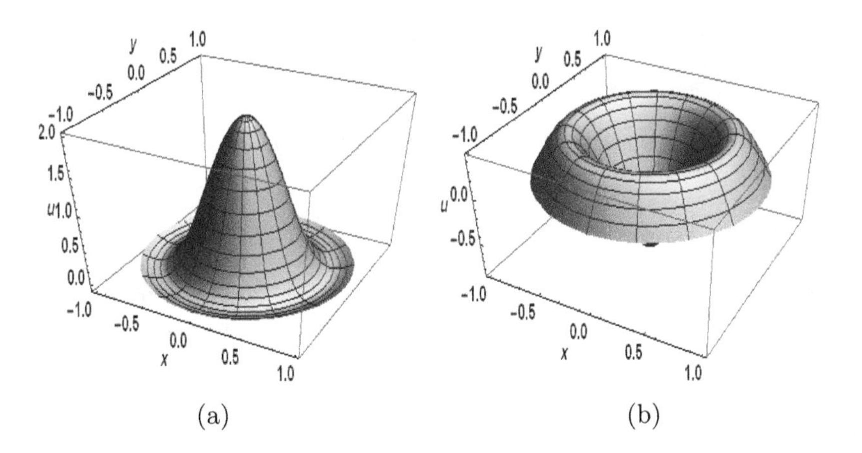

Fig. 6.13 Deformations of membrane at $t = 0$ and $t = 0.6$, obtained for $a = c = 1$.

Problem.

(1) At the boundary of a thin sector-shaped plate $r < a$, $0 < \varphi < \alpha$ the following temperature field is maintained:

$$u = \begin{cases} f(\varphi) \text{ as } r = a; \\ 0 \text{ as } \varphi = 0 \text{ and } \varphi = \alpha. \end{cases}$$

What configuration will have the temperature field in the limiting case $t \to +\infty$?

(2) Find the stationary velocity field in the incompressible fluid, appearing when the fluid passes an infinite cylindrical surface of a radius a, whose symmetry axis is perpendicular to the flow. Use the coordinate system whose OZ axis is parallel to the symmetry axis of the cylindrical surface and assume that velocity field at long distances from the cylindrical surface approaches a constant velocity $\vec{V}$ directed along one of the axes (for example along the axis OX).

Hints. The stationary velocity field can be described by means of the potential function $\varphi(x, y)$ (in the cartesian system $V_x = \partial_x \varphi(x, y)$, $V_y = \partial_y \varphi(x, y)$), being the solution of the Laplace equation $\Delta\varphi(x, y) = 0$. The velocity field in the polar coordinates is described by the formula

$$V_r = \frac{\partial \varphi}{\partial r}, \qquad V_\theta = \frac{1}{r} \frac{\partial \varphi}{\partial \theta}.$$

The boundary conditions are determined by the value of the liquid velocity at infinity and the condition of liquid adhesion to the cylinder surface, which formally means that the velocity field on the surface of the cylinder is zero.

6.3.3 *Solving linear evolutionary equations for three spatial variables*

The scheme for solving three-dimensional problems within the method of separating variables is similar to those used for a smaller number of spatial variables. There are, however, some differences that require comments.

We will start by considering the model describing the wave propagation on the surface of the ocean evenly covering the surface of a planet with parameters identical to those of the Earth [Dubin (2003)]. Suppose the planet's surface is a perfect sphere with a radius R. Given this assumption, it is natural to solve the problem in the spherical coordinates

$$x = \rho\cos\varphi\sin\theta, \quad y = \rho\sin\varphi\sin\theta, \quad z = \rho\cos\theta.$$

The ocean depth in these coordinates is given by the formula $d(t,\varphi,\theta) = d_0 + h(t,\varphi,\theta,\rho = R)$, where d_0 is the average depth of the ocean. We also assume that $d_0/R \ll 1$, so that the *theory of shallow water* can be applied, in which for the velocity of propagation of the surface wave c is described by the formula $c = \sqrt{gd_0}$, where $g \approx 9.8\,m/sec^2$ is the gravitational force acceleration. Thus, the function h describing the local change in the position of the liquid surface relative to the liquid level in the equilibrium satisfies the equation:

$$h_{tt} = c^2\Delta h, \qquad \Delta = \left(\frac{\partial^2}{\partial x^2} + \frac{\partial^2}{\partial y^2} + \frac{\partial^2}{\partial z^2}\right).$$

In spherical system the Laplace operator takes the form [Korn and Korn (1968)]:

$$\Delta = \frac{1}{\rho^2}\left[\frac{\partial}{\partial\rho}\left(\rho^2\frac{\partial}{\partial\rho}\right) + \frac{1}{\sin\theta}\frac{\partial^2}{\partial\theta}\left[\sin\theta\frac{\partial}{\partial\theta}\right] + \frac{1}{\sin^2\theta}\frac{\partial^2}{\partial\varphi^2}\right].$$

In case of the low-amplitude vibrations, it can be assumed that $\rho = R =$ const. Under this assumption, the function h will satisfy the following simplified equation:

$$h_{tt} = \frac{c^2}{R^2}\left[\frac{1}{\sin\theta}(h_\theta\,\sin\theta)'_\theta + \frac{1}{\sin^2\theta}h_{\varphi\varphi}\right]. \tag{6.88}$$

We use the ansatz $h = T(t)Y(\varphi,\theta)$, which, being inserted into (6.88) leads to the separation of variables:

$$\ddot{T} = -\lambda\frac{c^2}{R^2}T, \tag{6.89}$$

$$-\lambda Y = \left[\frac{1}{\sin\theta}(\sin\theta Y_\theta)'_\theta + \frac{1}{\sin^2\theta}Y_{\varphi\varphi}\right]. \tag{6.90}$$

Inserting into the last equation ansatz $Y = \Phi(\varphi)F(\theta)$, we get the following pair of equations

$$\frac{d^2\Phi}{d\varphi^2} + m^2\Phi = 0, \tag{6.91}$$

$$\frac{1}{\sin\theta}\frac{d}{d\theta}(\sin\theta F_\theta) - \frac{m^2}{\sin^2\theta}F = -\lambda F. \tag{6.92}$$

Solving the first of them, we get immediately that $\Phi = e^{\pm im\varphi}$, $m \in \mathbf{N}$. Using the change of independent variable $x = \cos\theta$ in the second equation, we get the Legendre equation discussed in Appendix C:

$$\frac{d}{dx}\left[(1-x^2)\frac{dF}{dx}\right] - \frac{m^2}{1-x^2}F = -\lambda F.$$

In case when $\lambda = \ell(\ell+1)$, ℓ is a natural number satisfying the condition $\ell \geq |m|$, the equation (6.92) possesses the bounded solution of the following form:

$$F = P_\ell^m(\cos\theta),$$

where $P_\ell^m(\cos\theta)$ are the Legendre's polynomials. Taking into account that $\lambda = l\,(l+1)$, we can present the equation (6.89) in the form

$$\ddot{T} + \ell(\ell+1)\frac{c^2}{R^2}T = 0.$$

The general solution to this equation is as follows:

$$T_\ell = C_1\cos\omega_\ell t + C_2\sin\omega_\ell t, \qquad \omega_\ell^2 = \frac{c^2\ell(\ell+1)}{R^2}.$$

Ultimately, the solution of the wave equation takes the form

$$h = \sum_{\ell=0}^{\infty}\sum_{m=-\ell}^{m=\ell}(A_{\ell,m}\cos\omega_\ell t + B_{\ell,m}\sin\omega_\ell t)Y_{\ell,m}(\theta,\varphi),$$

where $Y_{\ell,m} = P_\ell^m(\cos\theta)\,e^{im\varphi}$ (for symmetry reasons, it is sufficient to use only the function $\Phi = e^{+i\,m\,\varphi}$). Let us note, that the function $Y_{\ell,m}$ is described in the *Mathematica* package by the command SphericalHarmonicY[l, m, θ, ϕ].

The above formula can also be presented as

$$\begin{aligned} h = \sum_{\ell=0}^{\infty}\sum_{m=0}^{\ell} \{&(A_{\ell,m}\cos\omega_\ell t + B_{\ell,m}\sin\omega_\ell t)P_{\ell,m}(\cos\theta)\cos m\varphi \\ &+(C_{\ell,m}\cos\omega_\ell t + D_{\ell,m}\sin\omega_\ell t)P_{\ell,m}(\cos\theta)\sin m\varphi\}, \end{aligned} \tag{6.93}$$

where

$$A_{\ell,m} = \frac{\int_0^\pi d\theta \int_0^{2\pi} u(0,\theta,\varphi)\cos m\varphi P_\ell^m(\cos\theta)\sin\theta d\varphi}{\int_0^\pi d\theta \int_0^{2\pi} \cos^2 m\varphi (P_\ell^m(\cos\theta))^2 \sin\theta d\varphi};$$

$$B_{\ell,m} = \frac{\int_0^\pi d\theta \int_0^{2\pi} u_t(0,\theta,\varphi)\cos m\varphi P_\ell^m(\cos\theta)\sin\theta d\varphi}{\int_0^\pi d\theta \int_0^{2\pi} \cos^2 m\varphi (P_\ell^m(\cos\theta))^2 \sin\theta d\varphi};$$

$$C_{\ell,m} = \frac{\int_0^\pi d\theta \int_0^{2\pi} u(0,\theta,\varphi)\sin m\varphi P_\ell^m(\cos\theta)\sin\theta d\varphi}{\int_0^\pi d\theta \int_0^{2\pi} \sin^2 m\varphi (P_\ell^m(\cos\theta))^2 \sin\theta d\varphi};$$

$$D_{\ell,m} = \frac{\int_0^\pi d\theta \int_0^{2\pi} u_t(0,\theta,\varphi)\sin m\varphi P_\ell^m(\cos\theta)\sin\theta d\varphi}{\int_0^\pi d\theta \int_0^{2\pi} \sin^2 m\varphi (P_\ell^m(\cos\theta))^2 \sin\theta d\varphi}.$$

As an example, consider a spherical oscillations satisfying the following initial conditions:

$$u(t=0,\theta,\varphi) = P_\ell^0(\cos\theta), \qquad u_t(t=0,\theta,\varphi) = 0.$$

In this case the solution (6.93) takes the form

$$h=\sum_{\ell=0}^{\infty}\sum_{m=0}^{\ell}\{A_{\ell,m}\cos\omega_\ell t P_{\ell,m}(\cos\theta)\cos m\varphi + C_{\ell,m}\cos\omega_\ell t P_{\ell,m}(\cos\theta)\sin m\varphi\},$$

where

$$A_{\ell,m} = \begin{cases} 1 & \text{as } m=0, \\ 0 & \text{as } m\neq 0, \end{cases} \qquad C_{\ell,m} = \begin{cases} 1 & \text{as } m=0, \\ 0 & \text{as } m\neq 0. \end{cases}$$

So the solution can be presented as

$$h = P_\ell^0(\cos\theta)\cos\omega_\ell t. \tag{6.94}$$

It describes the azimuth symmetric natural vibration of the surface of a liquid on a sphere (see file PM2_8.nb).

If we pose another initial conditions, namely,

$$u(0) = P_\ell^m(\cos\theta)\cos m\varphi, \qquad u_t(0) = \omega_\ell P_\ell^m(\cos\theta)\sin m\varphi,$$

then the solution will be as follows:

$$u = P_\ell^m(\cos\theta)\cos(m\varphi - \omega_\ell t), \qquad \omega_\ell = c\sqrt{\ell(\ell+1)}.$$

The above function describes another kind of oscillations called *equatorial.*

In order to illustrate the above motions, consider the oscillation of a sphere of unit radius ($R = 1$) centered at the origin. In the spherical system, the vibrations of the sphere will be described by the following formula:

$$u = \rho(\varphi,\theta)\{\sin\theta\cos\varphi, \sin\theta\sin\varphi, \cos\theta\}, \tag{6.95}$$

where in case of elliptical traveling waves the following formula holds

$$\rho(\varphi, \theta) = (R + \varepsilon_1 \text{LegendreP}[l_1, m_1, \cos\theta] \cos[2\varphi - \omega_\ell t]),$$

while in the case of the azimuthal oscillations the formula is as follows:

$$\rho(\varphi, \theta) = (R + \varepsilon_2 \cos[\omega_\ell t] \text{SphericalHarmonicY}[l_2, m_2, \theta, \varphi]).$$

Note that in the *Mathematica* package the symbols LegendreP and SphericalHarmonicY are employed for the description of the associated Legendre and spherical functions, correspondingly. The parameters $\varepsilon_{1,2}$ in the above formula describe the amplitudes of the disturbances.

Let us consider the above problem at the following parameter values: $\varepsilon_1 = 0.1$, $l_1 = m_1 = 2$, $\varepsilon_2 = 0.5$, $l_2 = 3$, $m_2 = 0$. In both cases, $\omega_\ell t = 2\pi n/100$ and $n \in \mathbf{N}$. To visualize the nature of the vibrations, consider the dynamics of the behavior of the section $y = 0$ (Fig. 6.14(a)). The shape of the boundary of the section at $\varphi = 0$ is described by the formula (6.95). Figure 6.14(b) shows the deformation of the section caused by the propagation of an elliptical wave. Figure 6.14(c), in turn, illustrates deformations caused by the propagation of azimuthal spherical vibrations. In this case, the six fragments of undisturbed circumference (line 1) are subjected to irregular deformations and create pear-shaped figures (one such figure, corresponding to the section 3 of the panel c, is shown in Fig. 6.14(d)).

Example 6.13. Let us find the solution to the heat transport equation

$$T_t = \chi \Delta T$$

inside the sphere of the radius R, assuming that the solution satisfies the initial condition $T(0, x, y, z) = T_0 = \text{const}$ and the boundary condition $T(t, x^2 + y^2 + z^2 = R^2) = T_1$.

Due to the geometry of the area, the equation should be solved in spherical coordinates $x = r\cos\phi\sin\theta$, $y = r\sin\phi\sin\theta$, $z = r\cos\theta$, while the symmetry of the initial conditions suggests that the solution will depend only on the variable r. So, the problem reduces to finding the solution of the equation

$$T_t = \chi \frac{1}{r^2} \frac{\partial}{\partial r} r^2 \frac{\partial T}{\partial r}, \tag{6.96}$$

satisfying the conditions $T(0, r) = T_0$, $T(t, R) = T_1$. To simplify the problem, we make a substitution $T(t, r) = u(t, r) + T_1$. The function $u(t, r)$ satisfies the problem

$$u_t = \chi \frac{1}{r^2} \frac{\partial}{\partial r} r^2 \frac{\partial u}{\partial r}, \qquad u(0, r) = T_0 - T_1, \qquad u(t, R) = 0. \tag{6.97}$$

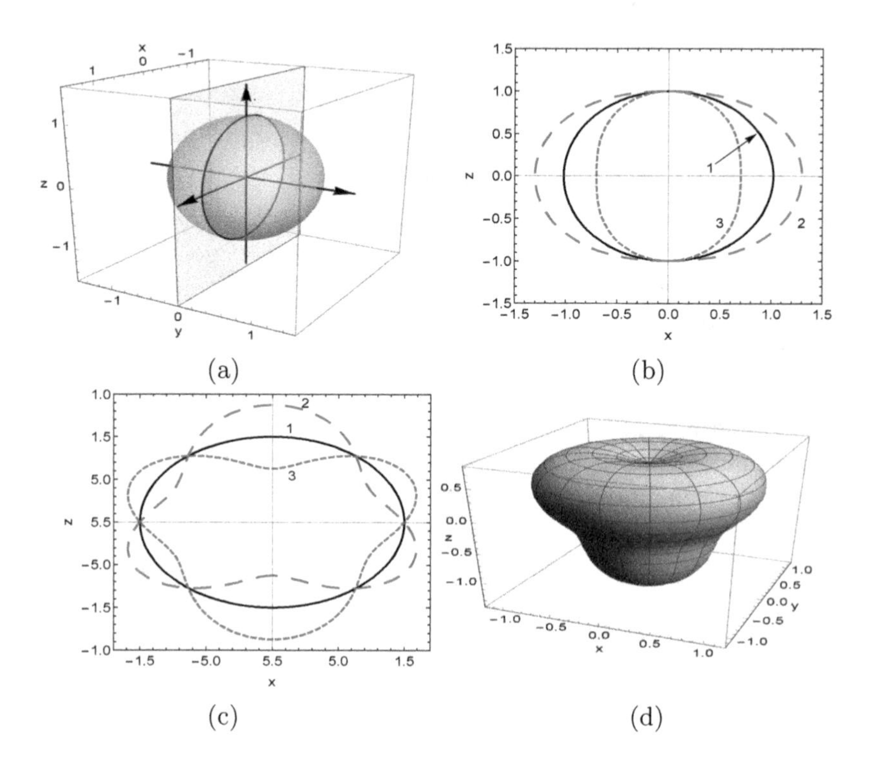

Fig. 6.14 Panel (a): the projection of the ball onto the plane $y = 0$. Panel (b): an elliptical wave rotating about the Oz axis, which causes the deformation of the section (line 1) shown on the panel (a) within the extreme values (lines 2 and 3). Panel (c): spherical azimuthal oscillations that are reflected in the oscillation of the section (line 1) in such a way that the lines 2, 3 turn one into another. Panel (d): the shape of the deformed sphere which corresponds to the line 3 on the panel (c).

Using the ansatz $u = F(t)W(r)$, we get the separable equations:

$$\dot{F} + \lambda F = 0, \qquad W'' + \frac{2}{r}W' + \frac{\lambda}{\chi}W = 0.$$

The substitution $W = Y/r$, reduces the second equation to the following one:

$$Y'' + \frac{\lambda}{\chi}Y = 0.$$

Its general solution takes the form $Y = C_1 \cos\sqrt{\lambda/\chi}r + C_2 \sin\sqrt{\lambda/\chi}r$. Thus,

$$W = \frac{1}{r}(C_1 \cos\sqrt{\lambda/\chi}\,r + C_2 \sin\sqrt{\lambda/\chi}\,r).$$

In order that the solution be bounded as $r \to 0+$, it is necessary to put $C_1 = 0$. Then

$$W = C_2 \frac{\sin\sqrt{\lambda/\chi}\,r}{r}.$$

It appears from the boundary conditions, that $\lambda_n = \chi(\pi n/R)^2$, $n = 1, 2, \ldots$. So the solution takes the form

$$u = \sum_{n=1}^{\infty} A_n \frac{\sin\sqrt{\lambda_n/\chi}\,r}{r} e^{-\lambda_n t}, \quad A_n = \int_0^R u(0,r) r \sin\sqrt{\lambda_n/\chi} r dr. \quad (6.98)$$

Let us analyze the solution to the equation (6.96), assuming that $T_0 = 20$; $T_1 = 100$; $R = 1$; $\chi = 0.01$. The description of the temperature field $T(t,r) = T_1 + u(t,r)$ along the radius of the sphere is carried out using the series (6.98) limited to 200 components. The calculation of the function $T(t,r)$ is presented in the cell 6.6 (and in more detail in the file PM2_8.nb).

Cell 6.6.

In[1]:
```
T0 = 20; T1 = 100; R = 1; χ = 0.01; M = 200;
```
In[2]:
```
A[n_] = (T0 - T1) Simplify[Integrate[r Sin[n Pi r/R], {r, 0, R}]/
Integrate[Sin[n Pi r/R]^2, {r, 0, R}]];
```
In[3]:
```
T[r_, t_] =
T1 + Sum[A[n] (Sin[(n Pi r/R)]/r) Exp[- χ (n Pi/R)^2 t], {n, 1,
M}];
```

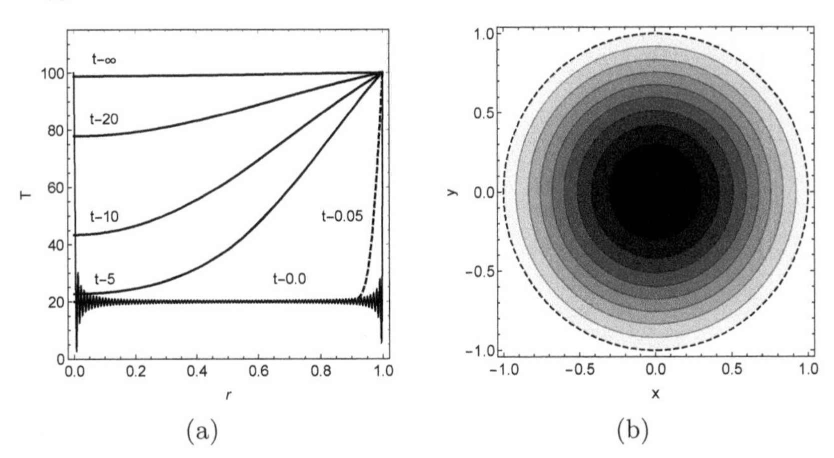

(a) (b)

Fig. 6.15 Panel (a): temperature field $T(t,r)$ at different moments of time. Panel (b): temperature field $T(t, \sqrt{x^2+y^2})$ in section $z = 0$ at $t = 10$.

Figure 6.15 shows the temperature fields along the radius of the sphere for specific values of the parameter t. It is clearly seen in Fig. 6.15 that the finite series approximates the initial condition $T(0,r) = T_0 + (T_1 - T_0)\mathrm{H}[r - R]$, $r \in [0, R]$ having a discontinuity of the first kind, not very accurately as there are oscillations at the ends of the interval. As $t > 0$ increases, the oscillations disappear and the temperature profile becomes smooth. The analysis of the graphs shows that the temperature inside the sphere increases over time, approaching the value of T_1 at each point.

Chapter 7

Application of numerical methods for solving partial differential equations

7.1 Finite-difference method

Finite-difference method is based on approximation of the differential operators. Using the definition of the derivative of the differentiable function $u(x)$ at the point x_i, we can approximate it as follows:

$$\left(\frac{du}{dx}\right)_{x=x_i} \approx \frac{u(x_i+h)-u(x_i)}{h}. \tag{7.1}$$

It is evident that the r.h.s. of (7.1) tends to the derivative of $u(\cdot)$ at the point x_i as $h \to 0$. In the similar way we can construct another approximations:

$$\left(\frac{du}{dx}\right)_{x=x_i} \approx \frac{u(x_i)-u(x_i-h)}{h} \quad \text{or} \quad \left(\frac{du}{dx}\right)_{x=x_i} \approx \frac{u(x_i+h)-u(x_i-h)}{2h}. \tag{7.2}$$

The general procedure for obtaining approximation with a higher accuracy is based on the use of the Taylor series expansion:

$$\frac{u(x_i+h)-u(x_i)}{h} = u'(x_i) + O(h), \tag{7.3}$$

$$\frac{u(x_i)-u(x_i-h)}{h} = u'(x_i) + O(h), \tag{7.4}$$

$$\frac{u(x_i+h)-u(x_i-h)}{2h} = u'(x_i) + O(h^2). \tag{7.5}$$

The last approximation gives an approximation of the order two. Finite-difference analog of derivative in the formula (7.5) is introduced with the help of the *three-point template* (it is also possible to obtain a more precise approximation based on a multi-point template). In order to approximate the second derivative, it is possible to use, in particular the following expression:

$$\left(\frac{d^2u}{dx^2}\right)_{x=x_i} \approx \frac{u'(x_i+h)-u'(x_i)}{h} = \frac{u(x_i+h)-2u(x_i)+u(x_i-h)}{h^2} \tag{7.6}$$
$$= u''(x_i) + O(h^2),$$

which delivers the second-order approximation base on the three-point template.

One of the methods enabling to approximate derivatives by their finite-difference analogs is based on employment of the interpolating Lagrange polynomials

$$P(x) = \sum_{j=0}^{k} u_j \ell_j(x), \qquad \ell_j(x) = \prod_{i=0, i\neq j}^{k} \frac{x-x_i}{x_j-x_i},$$

where $u_j(x_j)$ is the value of the function at the j–th node of the mesh $\{x_j\}_{j=1}^N$. If we take a mesh that is uniform and symmetrical about x_j, that is $x_j \pm k\,h$, $k = 1.2, \ldots, p$, create a polynomial $P(x) = \sum_{k=-p}^{p} u_{j+k}\ell_{j+k}(x)$ and differentiate it as many times as necessary, we will achieve the desired approximation of the derivative. For example, to approximate the first derivative on a five-point template using the method specified above, one can use the following *Mathematica* command:

Cell 7.1.

In[1]:

```
Simplify[D[ InterpolatingPolynomial[ Table[{ Subscript[x, i] + k h, u[Subscript[x, i +
k]]}, {k, -2, 2}], z], {z, 3}] /. z → Subscript[x, i]]
```

Out[1]:

$$\frac{-u[x_{i-2}]+2u[x_{i-1}]-2u[x_{i+1}]+u[x_{i+2}]}{2h^3}$$

Let us apply the method of discretization for finding the solution to the initial value problem

$$u_{tt} = c^2 u_{xx}, \quad x \in [a, b], \quad u(0,x) = \varphi(x), \quad u_t(0,x) = \psi(x), \tag{7.7}$$

satisfying the boundary conditions

$$u(t,a) = g_1(t), \qquad u(t,b) = g_2(t).$$

Let us introduce on the set of independent variables (t,x) a uniform mesh $t = t_k$ and $x = x_j$, $k = 1, \ldots,$ $j = 1, 2, \ldots, N+1$ in such a way that $x_1 = a$, $x_{N+1} = b$. Thus, $t_k = \tau(k-1)$, $x_j = a + (j-1)h$, where $h = (b-a)/N$ and τ are steps of the spatial and temporal variables, correspondingly.

Thus, functions and their derivatives at any point (t, x) are approximated by difference terms' values, taken in the nodes (t_k, x_j). Depending on the methods of approximation of derivatives, we distinguish *explicit* and *implicit* schemes.

Explicit scheme. Approximating the second derivative in the wave equation by the three-point template, we get the explicit scheme

$$U_{k+1,j} - 2U_{k,j} + U_{k-1,j} = r^2(U_{k,j+1} - 2U_{k,j} + U_{k,j-1}),\\ k = 1, 2, \dots; \ j = 2, 3, \dots, N, \tag{7.8}$$

where $r = c\tau/h > 0$. The system obtained should be solved with respect to the variables $U_{k+1,j}$ corresponding to the highest time layer. Doing this, we get:

$$U_{k+1,j} = 2U_{k,j} - U_{k-1,j} + r^2(U_{k,j+1} - 2U_{k,j} + U_{k,j-1}),\\ k = 1, 2, \dots; \ j = 2, 3, \dots, N.$$

The scheme is called explicit, because the value of $U_{k+1,j}$ on the highest time layer can be presented as a function of the parameter belonging to the lower time layers. The values $U_{k,1}$ and $U_{k,N+1}$ are derived from the boundary conditions. The values of $U_{1,j}$ and $U_{2,j}$ are derived, using the initial conditions and the approximation of u_t:

$$U_{1,j} = \varphi(x_j), \qquad U_{2,j} = U_{1,j} + \psi(x_j)\tau.$$

Let us discuss the stability of the numerical scheme. Among the various methods used, we choose the *spectral approach* that allows us to study the behavior of small disturbances of a certain type and reject unstable schemes with inappropriate solutions resulting from the discretization of the initial problem.

Let us consider the following *test solutions* of the system (7.8):

$$U_{k,j} = U_0 q^k e^{i\theta j}, \tag{7.9}$$

where U_0 is the amplitude of perturbation, $i = \sqrt{-1}$, $q = q(\theta)$ is a complex number. It is possible to derive from (7.8) the functional dependence of $q(\theta)$, called the *dispersive relation.* Indeed, after simple algebraic transformations, we get the formula

$$q - 2 + \frac{1}{q} = r^2\left(e^{i\theta} - 2 + e^{-i\theta}\right),$$

which is equivalent to the following quadratic equation:

$$q^2 - 2q\left(1 - 2r^2 \sin^2 \frac{\theta}{2}\right) + 1 = 0.$$

Since q determines the temporal evolution of the perturbation, the inequality $|q| \leq 1$ should be satisfied, because otherwise the solution will grow exponentially fast over time.

Solving the quadratic equation, we get:

$$q = 1 - 2r^2 \sin^2 \frac{\theta}{2} \pm 2r \sin \frac{\theta}{2} \sqrt{r^2 \sin^2 \frac{\theta}{2} - 1}.$$

It can be shown that the inequality

$$q = 1 - 2r^2 A^2 - 2rA\sqrt{r^2 A^2 - 1} < -1, \qquad A = \sin \frac{\theta}{2}$$

takes place. Indeed,

$$-2r^2 A^2 - 2rA\sqrt{r^2A^2 - 1} < -2 \Rightarrow rA\sqrt{r^2A^2 - 1} > 1 - r^2A^2.$$

The last inequality holds true when $r\,A > 1$. Going back to the initial expression, we get the inequality $q < -1$. The restriction $r\,A > 1$ means that $\sin \frac{\theta}{2} > 1/r$. So we can see that the inequality $r > 1$ implies the existence of harmonics (θ) which satisfy the inequality $|q| > 1$, and the corresponding perturbation unlimitedly grows over time.

Contrary, if $|r| < 1$, then $|r|^2 A^2 < A^2 \leq 1$ and the expression under the root is negative,

$$q = 1 - 2r^2 A^2 \pm 2rA\sqrt{1 - r^2 A^2} i \in \mathbb{C}$$

and $|q| = \sqrt{(1 - 2r^2A^2)^2 + 4r^2A^2(1 - r^2A^2)} = 1$. Hence the condition $r < 1$ ensures the spectral stability of the numerical scheme.

Implicit scheme. Let us consider the scheme

$$U_{k+1,j} - 2U_{k,j} + U_{k-1,j} = r^2(\sigma(U_{k+1,j+1} - 2U_{k+1,j} + U_{k+1,j-1}) + (1-\sigma)(U_{k-1,j+1} - 2U_{k-1,j} + U_{k-1,j-1})), \quad k = 1, 2, \ldots;\ j = 2, 3, \ldots, N, \tag{7.10}$$

assuming that $0 < \sigma \leq 1$. The system (7.10) cannot be solved with respect to $U_{k+1,j}$, and therefore in this case (and other similar cases) we say that the scheme is *implicit.*

We can also consider another implicit scheme, namely,

$$U_{k+1,j} - 2U_{k,j} + U_{k-1,j} = r^2(U_{k+1,j+1} - 2U_{k+1,j} + U_{k+1,j-1}), \quad k = 1, 2, \ldots;\ j = 2, 3, \ldots, N. \tag{7.11}$$

Inserting the test solution (7.9) into this scheme and finding $q = q(\theta)$, we get:

$$q = 1 \pm 2r \sin \theta/2 i \rightarrow |q| = \sqrt{1 + 4r^2 \sin^2 \theta/2} > 1.$$

From this appears that the scheme is unstable at arbitrary value of r.

Using the boundary conditions we obtain that $U_{k,1} = g_1(t_k)$, $U_{k,N+1} = g_2(t_k)$. Initial conditions imply the equalities $U_{0,j} = \varphi(x_j)$, $U_{1,j} = U_{0,j} + \tau\psi(x_j)$. The last expression appears from the approximation of the derivative of u with respect to t.

Evolution of the initial perturbation

$$u(x,0) = \begin{cases} 1 + \cos(x), & \text{as} \quad |x| \leq \pi, \\ 0, & \text{otherwise} \end{cases}$$

approximated by means of the explicit scheme at $\tau = 0.1$, $c = 0.5$, and $h = 0.075$ is shown in Fig. 7.1(a). It is seen that the initial perturbation splits into two identical impulses moving with the speeds $\pm c$ into opposite directions and evolving in self-similar mode without the change of the shape. This fully corresponds to the analytical description of the same problem. It can also be said that in this case the numerical scheme is stable.

The time evolution of the same disturbance, calculated by means of the implicit scheme (7.11) (Fig. 7.1(b)) shows the split of the initial profile into two waves, whose amplitudes do not maintain a constant values in the course of the evolution. This indicates that the scheme is unstable. The numerical solution to the wave equation satisfying the initial conditions $u(0,\, x) = [1 + \cos(x - x_0)]H[x + \pi - x_0] * H[\pi + x_0 - x]$ (where $H(\cdot)$ is the Heaviside function), $u_t(0,\, x) = 0$ by means of the explicit numerical

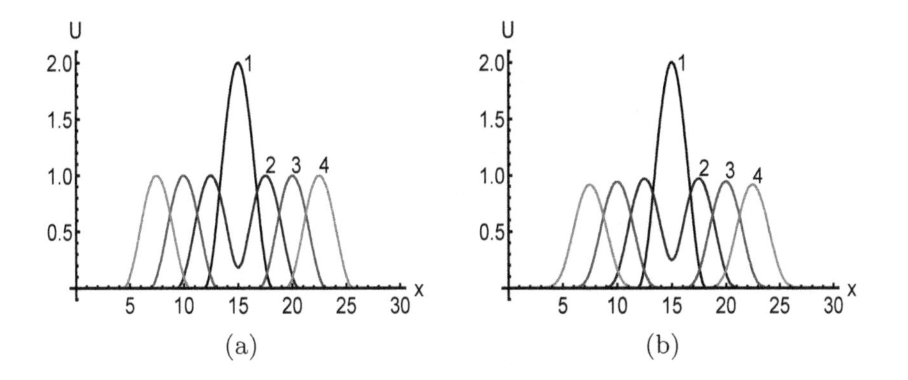

Fig. 7.1 Numerical solution of the initial value problem (7.7) with $\varphi(x) = 1 + \cos(x)$, $\psi(x) = 0$. Panel (a): temporal evolution of initial disturbance obtained by applying the explicit scheme at $r = 0.67 < 1$; panel (b): solving the same problem using the implicit scheme. The plot 1 presents the initial perturbation $u(0,\, x) = \varphi(x)$; the plot 2 presents the solution for $t = 50\,\tau$; the plot 3 corresponds to $t = 100\tau$; the plot 4 corresponds to $t = 150\tau$. The following parameters' values were used in the numerical experiments: $\tau = 0.1$, $c = 0.5$, $h = 0.075$.

scheme, as well as the visualization of the temporal evolution of the initial perturbation can be found in the file PM2_9.nb.

Since any numerical scheme can be treated as a discrete analog of the corresponding continuous problem, a question arises: whether (and when) the numerical scheme approximates the solution of the initial problem. Obtaining the answer on this question in general case is not easy, yet it can be done for most of the linear problems.

Let us now address the study of the discrete analog of the heat transport equation

$$U_t = aU_{xx}.$$

One of possible numerical schemes approximating this equation takes the following form:

$$U_{k+1,j} - U_{k,j} = r(U_{k,j+1} - 2U_{k,j} + U_{k,j-1}), \quad k = 1, 2, \ldots; \; j = 2, 3, \ldots, N,$$

where $r = a\tau/h^2$. Inserting the test solution $U_{k,j} = U_0 q^k e^{j\theta i}$ to this system, we get the following dispersive relation:

$$q = 1 - 4r \sin^2 \theta/2.$$

From this appears that $|q| \leq 1$, provided that $0 < r < 1/2$.

The stability of the schema requires the fulfillment of the condition $-1 \leq q \leq 1$. Then $0 \leq r \sin^2 \theta/2 \leq 1/2$.

The condition $r < 1/2$ ensures the fulfillment of both of the inequalities for any value of the parameter θ.

The general conclusion is that explicit schemes are more easy to be implemented, but one should be careful when using this type of the scheme, because the fulfillment of inequality $|q| > 1$ serves as the sufficient condition for the instability of the scheme, while the inequality $|q| \leq 1$ provides only the necessary condition for the spectral stability.

7.2 The method of lines

The method of lines can be considered as an analog of the finite differences method. Below we describe the algorithm, enabling to solve the problem of finding the approximate solution to the non-homogeneous wave equation

$$u_{tt} = u_{xx} + f(x,t), \qquad x \in [0, L], \; t \geqslant 0, \tag{7.12}$$

satisfying the initial conditions

$$u(0, x) = \varphi_1(x), \qquad u_t(0, x) = \varphi_2(x)$$

and the boundary conditions

$$u(t,0) = \psi_1(t), \quad u(t,L) = \psi_2(t).$$

Let $x_k = kh$, $k = 0, 1, \ldots, n+1$, $h = L/(n+1)$ be a set of the parallel straight lines, and $U_k(t) \approx u(x_k, t)$. Using the second-order approximations for the derivatives in (7.12), we get the following system of second-order ODEs

$$\begin{aligned} U_k'' &= c^2 \frac{U_{k+1} - 2U_k + U_{k-1}}{h^2} + f(x_k, t), \\ U_0 &= \psi_1(t), \qquad U_{n+1} = \psi_2(t) \end{aligned} \tag{7.13}$$

satisfying the initial conditions $U_k(0) = \varphi_1(x_k)$, $U_k'(0) = \varphi_2(x_k)$. Let us note, that the scheme (7.13) approximates the equation (7.12) up to $O(h^2)$. This accuracy can be increased by employment of the Taylor decomposition, since the following relation holds true (see [Berezin and Zhidkov (1965)], p. 583):

$$\begin{aligned} &u(x_{k+1}, t) - 2u(x_k, t) + u(x_{k-1}, t) \\ &= \frac{5h^2}{6} u_{xx}(x_k, t) + \frac{h^2}{12} \left[u_{xx}(x_{k+1}, t) + u_{xx}(x_{k-1}, t) \right] + O(h^6). \end{aligned} \tag{7.14}$$

Taking into account that

$$u_{xx}(x_k, t) = u_{tt}(x_k, t) - f(x_k, t) \equiv U_k''(t) - f_k(t),$$

where $f_k(t) = f(x_k,\, t)$, we get the following system:

$$\begin{aligned} &\frac{5}{6} U_k'' + \frac{1}{12} \left[U_{k+1}'' + U_{k-1}'' \right] - \frac{1}{h^2} \left[U_{k+1} - 2U_k + U_{k-1} \right] \\ &= \frac{5}{6} f_k(t) + \frac{1}{12} \left[f_{k+1}(t) + f_{k-1}(t) \right], \quad k = 1, 2, \ldots, n. \end{aligned} \tag{7.15}$$

This system approximates the initial equation up to $O(h^4)$.

Problem.

Consider the initial value problem

$$u_{tt} = \frac{1}{4} u_{xx},$$

$$u(x,0) = \begin{cases} 1 + \cos(x - x_0), \text{ as } |x - x_0| \le \pi \\ 0, \qquad\qquad \text{elsewhere} \end{cases},$$

$$u_t(x,\, 0) = 0.$$

Study numerically the evolution of the initial perturbation, using the finite-difference scheme on the segment $0 < x < L = 30$. Compare the results obtained with the solution of the same problem presented in the files PM2_9.nb and PM2_10.nb.

7.3 Galerkin method

Let us consider the equation [Dubin (2003)]

$$T_t = \chi T_{xx} + S(x,t) \quad x \in [0,\ L], \quad t \geq 0 \tag{7.16}$$

together with the initial condition

$$T(x,0) = T_0(x) \tag{7.17}$$

and the boundary conditions

$$T(0,t) = T_1(t), \qquad T(L,t) = T_2(t). \tag{7.18}$$

We can also consider the condition

$$T_x(L,t) = 0, \tag{7.19}$$

posed on the right edge.

Let us use the ansatz $T(x,t) = u(x,t) + Y(x,t)$, choosing the function $u(t,\ x)$ in such a way that conditions $u(0,t) = T_1(t)$ and $u(L,t) = T_2(t)$ be fulfilled. With such a choice, the function Y will satisfy the boundary conditions $Y(0,t) = Y(L,t) = 0$ and the initial value problem

$$\begin{aligned} &Y_t = \chi Y_{xx} + S(t,x) - u_t + \chi u_{xx}, \\ &Y(x,0) = T_0(x) - u(x,0). \end{aligned} \tag{7.20}$$

The facilitation associated with such a substitution consists in that it is easier to work with the base functions attaining zero at the boundary of the area.

In the Galerkin method the solution of the equation (7.20) is sought in the form

$$Y = \sum_{n=0}^{M} c_n(t) v_n(x), \tag{7.21}$$

where $u(x,t)$ is any function satisfying the boundary conditions, $c_n(t)$ are *the Fourier coefficients* to be defined, $\{v_n(x)\}_{n=1,2,\ldots}$ is a set of functions nullifying on the boundary of the domain, and forming the orthogonal basis on the set of the functions $C^k[0,\ L]$.

Inserting the ansatz (7.21) into the formula (7.16), we get the equation

$$\sum_{n=0}^{M} c_n'(t) v_n = \chi \left(u_{xx} + \sum_{n=0}^{M} c_n(t)(v_n)_{xx} \right) + S(t,x) - u_t + \chi u_{xx}.$$

Multiplying both sides by v_k, $k = 1, \dots, M$, integration over the set $x \in [0, L]$ and using the orthogonality conditions, we obtain the following system of ODEs:

$$\sum_{n=0}^{M} c_n'(t) \int_0^L v_n v_k dx = \chi \sum_{n=0}^{M} c_n(t) \int_0^L (v_n)_{xx} v_k dx + \int_0^L (S - u_t + \chi u_{xx}) v_k dx. \tag{7.22}$$

The functions we are looking for should satisfy the initial conditions associated with (7.17). For the equation (7.16) and the boundary conditions (7.18), one can make the following choice:

$$u(x,t) = T_1(t) + \frac{x}{L}\left(T_2(t) - T_1(t)\right), \quad v_n = \sin(\pi n x / L).$$

In case of the boundary condition (7.19) the choice will be as follows:

$$u(x,t) = T_1(t), \quad v_n = x^n \left(1 - \frac{n}{n+1}\frac{x}{L}\right).$$

Problem. Using the Galerkin method, construct a numerical scheme for the equation

$$u_t = \chi u_{xx}, \qquad x \in [0, L],$$

describing the heat transport in the thin wire of the length L. Take the heat transport coefficient $\chi = 1/8$. Assume that $u(0, x) = 1$, $u(t, 0) = 1 + \sin t$, and that both the right edge and the lateral surface are insulated [Dubin (2003)].

7.4 Finite elements method

The finite element method, also known as the Petrov-Galerkin method [Kythe (1997)], shares some features with the Galerkin method. In this method, the approximate solution is sought in the form

$$u_h(t,x) = \sum_{j=1}^{N} U_j(t)\phi_j(x), \tag{7.23}$$

where $\{\phi_j(x)\}_{j\in J}$ is a set of *trial functions.*

Approximation of nonlinear functions within the classical Galerkin method causes considerable difficulties. An alternative approximation method was developed in the article [Christie *et al.* (1981)]. Within this method, he powers $u(t,x)^\mu$ are approximated as follows:

$$(u_h(t,x))^\mu = \sum_{j=1}^{N} (U_j(t))^\mu \phi_j(x).$$

The trial function $\phi_j(x)$ is chosen such that it is equal to zero everywhere except in the small vicinity of the nodal point x_j, while at the nodal point itself it is equal to one. An example of a trial function is the so-called *hat function* (see Fig. 7.2(a)) defined as follows:

$$\phi_j(x) = \begin{cases} 1 + \frac{x-x_j}{h} \text{ if } x \in [x_{j-1}; x_j] \\ 1 - \frac{x-x_j}{h} \text{ if } x \in [x_j; x_{j+1}] \\ 0 \text{ elsewhere.} \end{cases} \tag{7.24}$$

The search function $U_j(t)$ can be obtained by means of the *variational method.* Within this method, we substitute the formula (7.23) into the equation under consideration, multiply the result by the *test function* $\psi_j(x)$ and then calculate the integral of this expression on the segment $[a, b]$. Note that if the test functions coincide with the trial functions, then the described procedure will not differ from the Galerkin method [Marchuk (1982)], but if the sets $\{\psi_j(x)\}_{j\in J}$ and $\{\phi_j(x)\}_{j\in J}$ are not identical, then arbitrariness in the selection of the test functions allows to increase the accuracy of the approximation.

The choice of the test function is related to the order N of the highest partial derivative with respect to x in the equation under consideration, since the test function should be N times continuously differentiable. For example, if the highest derivative of x is of the order 2, i.e. the equation contains u_{xx}, then $\psi_j(x)$ is constructed as a third order polynomial (see Fig. 7.2(b)):

$$\psi_j(x) = \frac{1}{h^3}$$
$$\begin{cases} (x - x_{j-2})^3, & x \in [x_{j-2}; x_{j-1}] \\ h^3 + 3h^2(x - x_{j-1}) + 3h(x - x_{j-1})^2 - 3(x - x_{j-1})^3, & x \in [x_{j-1}; x_j] \\ h^3 + 3h^2(x_{j+1} - x) + 3h(x_{j+1} - x)^2 - 3(x_{j+1} - x)^3, & x \in [x_j; x_{j+1}] \\ (x_{j+2} - x)^3, & x \in [x_{j+1}; x_{j+2}] \\ 0, & \text{elsewhere.} \end{cases} \tag{7.25}$$

Like the trial function, the test function $\psi_j(x)$ is chosen such that it nullifies everywhere except a small vicinity of the nodal point x_i.

Example 7.1. Let us consider the following equation

$$u_t + (u^2)_x = \nu u_{xx}.$$

The equation defining $U_j(t)$ in this case will take the form

$$((u_h)_t, \psi_j(x)) + ((u_h)_x^2, \psi_j(x)) - \nu((u_h)_{xx}, \psi_j(x)) = 0, \tag{7.26}$$

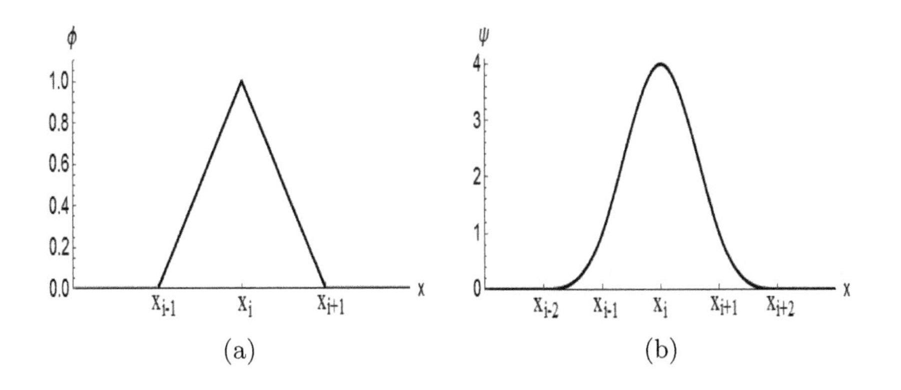

(a) (b)

Fig. 7.2 Profiles of the functions $\phi_j(x)$ (a) and $\psi_j(x)$ (b).

where (f, g) denotes the scalar product $\int_a^b f(x)\, g(x)\, dx$. Note that the last term should be presented in a different form, because the function u_h is not necessarily differentiable with respect to x. Integrating by parts this expression, we can give it the proper meaning:

$$\int_a^b (u_h)_{xx}\psi_j(x)dx = \int_a^b u_h(\psi_j(x))_{xx}dx.$$

This is why we require the test function to be sufficiently smooth.

Let us consider a procedure that allows us to determine the first term in the equation (7.26):

$$((u_h)_t, \psi_j(x)) = \int_{x_{j-2}}^{x_{j+2}} \psi_j(x) \sum_{s=j-2}^{s=j+2} \dot{U}_s(t)\phi_s(x)dx = \sum_{s=j-2}^{s=j+2} \dot{U}_s(t)a_s,$$

where

$$a_{j-2} = \int_{x_{j-2}}^{x_{j-1}} \psi_j(x)\phi_{j-2}(x)dx = \frac{h}{20},$$

$$a_{j-1} = \int_{x_{j-2}}^{x_j} \psi_j(x)\phi_{j-1}(x)dx = \frac{26h}{20},$$

$$a_j = \int_{x_{j-1}}^{x_{j+1}} \psi_j(x)\phi_j(x)dx = \frac{66h}{20},$$

$$a_{j+2} = a_{j-2}, \qquad a_{j+1} = a_{j-1}.$$

So, the first term in (7.26) changes into the expression

$$\frac{h}{20}\left(\dot{U}_{j-2} + 26\dot{U}_{j-1} + 66\dot{U}_j + 26\dot{U}_{j+1} + \dot{U}_{j+2}\right).$$

The remaining terms are calculated in the similar way.

Chapter 8

Some completely integrable nonlinear models

8.1 Introduction

It was shown in the previous sections, that there is a large number of methods that allow for solving various initial and boundary value problems for linear PDEs. The situation is different with nonlinear PDEs, which can be integrated only in exceptional cases. Nevertheless, there is one distinguished class of nonlinear evolutionary equations, called completely integrable, for which methods enabling to integrate wide families of initial and boundary value problems have been developed. Several of these models, relevant to the continuum models discussed earlier, will be analyzed in this chapter. These models possess solutions that have no analogs among the solutions of linear equations. Such solutions, while being interesting in their own right, can also shed light on what can be expected from solutions to more realistic continual media models.

8.2 The simplest wave equations. Influence of nonlinearity and dispersion on the evolution of perturbations

When analyzing linear homogeneous partial differential equations $F(\partial_x,\, \partial_t)u(t,\, x) = 0$ describing plane waves, where $F(v,\, w)$ is an algebraic function, one can get convinced that these equations support the solution of the form

$$u = A\, e^{i(\omega t - kx)}. \tag{8.1}$$

This solution describes so called *monochromatic wave*. The parameter ω is called the *wave frequency*, the parameter k is called the *wave number*, while A is the wave amplitude. Substituting the solution of this form into the wave equation, we obtain a certain algebraic equation $\tilde{F}(k,\, \omega) = 0$, which determines the *dispersion relation*, i.e. the relationship between the wave number and the frequency. For the simplest wave equation

$$u_t + c u_x = 0 \tag{8.2}$$

the dispersion relation takes the form $\omega - ck = 0$.

Using the dispersion relation, two important wave characteristics that we mentioned earlier can be determined, namely, the *phase velocity* $v_{ph} = \omega/k$ and the *group velocity* $v_g = d\omega/dk$. In the case of the wave equation (8.2) $v_{ph} = v_g = c =$ const. Phase velocity characterizes the movement of a monochromatic wave, while the group velocity determines the speed of propagation of a wave pack, which is associated with the time evolution of any (not necessarily monochromatic) initial disturbance. As it is shown in the theory of Fourier series and Fourier integrals, each smooth function with a compact support can be represented as a sum of harmonic functions similar to (8.1), while more general non-compact perturbations, under certain conditions, can be presented in the form of a Fourier integral of such functions. When v_{ph} and (or) v_g depend on k, then each initial perturbation will be deformed in the course of evolution due to the difference in the velocities of individual harmonics. This phenomenon is called *dispersion*. We will discuss the consequences of this effects using the model equations.

Remark. Passing in (8.2) to the variables $\xi = x - ct$, $\eta = x + ct$, we get the equation $u_\eta = 0$, having the solution $u = f(\xi) = f(x - ct)$, where $f(\cdot)$ is arbitrary smooth function. Solutions of this type describe waves moving with velocity c without change of their form, which is characteristic for solutions of dispersionless equations.

One of the simplest equation in which the dispersion effects occur has the following form [Bhatnagar and Prasad (1970)]:

$$u_t + u_x + u_{xxx} = 0. \tag{8.3}$$

Inserting the ansatz (8.1) into this equation, we obtain the relations

$$\omega = k(1 - k^2) \Rightarrow v_{ph} = 1 - k^2, \qquad v_{gr} = 1 - 3\,k^2.$$

Thus, the velocity of propagation of the non-monochrome disturbance differs from the velocity of propagation of the monochromatic wave. Moreover, a particular monochrome wave moves with a speed dependent on the wavenumber k. The combination of these properties leads to a clear dispersion effect, as shown in Fig. 8.1. This figure shows the evolution of the initial perturbation

$$u(0,\,x) = \varphi(x) = \begin{cases} \frac{1}{2}\left[1 + \cos\frac{\pi}{10}(x + 140)\right], & \text{as} \quad |x + 140| < 10, \\ 0, & \text{elsewhere} \end{cases} \tag{8.4}$$

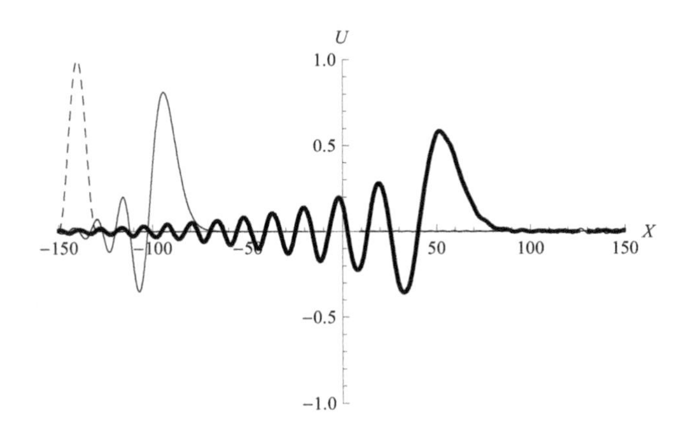

Fig. 8.1 Profiles of the solutions of the initial value problem (8.3), (8.4). Dashed line: $t = 0$; thin solid line: $t = 50$; bold line: $t = 200$.

obtained by means of the numerical solution of the initial value problem (8.3), (8.4).

We will return to the analysis of the equation (8.3) and its nonlinear generalization later on, but now we address the Hopf equation (5.17), which is the simplest nonlinear equation describing wave motions. Solutions of this equation exhibit properties, in a sense, opposite to the properties of the simplest dispersive wave equation, as will be shown below. A certain analogy can be seen between the simplest wave equation (8.2) and the Hopf equation, in which the parameter u plays formally the role of a phase velocity. Guided by this observation, we postulate that the solution of the Cauchy problem

$$u_t + u\,u_x = 0, \qquad u(0,\,x) = \phi(x) \tag{8.5}$$

takes the form

$$u(t,\,x) = \phi(x - u\,t). \tag{8.6}$$

To make sure that the function (8.6) satisfies the Hopf equation, it is necessary to calculate its partial derivatives with respect to both independent variables. Using the rule of differentiation of an implicit function, we get:

$$\tfrac{\partial}{\partial t} f(x - u\,t) = -\dot{f}(z)\Big|_{z=x-u\,t}(u + t\,u_t), \Longrightarrow \tfrac{\partial\,u}{\partial\,t} = -u\,\tfrac{\dot{f}(z)}{1+t\,\dot{f}(z)}$$

$$\tfrac{\partial}{\partial x} f(x - u\,t) = \dot{f}(z)\Big|_{z=x-u\,t}(1 - t\,u_x), \Longrightarrow \tfrac{\partial\,u}{\partial\,x} = \tfrac{\dot{f}(z)}{1+t\,\dot{f}(z)}.$$

Inserting these formulas to the equation (5.17), we obtain the identity. The fulfillment of the initial condition is obvious.

The formula (8.6) describes an implicit solution to the Hopf equation, and in the general case, the explicit expression of the solution is not available. Therefore, most often the solution in the specified form is used for graphical visualization, which can be carried out in the *Mathematica* package using the command

```
ContourPlot[{u-f[x-u*0] == 0, u-f[x-u*t] == 0}, {x, -1, 3}, {u, 0, 1}]
```

allowing to present the disturbance profile at time $t > 0$ against the background of the initial disturbance. However, for specially selected initial conditions, especially for those cases when the initial disturbance is described by a continuous piecewise linear function, such a solution can be constructed analytically.

Example 8.1. Let us consider triangle-shaped initial disturbance:

$$u(0,\,x) = \varphi(x) = \begin{cases} x, \text{ as } 0 < x < 1, \\ 2 - x, \text{ as } 1 < x < 2, \\ 0 \text{ elsewhere.} \end{cases} \tag{8.7}$$

According to the formula (8.6), solution of the Hopf equation satisfying the initial conditions (8.7) is given by the following formula:

$$u = \begin{cases} x - u\,t, \text{ as } 0 < x - u\,t < 1, \\ 2 - (x - u\,t), \text{ as } 1 < x - u\,t \le 2, \\ 0 \text{ elsewhere.} \end{cases}$$

This expression can be solved with respect to the variable u:

$$u = \begin{cases} \frac{x}{1+t}, \text{ as } 0 < x < 1 + t, \\ \frac{2-x}{1-t}, \text{ as } \begin{cases} 1 + t < x \le 2, \text{ and } 0 < t < 1, \\ 2 \le x < 1 + t, \text{ and } t > 1, \end{cases} \\ 0 \text{ elsewhere.} \end{cases} \tag{8.8}$$

Figure 8.2 (left panel) shows the characteristic feature of the evolution of the initial disturbance connected with the fact that the amplitude u of the initial perturbation plays the role of the local velocity, so that the point at which the amplitude is larger moves at the highest speed. This leads to an increase in the steepness of the front of the wave moving to the right and, as a result, to the formation of the discontinuity of the first type at some moment of time t_{cr} (in the given specific example $t_{cr} = 1$), and then to the formation of multivalued solution. Such a solution occurs to be unstable and is not implemented in practice, unlike the *shock wave* solution, which for the times $t \geq t_{cr}$ is described by the formula

$$u = \begin{cases} \frac{x}{1+t}, & \text{as } 0 < x < x_f(t), \\ 0 & \text{elsewhere}, \end{cases} \tag{8.9}$$

where $x_f(t)$ defines the position of the shock wave front. We can determine the function $x_f(t)$ writing the Hopf equation in the following equivalent form:

$$\frac{d}{dt} u\Big|_{\Gamma} = 0, \qquad \Gamma: \quad \frac{dx}{dt} = u,$$

where $\frac{d}{dt} = \frac{\partial}{\partial t} + \frac{dx}{dt}\frac{\partial}{\partial x} = \frac{\partial}{\partial t} + u\frac{\partial}{\partial x}$ denotes the substantial derivative (introduced in Section 5.2). From the above formula in an elementary way follows the *conservation law*

$$\int_{\mathbb{R}} u(t, x)\, dx = \text{const}\,.$$

In particular, in the case of the solution (8.9) the following formula holds true:

$$\int_0^{x_f(t)} \frac{x}{1+t}\, dx = \int_{\mathbb{R}} \varphi(x)\, dx = 1.$$

From this we get, after elementary calculations, that $x_f(t) = \sqrt{2\,(1+t)}$, hence at $t > t_{cr} = 1$ the shock wave solution is given by the formula

$$u = \begin{cases} \frac{x}{1+t}, & \text{as } 0 < x < \sqrt{2\,(1+t)}, \\ 0 & \text{elsewhere}. \end{cases}$$

The shock wave solutions are presented on the right panel in Fig. 8.2.

Example 8.2. Let us consider the propagation of the shock wave associated with the step-like initial disturbance

$$u(0, x) = \varphi(x) = \begin{cases} u_2 & \text{as } x < 0, \\ u_1 & \text{as } x \geq 0 \end{cases}$$

(we assume that $u_2 > u_1 \geq 0$).

In the course of evolution, the discontinuity moves to the right with constant velocity $V > 0$ and the solution takes the form

$$u(t, x) = \begin{cases} u_2 & \text{as } x < V\,t, \\ u_1 & \text{as } x \geq V\,t. \end{cases} \tag{8.10}$$

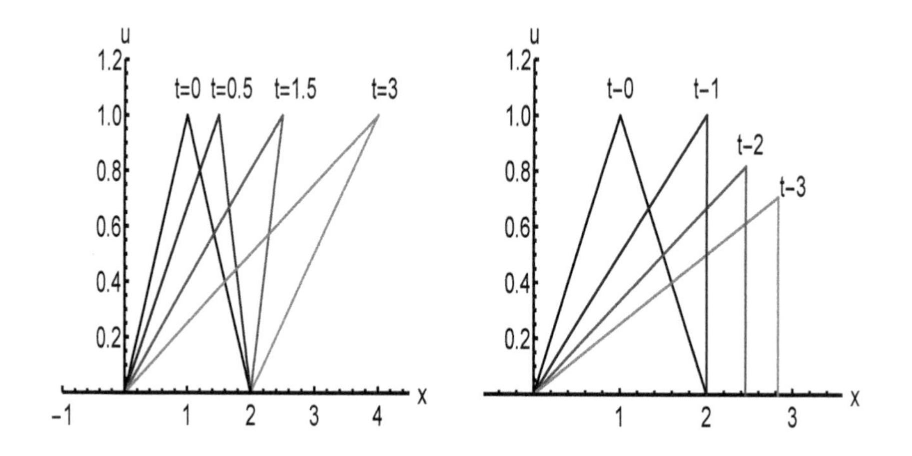

Fig. 8.2 Profiles of solution of the initial value problem (5.17)–(8.7) corresponding to the moments of time $t = 0;\ 0.5;\ 1.5;\ 3.0$. Left panel: formation of an unstable multivalued solution described by the formula (8.8); right panel: solution that describes propagation of the shock wave created under the same initial conditions.

It turns out that the constant V is uniquely defined by the boundary conditions $u_2 = \lim_{x \to -\infty} u(t, x)$, $u_1 = \lim_{x \to +\infty} u(t, x)$.

Lemma 8.1. *The following formula holds true:*

$$V = \frac{u_1 + u_2}{2}. \tag{8.11}$$

Proof. The Hopf equation (5.17) can be rewritten in the form of *conservation law*

$$\frac{\partial F}{\partial t} + \frac{\partial G}{\partial x} = 0, \quad F = u(t, x), \quad G = \frac{u^2(t, x)}{2}. \tag{8.12}$$

From the preceding equation we get the integral identity

$$0 = \int\int_S \frac{\partial F}{\partial t} + \frac{\partial G}{\partial x} = \int_{ABCD} F\,dx - G\,dt,$$

where $ABCD$ is the rectangle symmetric with respect to the straight line being the locus of the coordinates of the front of shock wave in the plane (t, x) (see Fig. 8.3), S is the interior of the rectangle $ABCD$. Pulling down to zero the sides AD and BC, and keeping AB and DC constant, we get:

$$\int_{AB} u\,dx - \frac{u^2}{2}dt = \int_{DC} u\,dx - \frac{u^2}{2}dt + O(|AD|).$$

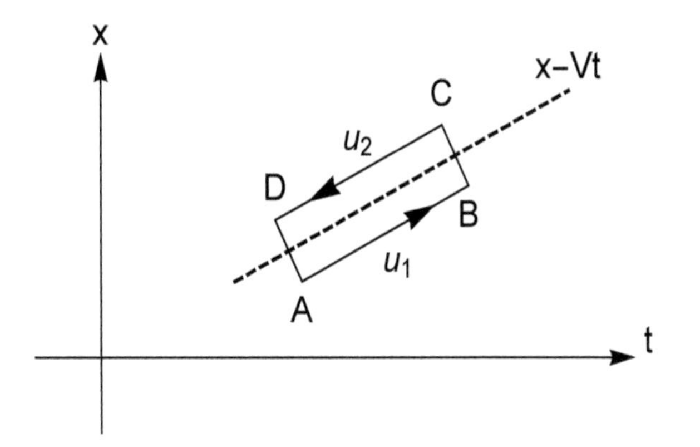

Fig. 8.3 Trajectory of the front of shock wave.

Taking into account equality $dx = V\,dt$ and that, in accordance with the initial configuration, u is identical with u_1 before the front of the shock wave, and attains the value u_2 behind the shock wave, we get:

$$\left(u_2\,V - \frac{u_2^2}{2}\right) = \left(u_1\,V - \frac{u_1^2}{2}\right) + O(|AD|).$$

In the limiting case when both $|AD| = |BC|$ tend to zero, we get the formula (8.11).

8.3 Further examples of generalized solutions of the Hopf equation

Example 8.3. Let us find the solution of the initial value problem (8.5) with

$$\phi(x) = \begin{cases} 0, & \text{when} \quad x < 0, \\ a > 0, & \text{when} \quad x > 0. \end{cases}$$

The formula (8.6) gives us the implicit solution of the problem, and solving it with respect to u, we get the solution on the whole real line, excluding the interval $[0,\, a\,t]$:

$$u(t,\, x) = \begin{cases} 0, & \text{when} \quad x < 0, \\ a, & \text{when} \quad x > a\,t. \end{cases}$$

To determine the values of the function $u(t,\, x)$ for $x \in [0,\, a\,t]$, let us notice that the jump of the initial data $\phi(x)$ can be characterized by the continuous parameter $\mu \in [0, a]$, and this segment defines the range of the values of the function $u(0, x)$ in the singular point $x = 0$. As it was mentioned earlier, $u(t, x)$ in the argument of the function (8.6) plays the rule of velocity. Thus, the point of the jump characterized by the value μ at time $t > 0$ will be on the right in the distance μt, and the complete solution of the problem will take the form:

$$u(t, x) = \begin{cases} 0 & \text{when} \quad x < 0, \\ \frac{x}{t}, & \text{when} \quad 0 \leq x \leq a\,t, \\ a & \text{when} \quad x > a\,t. \end{cases} \tag{8.13}$$

Example 8.4. Now that we know the solution of the Cauchy problem for both of the step functions, we can solve the Cauchy problem (8.5) with

$$\phi(x) = \begin{cases} 0 & \text{when} \quad x < 0, \\ 1 & \text{when} \quad 0 \leq x \leq 1, \\ 0 & \text{when} \quad x > 0. \end{cases}$$

Solution of the problem can be divided into two steps. For $0 < t < t^*$ the right discontinuity moves with the velocity $v_r = \frac{1}{2}$, while the left discontinuity turns into inclined segment that catches up the step with the velocity $v_l = 1$, see Fig. 8.4, left panel. It is easily calculated that at $t^* = 2$ the function u will look as the rectangular triangle. For $t > t^*$ the solution is described by the formula (see Fig. 8.4, right panel):

$$u(t, x) = \begin{cases} 0 & \text{when} \quad x < 0, \\ \frac{x}{t}, & \text{when} \quad 0 \leq x \leq x_f(t) \\ 0 & \text{when} \quad x > x_f(t), \quad t > 2. \end{cases} \tag{8.14}$$

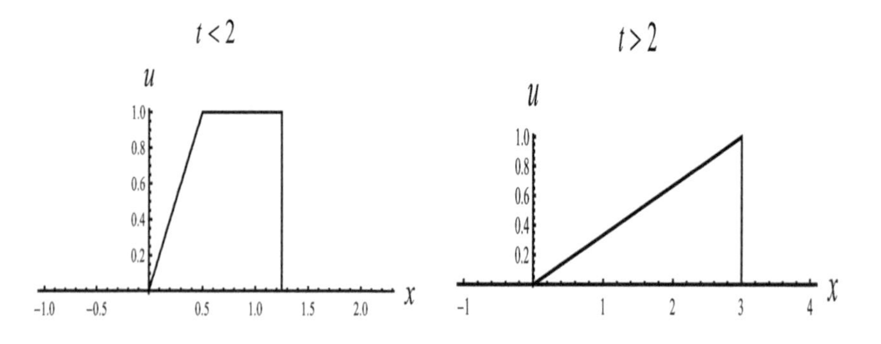

Fig. 8.4 Evolution of square-like disturbance.

The coordinate $x_f(t)$ of the moving front can be found using the fact that the Hopf equation can be written in the form (8.12). Integrating this equation with respect to $x \in \mathbb{R}$, and taking into account that $u(t, x) \to 0$ as $|x| \to \infty$, we obtain that

$$\int_{\mathbb{R}} u(t, x)\, dx = \text{const}.$$

So we have the equalities:

$$1 = \int_{-\infty}^{+\infty} u(0, x) dx = \int_0^1 dx = \int_0^{x_f(t)} \frac{x}{t} dx = \frac{x_f(t)^2}{2t} \quad \text{as} \quad t > 2.$$

So, $x_f(t) = \sqrt{2t}$.

Example 8.5. Now let us consider the problem in a sense opposite to the previous one, namely, we are going to solve the Cauchy problem (8.5) with

$$\phi(x) = \begin{cases} 1 & \text{when} \quad x < 0, \\ 0 & \text{when} \quad 0 \le x \le 1, \\ 1 & \text{when} \quad x > 1. \end{cases}$$

It is evident, that the left step-like function will catch up the right one with the velocity $v_l = 1/2$, while the right step will turn into the inclined segment whose right edge moves with the velocity $v_r = 1$, see Fig. 8.5, left panel. At $t = 2$ the step function will attain the foot of the right disturbance (Fig. 8.5, middle left panel). At $t > 2$ the solution will have the form (Fig. 8.5, right left panel):

$$u(t, x) = \begin{cases} 1 & \text{when} \quad x < x_L(t), \\ \frac{x}{t}, & \text{when} \quad x_L(t) \le x \le t \\ 1 & \text{when} \quad x > t. \end{cases} \tag{8.15}$$

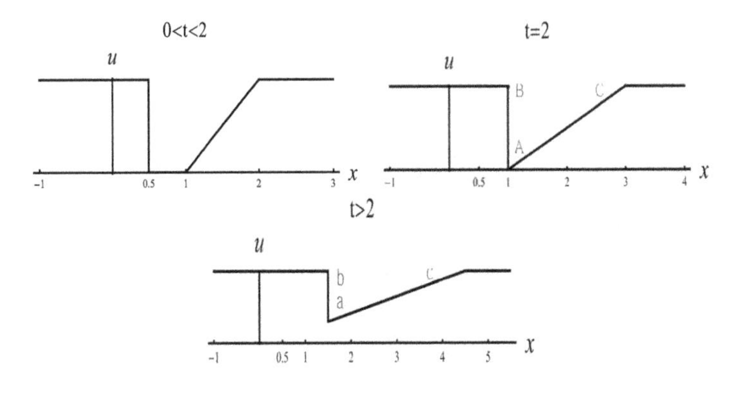

Fig. 8.5 Evolution of a square perturbation.

We find the function $x_L(t)$ from the following consideration. First of all, from the conservation law applied to the "hole" in the constant solution $u = 1$ appears that the areas of the triangles ABC and abc are the same that is

$$\frac{(1-u_L)(t-x_L)}{2} = 1. \tag{8.16}$$

From the other hand, we have the equality

$$\dot{x}_L(t) = \frac{1+u_L}{2}, \qquad x_L(2) = 1. \tag{8.17}$$

Expressing u_L from the equality (8.16), inserting it into (8.17) and taking into account the initial condition, we get the Cauchy problem

$$\dot{x}_L = \frac{1}{x_L - t} + 1, \qquad x_L(2) = 1,$$

which is solved by the substitution $w = x_L - t$. Finally we obtain the following solution:

$$x_L(t) = t - \sqrt{2t-3}, \qquad t \geq 2.$$

Problem.

Find the solution to the initial value problems (8.5) with the functions $\phi(x)$ taken as in the Figs. 8.6(a), (b), (c).

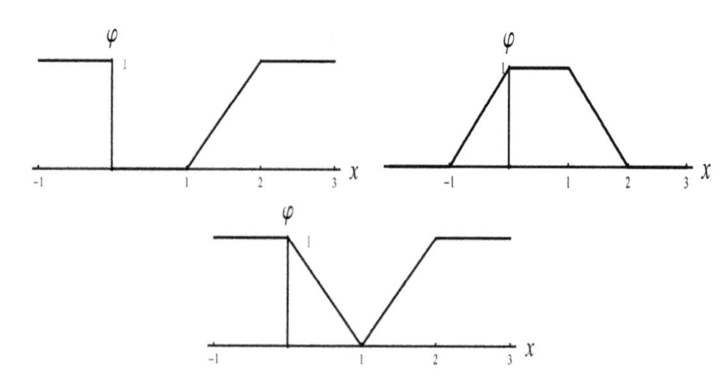

Fig. 8.6 Profiles of disturbances that are proposed to be used in the Exercise 8.3 as the Cauchy data.

8.4 The Burgers equation

Let us now consider a dispersion relation that differs from that associated with the equation (8.2) by adding a square term:

$$\omega = ck + i\nu k^2, \qquad \nu > 0. \tag{8.18}$$

This relationship corresponds to the equation

$$u_t + c\,u_x = \nu u_{xx}, \tag{8.19}$$

which coincides with the linearization of the Burgers equation (5.29). Thus, the Burgers equation can be derived from the dispersion relation (8.18) by resorting to the trick used in the case of the linear wave equation (8.2). The Burgers equation has many interesting properties. It belongs to the class of the *completely integrable* equation in the sense that the Cole-Hopf transformation

$$u(t,\, x) = -2\,\nu\partial_x\, \log\, \Phi(t,\, x) \tag{8.20}$$

connects solutions of the Burgers equation with solutions of the heat transport equation.

$$\Phi_t - \nu\,\Phi_{xx} = 0. \tag{8.21}$$

Indeed, taking partial derivatives of the r.h.s of the formula (8.20) and inserting it into the equation (5.29), we can obtain the following equation:

$$\left[\Phi\,\Phi_x - \Phi^2\frac{\partial}{\partial\, x}\right]\left[\Phi_t - \nu\,\Phi_{x\,x}\right] = 0.$$

Thus, the substitution (8.20) turns the Burgers equation into the identity if the function Φ is a solution to the equation (8.21).

The possibility of obtaining analytical solution of this equation makes it, on the one hand, an important tool for testing numerical schemes, and, on the other hand, opens the possibility of examining the dependence of solutions on the viscosity coefficient.

The numerical study of the Burgers equation will be based on the finite differences method. We introduce on a certain rectangle belonging to $\mathbb{R}^2$ a uniform mesh and make the substitution $u(t_k, x_j) = U_{k,j}$, where $t_{k+1} - t_k = \tau = \text{const}$, $x_{j+1} - x_j = h = \text{const}$. Approximating the first and the second derivative, we obtain the following scheme:

$$\frac{U_{k+1,j} - U_{k,j}}{\tau} + U_{k,j}\frac{U_{k,j+1} - U_{k,j-1}}{2h} = \nu\frac{U_{k,j+1} - 2U_{k,j} + U_{k,j-1}}{h^2}. \tag{8.22}$$

This scheme can be presented in the form

$$U_{k+1,j} = U_{k,j} + \tau\left(-U_{k,j}\frac{U_{k,j+1}-U_{k,j-1}}{2h} + \nu\frac{U_{k,j+1}-2U_{k,j}+U_{k,j-1}}{h^2}\right).$$

Let us analyze the stability of the scheme, or rather its linearization, by replacing $U_{k,j}$ in the second term of the r.h.s. by $U^0_{k,j} = \text{const}$. Using the ansatz $U_{k,j} = \bar{U}q^k e^{i\varphi j}$, we get the dispersion relationship

$$q = 1 - 4\frac{\tau\nu}{h^2}\sin^2(\varphi/2) - iU^0_{k,j}\frac{\tau}{h}\sin\varphi.$$

The stability condition $|q| \le 1$ can be presented as

$$|q| = (1 - 4r\sin^2(\varphi/2))^2 + \left(U^0_{k,j}\frac{\tau}{h}\right)^2\sin^2\varphi \le 1, \tag{8.23}$$

where $r = \tau\nu/h^2$. It is easily seen that the condition $\sin\varphi = 0$, implies the equality $\sin(\varphi/2) \in \{0, \pm 1\}$ and the relation $|q| = (1-4r)^2 \le 1$. Hence

$$0 < r = \frac{\tau\nu}{h^2} \le \frac{1}{2}.$$

Let us rewrite the inequality (8.23) in the form

$$-2r + 4r^2\sin^2\varphi/2 + (U^0_{k,j}\tau/h)^2\cos^2\varphi/2 \le 0.$$

If we put $r = 1/2$, then the following inequality will be fulfilled:

$$U^0_{k,j}\frac{\tau}{h} \le 1.$$

Thus, unlike the linear case, the necessary condition for the stability of the scheme depends on the parameter $U^0_{k,j}$ used to linearize the equation, and therefore, when applying an explicit scheme to the Burgers equation, one should be careful and, in particular, make the size of the time step dependent on the maximal amplitude of the initial disturbance.

8.4.1 *Traveling wave solution to the Burgers equation, describing viscous shock wave*

Let us consider the problem

$$u_t + u\,u_x = \nu\,u_{xx}, \quad u(-\infty, x) = u_2, \quad u(+\infty, x) = u_1, \tag{8.24}$$
$$\frac{\partial u}{\partial x}(t, \pm\infty) = 0, \quad u_2 > u_1 \ge 0.$$

We are looking for the solution to (8.24) taking the form

$$u(t, x) = U(z), \qquad z = x - s\,t, \quad s = \text{const}. \tag{8.25}$$

Inserting (8.25) into (8.24), we get the following ODE:

$$\frac{d}{d\,z}\left\{\frac{U^2}{2} - s\,U - \nu\,\dot{U}\right\} = 0.$$

Integrating it once, we get:

$$\frac{U^2}{2} - s\,U - \nu\,\dot{U} = C.$$

It appears from the boundary conditions that

$$\frac{u_2^2}{2} - s\,u_2 = C = \frac{u_1^2}{2} - s\,u_1.$$

From this we get

$$s = \frac{u_2 + u_1}{2}.$$

Note that the velocity s coincides with that obtained when describing the evolution of the step function within the Hopf equation. Inserting s into the first equality, we express the constant of integration as

$$C = -\frac{u_1\,u_2}{2}.$$

So the initial problem is reduced to solving the equation

$$\nu\dot{U} = \frac{1}{2}\,U(U - 2\,s) + \frac{u_1\,u_2}{2},$$

which can be presented in the form

$$\frac{d\,U}{U(U - 2\,s) + u_1\,u_2} = \frac{1}{2\,\nu}\,d\,z.$$

Integrating this equation we obtain:

$$\frac{1}{u_2 - u_1}\,\log\frac{|U - u_2|}{|U - u_1|} = \frac{1}{2\,\nu}\,(z - z_0).$$

Assuming in addition that the solution we are looking for is monotone, that is, $u_1 < U < u_2$, we can present the above equality as

$$u_2 - U = (U - u_1)\,\exp\lambda\,(z - z_0), \qquad \lambda = \frac{u_2 - u_1}{2\,\nu}.$$

Performing simple algebraic transformations, we finally obtain the following solution:

$$u(t,\,x) = \frac{u_2 + u_1\,e^{\lambda\,(z - z_0)}}{1 + e^{\lambda\,(z - z_0)}}. \tag{8.26}$$

The formula (8.26) describes a so-called *viscous shock wave*. The next question is how to define the *effective width* of the shock wave. Having

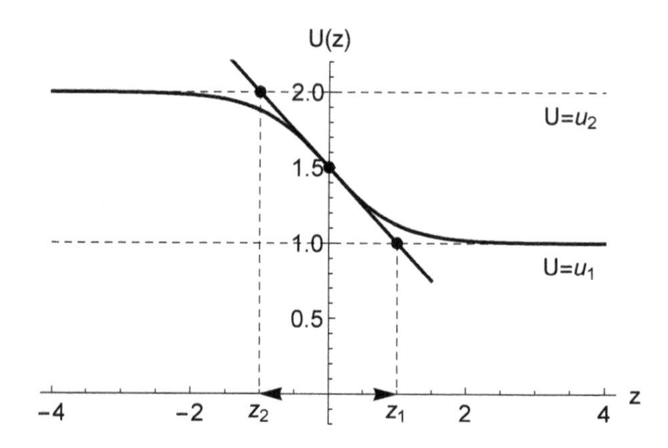

Fig. 8.7 Determination of the effective width of a viscous shock wave.

determined in some reasonable way the effective width, we thereby obtain a formula that makes it possible to analyze the dependence of the solution on the parameters and the closeness of the solution (8.26) to the shock-wave solution of the Hopf equation.

The effective width can be determined in many ways which generally lead to qualitatively identical results. We will define the width as follows (see Fig. 8.7):

- at first we find the point of inflection of the graph of the function $U(z)$;
- then we determine the points of intersection of the straight line tangent to the inflection point with the asymptotes $u = u_2$ and $u = u_1$ (we denote these points as (u_2, z_2) and (u_1, z_1), respectively);
- finally we define the effective width as $\Delta = z_1 - z_2$.

Calculating the second derivative of $U(z)$, we get:

$$U''(z) = \lambda\,(u_1 - u_2)e^{\lambda(z-z_0)}\frac{e^{\lambda(z-z_0)} - 1}{\left(1 + e^{\lambda(z-z_0)}\right)^3}. \tag{8.27}$$

For $z_0 = 0$, the inflection point will have the coordinate $z = 0$.

The straight line tangent to the inflection point is described by the formula

$$L: \qquad \frac{u_1 + u_2}{2} - \frac{(u_2 - u_1)^2}{2\,\nu}\,z.$$

The points of intersection of L with the asymptotes u_1, u_2 are, respectively, $z_{1,2} = \pm \frac{4\,\nu}{u_2 - u_1}$, and therefore

$$\Delta = \frac{8\,\nu}{u_2 - u_1}. \tag{8.28}$$

We conclude from this that the effective width is directly proportional to the viscosity coefficient and inversely proportional to the difference $u_2 - u_1 > 0$. In the limiting case $\nu \to 0+$, the solution (8.26) turns into the solution of the Hopf equation, describing the propagation of the shock wave. So one can see that the viscosity causes widening of the shock front. Otherwise, the solutions (8.26) and (8.10) are identical.

The numerical solution of the viscous shock problem is presented in the file PM2_11.nb, in which the scheme (8.22) is used. The file presents the numerical solution of the Burgers equation with the function (8.26) taken as the Cauchy's data. As one can see, the initial disturbance moves without changing the shape, which serves as the element of the proof of the stability of the solution describing the viscous shock wave.

8.4.2 *Exact solutions of the Burgers equation*

Using the Cole-Hopf transformation, solution to any initial value problem for the Burgers equation can be obtained. This is apparent from the following statement.

Lemma 8.2. *The Cole-Hopf transformation (8.20) connects the solution to the Cauchy problem*

$$\begin{aligned} u_t + u\,u_x &= \nu\,u_{xx}, \\ u(0,\,x) &= F(x), \end{aligned} \tag{8.29}$$

where $F(x)$ is any bounded continuous function, with the solution of the initial value problem

$$\begin{aligned} \Phi_t - \nu\Phi_{xx} &= 0, \\ \Phi(0,\,x) &= C_1 \exp\left[-\frac{1}{2\,\nu}\int_0^x F(\xi)\,d\,\xi\right]. \end{aligned} \tag{8.30}$$

Proof. The initial conditions of the problems (8.29) and (8.30) are connected with the help of the following formula

$$u(0,\,x) = -2\,\nu\left[\log\Phi(0,\,x)\right]_x.$$

Hence

$$d\,\log\Phi(0,\,x) = -\frac{F(x)}{2\,\nu}\,d\,x.$$

Integrating we obtain

$$\log\Phi(0,\,x) - \log C_1 = -\frac{1}{2\,\nu}\int_0^x F(x)\,d\,x,$$

or

$$\Phi(0,\,x) = C_1\,\exp\left[-\frac{1}{2\,\nu}\int_0^x F(\xi)\,d\,\xi\right].$$

Lemma 8.3. *If $\theta(x)$ is a bounded continuous function, then the solution of the Cauchy problem*

$$\Phi_t - \nu\Phi_{xx} = 0,$$
$$\Phi(0,x) = \theta(x)$$

is given by the formula

$$\Phi(t,\,x) = \frac{1}{\sqrt{4\,\pi\,\nu\,t}}\int_{-\infty}^{+\infty}\theta(\xi)\,\exp\left[-\frac{(x-\xi)^2}{4\,\nu\,t}\right]\,d\,\xi. \qquad (8.31)$$

The proof of this statement, which can be found, e.g. in [Vladimirov (2007)], is elementary.

Corollary 8.1. *Solution of the initial value problem (8.29) is given by the formula*

$$u(t,x) = \frac{\int_R \frac{x-\xi}{t}\,e^{-\frac{f(\xi,\,t,\,x)}{2\nu}}\,d\,\xi}{\int_R e^{-\frac{f(\xi,\,t,\,x)}{2\nu}}\,d\,\xi}, \qquad (8.32)$$

where

$$f(\xi,\,t,\,x) = \int_0^\xi F(z)\,d\,z + \frac{(x-\xi)^2}{2\,t}.$$

Example 8.6. The possibility of solving arbitrary initial value problem for the Burgers equation (including for small values of the parameter ν) opens up the possibility of easy comparison of the previously obtained shock wave solutions supported by the Hopf equation with the solution of the corresponding initial value problem supported by the Burgers equation. Consider as an example the following one:

$$u_t + u\,u_x = 0, \qquad u(0,\,x) = \phi(x),$$

where

$$\phi(x) = \begin{cases} 2 & \text{when} \quad x < 0, \\ 0 & \text{when} \quad 0 \le x \le 2, \\ 2 & \text{when} \quad x > 2. \end{cases} \qquad (8.33)$$

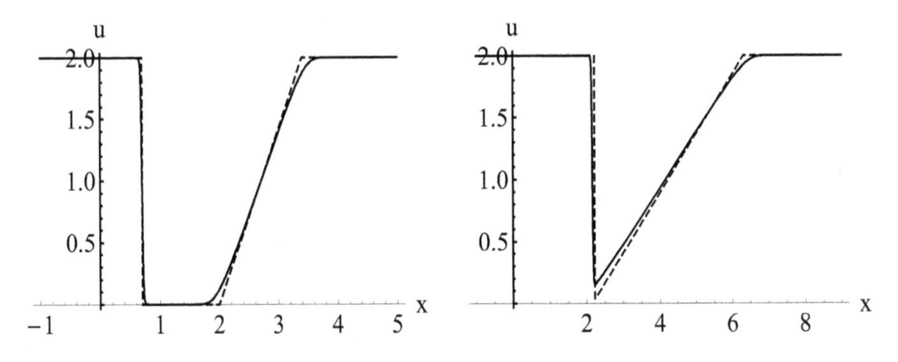

Fig. 8.8 Comparison of the evolution of the initial perturbation (8.33) supported by the Hopf equation (dashed line) and the Burgers equation with $\nu = 0.01$ (solid line). Left panel: $t = 0.7$; right panel: $t = 2.15$.

The problem is solved in full analogy with how it was done in Example 8.5. The left step-like function overtakes the right one at $t = 2$, and the solution profile becomes as follows:

$$u(t,\, x)|_{t=2} = \begin{cases} 2 \quad \text{when} \quad x < 2, \\ \frac{x-2}{2} \quad \text{when} \quad 2 \leq x \leq 6, \\ 2 \quad \text{when} \quad x > 6. \end{cases} \tag{8.34}$$

For $t > 2$ the solution takes the form

$$u(t,\, x)|_{t>2} = \begin{cases} 2 \quad \text{when} \quad x < x_L(t), \\ u_L(t) + \Delta(x) \quad \text{when} \quad x_L(t) \leq x \leq 2(1+t), \\ 2 \quad \text{when} \quad x > 2(1+t), \end{cases} \tag{8.35}$$

where

$$x_L(t) = 2\left[(1+t) - \sqrt{t+2}\right],\; u_L(t) = 2\left[1 - \frac{2}{\sqrt{t+2}}\right],\; \Delta = \frac{2\left[x - x_L(t)\right]}{t+2}.$$

Next, in order to obtain the solution of the Burgers equation satisfying the initial condition (8.34) we use the formula (8.32) with $\nu = 0.01$. The solutions of both equations corresponding to $t = 4$ are shown in Fig. 8.8. The figure shows that the solutions are close to each other, in particular, the speeds of the movements of the rear parts of disturbance are almost the same. However, discrepancies appear in those areas in which the solution of the Hopf equation is not smooth.

It is also interesting to consider the problem inverse to that formulated in Lemma 8.3, namely, how can we determine the solution to the linear transport equation corresponding to known solution of the Burgers equation. When solving this problem, one should take into account peculiarities, which we discussed on a specific example.

Example 8.7. Let us find the solution to the equation $\Phi_t - \nu\Phi_{xx} = 0$ corresponding to the function

$$u(t,\, x) = \frac{x}{t}, \tag{8.36}$$

being simultaneously the solution to the Burgers and Hopf equations.

The solution of the Burgers equation is related to the sought solution of the transport equation by the formula (8.20). Integrating the equation

$$\frac{x}{t} = -2\,\nu\frac{\partial}{\partial\, x}\log\Phi(t,\, x), \tag{8.37}$$

we obtain:

$$\Phi(t,\, x) = C(t)\exp\left[-\frac{x^2}{4\nu t}\right]. \tag{8.38}$$

It turns out that the parameter $C(t)$ in the above formula cannot be arbitrary. Substituting the formula (8.38) we obtain the equation

$$\dot{C}(t) + \frac{x^2}{4\nu t^2}C(t) = \nu C(t)\left[\frac{x^2}{4\nu^2 t^2} - \frac{1}{2\nu t}\right].$$

Solving the equation with respect to $C(t)$ and inserting the result into (8.38), we obtain the function

$$\Phi(t,\, x) = \frac{C_0}{\sqrt{t}}\exp\left[-\frac{x^2}{4\nu t}\right], \qquad C_0 = \text{const},$$

which, on one hand, satisfies the heat transport equation, and on the other hand, is related to the function $u(t,\, x) = x/t$ by means of the formula (8.37).

8.5 Korteweg-de Vries (KdV) equation and its modifications

8.5.1 *Introductory remarks*

In the previous sections, we got acquainted with the simplest generalization of the linear dispersion relationship $\omega = k\,c$, replacing it with the dispersion relation

$$\omega = kc - \beta k^3 \tag{8.39}$$

for the particular case $c = \beta = 1$. This relationship in the general case corresponds to the equation of the form

$$u_t + c\,u_x + \beta u_{xxx} = 0. \tag{8.40}$$

Note that the numerical values of the parameters are not important since they can be changed using the *scaling* transformation [Barenblatt (2012)]. Therefore, each equation of the form (8.40) represents the same type of behavior, characterized in that the compact initial profile is intensely deformed in the course of evolution. Previously, we also considered the Hopf equation, solutions of which tend to increase the steepness and ultimately to the formation of a shock wave. If we combine in one equation the nonlinear term responsible for increasing the wave's steepness by changing c with u in (8.40), maintaining simultaneously the dispersion term, then both effects will be balanced, which results in the existence of stable nonlinear waves with very interesting properties. The canonical equation containing both of these effects is the celebrated Korteweg-de Vries (KdV) equation:

$$u_t + 12\,u\,u_x + u_{xxx} = 0. \tag{8.41}$$

As in the case of the equation (8.40), the value of the coefficient at the non-linear term is irrelevant, because, using the scaling transformation

$$t \to e^{\alpha}\,t, \quad x \to e^{\beta}\,x, \quad u \to e^{\gamma}u$$

with appropriate choice of the parameters α, β and γ, we can pass to the equation

$$u_t + a\,u\,u_x + b\,u_{xxx} = 0,$$

where $0 < a$, b are arbitrary.

An equation that can also be associated with a dispersion relationship similar to (8.39) is the so-called *modified KdV equation* (mKdV to abbreviate):

$$\partial_t\,u + 12\;u^p\,u_x + \;\partial_x^3\,u = 0, \qquad p \geq 2. \tag{8.42}$$

Among the solutions to the mKdV equation, the most interesting are *soliton solutions*, which we are looking for in the form $u = u(z)$, $z = x - st$, where $s > 0$. Inserting this ansatz into the equation (8.42), we get the ODE

$$-\,su' + 12\,u^p\,u' + u''' = 0. \tag{8.43}$$

Integrating equation (8.43), we obtain, under the additional assumption $u^{(k)} \to 0$, $k = 0, 1, \ldots$ as $|z| \to \infty$, the equation

$$-su + \frac{12}{p+1}u^{p+1} + u'' = 0,$$

which can be presented in the Hamiltonian form

$$u' = -w = -H_w, \qquad w' = -su + \frac{12}{p+1} u^{p+1} = H_u, \tag{8.44}$$

where

$$H(u,\, w) = \frac{w^2}{2} + \frac{12}{(p+1)(p+2)}\, u^{p+2} - \frac{s}{2}\, u^2.$$

So we deal with a system with one degree of freedom, whose potential energy is expressed by the formula

$$U_{pot} = \frac{12}{(p+1)(p+2)}\, u^{p+2} - \frac{s}{2}\, u^2.$$

Depending on parity of the parameter p, U_{pot} has one local minimum (when $p = 2\,n + 1$) or two local minima (when $p = 2\,n$) and one local maximum. Local minima correspond to periodic solutions, separated from another solutions by the separatrices bi-asymptotic to the saddle point, which correspond to soliton solutions. Recall that in Chapter 2 we've obtained the solution of the system (8.44) in the case $p = 1$. Now for any p we look for a solution in a similar form:

$$u = \frac{A}{\cosh^{\gamma}[B\,z]}. \tag{8.45}$$

Inserting (8.45) into (8.43), we get the equation

$$-\frac{A\,s}{\cosh^{\gamma}[B\,z]} + \frac{12}{p+1}\frac{A^{p+1}}{\cosh^{\gamma(p+1)}[B\,z]} + A\,\gamma\,B^2 \frac{\gamma\,\cosh^2[B\,z] - (\gamma+1)}{\cosh^{\gamma+2}[B\,z]} = 0,$$

which can be rewritten as an algebraic equation with respect to the powers of hyperbolic functions. In order to avoid over-determination, which leads to the existence of merely trivial solutions, we assume, that $\gamma\,(p+1) = \gamma+2$, thus "balancing" the highest powers of the function $\cosh[B\,z]$. From this we get that $\gamma = 2/p$. Next, we reduce the equation to a common denominator and equate the numerator of the resulting expression to zero. This way we obtain the equation

$$\left(A\,\gamma^2\,B^2 - A\,s\right)\cosh^2[B\,z] + \frac{12\,A^{p+1}}{p+1} - A\,\gamma\,B^2(\gamma+1) = 0.$$

Equating to zero the coefficient at $\cosh^2[B\,z]$ and the remaining terms, we obtain the system of algebraic equation

$$\begin{aligned} &A\,\gamma^2\,B^2 - A\,s = 0,\\ &\tfrac{12\,A^{p+1}}{p+1} - A\,\gamma\,B^2(\gamma+1) = 0. \end{aligned}$$

Solving the obtained algebraic system with respect to A and B, we finally get:

$$u = \left[\frac{s(p+1)(p+2)}{24}\right]^{\frac{1}{p}} \cosh^{-\frac{2}{p}}\left(\frac{\sqrt{s}\,p}{2}\,z\right). \tag{8.46}$$

Let us note that for $p = 1$ relation (8.46) coincides with the formula (2.26) describing soliton solution of the classical KdV equation.

The shape of the soliton resembles a bell (see Fig. 8.9). It moves from left to right with the speed which is proportional to its amplitude.

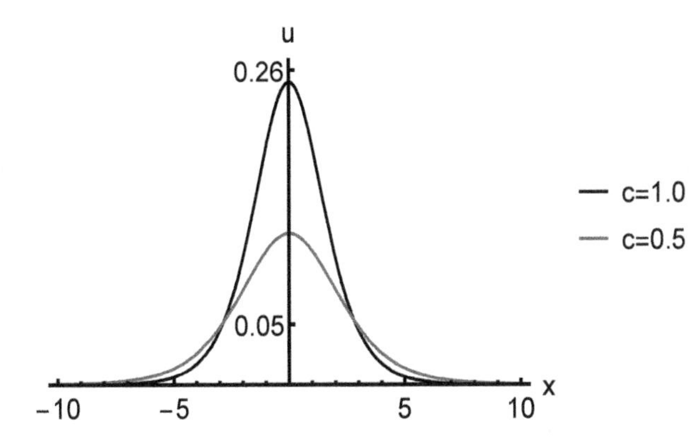

Fig. 8.9 Profiles of the soliton solutions obtained for $s = 1.0$ (upper curve) and $s = 0.5$ (lower curve).

8.5.2 *Scott Russell's discovery. Formal derivation of the KdV equation*

Solitons formed on the surface of the oceans as a result of underground earthquakes have undoubtedly been known to the inhabitants of some coastal areas since ancient times. They are manifested as long (ranging from several dozen to hundreds meters) waves on the water surface, which, when approaching the coast, often form a huge water hump, called the *tsunami wave*. Yet the first "documented" observation of soliton is owed to Scott Russell (1834). Here is his description of the phenomenon.

"I was observing the motion of a boat which was rapidly drawn along a narrow channel by a pair of horses, when the boat suddenly stopped — not so the mass of water in the channel which it had put in motion; it accumulated

round the prow of the vessel in a state of violent agitation, then suddenly leaving it behind, rolled forward with great velocity, assuming the form of a large solitary elevation, a rounded, smooth and well-defined heap of water, which continued its course along the channel apparently without change of form or diminution of speed. I followed it on horseback, and overtook it still rolling on at a rate of some eight or nine miles an hour, preserving its original figure some thirty feet long and a foot to a foot and a half in height. Its height gradually diminished, and after a chase of one or two miles I lost it in the windings of the channel. Such, in the month of August 1834, was my first chance interview with that singular and beautiful phenomenon which I have called the Wave of Translation".

Scott Russell was an engineer by profession. He was able to create a long wave on the water surface under laboratory conditions (Fig. 8.10 illustrates the experiment conducted by Scott Russell). He introduced the notion *solitary wave* and tried, unfortunately without success, to guess the form of the equation describing the soliton. He also asked many of his contemporaries for help. These consultations were also unsuccessful, because the soliton, being a significantly nonlinear object, did not fit within the framework of the theories of the time. The theoretical confirmation of Russell's observations was found only in 1895 in the works of Dutch scientists Korteweg and de Vries.

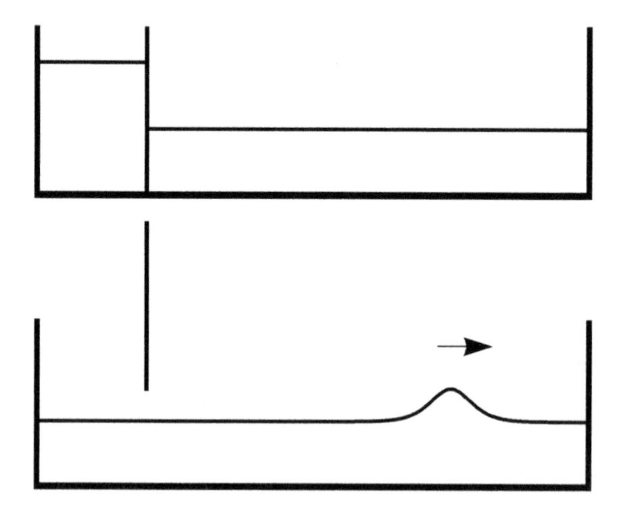

Fig. 8.10 Reconstruction of the solitary wave by Scott Russell.

In the 1950s, E. Fermi, J. Pasta and S. Ulam (FPU) dealt with the problem of energy distribution in a chain consisting of masses connected by springs (see Fig. 8.11). When the elastic force is linear, the kinetic energy delivered by a short impact of the one of the masses, divides evenly into all the degrees of freedom (modes) associated with all natural frequencies of the system. For a linear system, this effect, known as *ergodic hypothesis* is always true. To the great amazement of the researchers, numerical experiments with a long chain of balls (they used in the experiments chains consisting of about sixty balls) related to non-linear elastic forces did not confirm the ergodic hypothesis.

Fig. 8.11 Scheme of the experimental chain considered by Fermi, Pasta and Ulam.

The modeling system describing the FPU experiment can be presented as follows:

$$\ddot{Q}_n(t) = f(Q_{n+1} - Q_n) - f(Q_n - Q_{n-1}), \tag{8.47}$$

where Q_n is the deviation of n-th ball from the state of equilibrium,

$$f(z) = \gamma\, z + \alpha\, z^n,$$

$n > 1$ (we'll assume that $n = 2$). After an unsuccessful attempt to confirm the ergodic hypothesis in the nonlinear case, the system (8.47) began to take a special interest. In 1965, Kruskal and Zabussky searched for the continual analog of the equation (8.47). Their reasoning was as follows. Assuming that the distance a separating the balls at rest is very small, they equated the value $n \cdot a$, which approximates the coordinate of n-th ball, with the (continuous) parameter x by making the substitution

$$Q_n(t) = Q(t,\, n\, a) \equiv Q(t,\, x).$$

The coordinates of the balls adjacent to that with the coordinate x, are described with the help of the following formula:

$$Q_{n\pm 1} = Q(t,\, x \pm a) = e^{\pm a\, D_x} Q(t,\, x) = \sum_{k=0}^{\infty} \frac{(\pm\, a)^k}{k!} \frac{\partial^k}{\partial\, x^k} Q(t,\, x).$$

Under the above assumptions, the system (8.47) can be changed with the equation

$$\frac{\partial^2}{\partial\, t^2} Q(t,\, x) = f\left[\sum_{k=1}^{\infty} \frac{a^k}{k!} \frac{\partial^k}{\partial\, x^k} Q(t,\, x)\right] - f\left[-\sum_{k=1}^{\infty} \frac{(-a)^k}{k!} \frac{\partial^k}{\partial\, x^k} Q(t,\, x)\right].$$

From this appears the following equalities:

$$\begin{aligned}
Q_{tt} &= \gamma\left\{\tfrac{a}{1!}Q_x + \tfrac{a^2}{2!}Q_{xx} + \tfrac{a^3}{3!}Q_{xxx} + \tfrac{a^4}{4!}Q_{xxxx} + O(|a|^5)\right\}\\
&+\gamma\left\{-\tfrac{a}{1!}Q_x + \tfrac{a^2}{2!}Q_{xx} - \tfrac{a^3}{3!}Q_{xxx} + \tfrac{a^4}{4!}Q_{xxxx} + O(|a|^5)\right\}\\
&+\alpha\left[aQ_x + \tfrac{a^2}{2!}Q_{xx} + \tfrac{a^3}{3!}Q_{xxx} + \dots\right]^2\\
&+\alpha\left[-aQ_x + \tfrac{a^2}{2!}Q_{xx} - \tfrac{a^3}{3!}Q_{xxx} + \dots\right]^2\\
&+\gamma\left\{a^2Q_{xx} + \tfrac{a^4}{12}Q_{xxxx}\right\} + 2\alpha a^3Q_xQ_{xx} + O(|a|^5).
\end{aligned}$$

In the following step the scaling $Q = a\,P$ is used, leading to the equation

$$P_{tt} = \gamma\left(a^2P_{2x} + \frac{a^4}{12}P_{4x}\right) + 2\alpha\,a^4P_x\,P_{xx} \tag{8.48}$$

(here and henceforth the "irrelevant" tail is omitted). Differentiating the equation (8.48) over x, introducing new function $u(t,\,x) = P_x(t,\,x)$ we fall into the Boussinesq equation

$$\frac{\partial^2 u}{\partial t^2} = \frac{\partial^2}{\partial x^2}\left\{\gamma\left[a^2u + \frac{a^4}{12}u_{xx}\right] + \alpha a^4u^2\right\} \tag{8.49}$$

Remark. Like the KdV equation, the Boussinesq equation (8.49) possesses the soliton solutions, yet they are unstable.

Next we apply the scaling

$$u = \varepsilon^{\mu}u^1, \qquad \xi = \varepsilon^{p}(x - c\,t), \qquad \tau = \varepsilon^{q}\,t,$$

where $0 < \varepsilon \ll 1$ is a small parameter. This lead to the equation

$$\begin{aligned}
&\left\{\varepsilon^{2q}\tfrac{\partial^2}{\partial\tau^2} - 2\,c\,\varepsilon^{q+p}\tfrac{\partial^2}{\partial\xi\partial\tau} + c^2\,\varepsilon^{2p}\tfrac{\partial^2}{\partial\xi^2}\right\}u^1\\
&= \varepsilon^{2p}\tfrac{\partial^2}{\partial\xi^2}\left\{\gamma\left(a^2\,u^1 + \tfrac{a^4}{12}\varepsilon^{2p}\tfrac{\partial^2u^1}{\partial\xi^2}\right) + \alpha\,a^4\varepsilon^{\mu}[u^1]^2\right\}.
\end{aligned}$$

Now we put

$$p = \frac{1}{2}, \quad q = \frac{3}{2}, \quad c = a\sqrt{\gamma}.$$

With such a choice the term proportional to the second derivative over τ will be negligibly small compared to the remaining terms, so dropping it, we obtain the equation

$$2\,a\,\sqrt{\gamma}u^1_{\tau\xi} + \gamma\frac{a^4}{12}u^1_{\xi\xi\xi\xi} + \alpha\,a^4[u^1]^2_{\xi\xi} = 0.$$

Integrating it over the variable ξ, choosing the constant of integration to be equal to zero and dividing the resulting equation by $2\,\alpha\sqrt{\gamma}$, we get the equation

$$u^1_\tau + \frac{\sqrt{\gamma}a^3}{24}u^1_{\xi\xi\xi} + \frac{\alpha a^3}{\sqrt{\gamma}}u^1\,u^1_\xi = 0.$$

Making the substitution $W = u^1$ and then applying the scaling

$$T = e^\delta \tau, \qquad X = e^\sigma X,$$

where the parameters δ, σ satisfy the equations

$$e^{3\sigma-\delta} = \frac{24}{a^3\sqrt{\gamma}}, \qquad e^{\sigma-\delta} = \frac{6\sqrt{\gamma}}{\alpha\,a^3},$$

we finally get the standard KdV equation

$$W_T + 6\,W\,W_\xi + W_{\xi\xi\xi} = 0.$$

8.5.3 *Conservation laws of the KdV equation*

Let us consider the evolutionary equation

$$u_t = F(t,\,x,\,u,\,u_x...u_{mx}). \tag{8.50}$$

Definition 8.1. By the conservation law of $n-$th order related to (8.50) we mean a pair of functions $\rho(t,\,x,\,u,\,u_x...u_{nx})$, $\sigma(t,\,x,\,u,\,u_x...u_{nx})$ such that the equation

$$D_t\rho + D_x\sigma = 0 \tag{8.51}$$

turns into identity on virtue of the equation (8.50) and its differential consequences. Symbols D_t and D_x denote the total derivatives over the corresponding variables.

If in addition the function σ satisfies the condition

$$\lim_{|x|\to\infty} \sigma(t,\,x,\,u,\,u_x...u_{nx}) = 0,$$

then, integrating the equation (8.51) over the x, we obtain the equality

$$\frac{\partial}{\partial t}\int_{-\infty}^{+\infty}\rho\,d\,x = -\int_{-\infty}^{+\infty}D_x\,\sigma\,d\,x = -\lim_{M,N\to\infty}\sigma|_{x=-M}^{x=N} = 0.$$

Hence

$$\int_{-\infty}^{+\infty}\rho\,d\,x = \text{const}$$

on the solutions of the equation (8.50).

Examples of the conservation laws provide the gas dynamics equations which for one spatial variable can be written as

$$\frac{\partial}{\partial t}\rho\, u + \frac{\partial}{\partial x}\left(\rho\, u^2 + \frac{\beta}{n+2}\,\rho^{n+2}\right) = 0,$$
$$\frac{\partial}{\partial t}\rho + \frac{\partial}{\partial x}(\rho\, u) = 0.$$

From this we conclude that the integrals

$$\int_{-\infty}^{+\infty} \rho\, u\, d\, x; \qquad \int_{-\infty}^{+\infty} \rho\, d\, x$$

do not depend on t. The first integral expresses the law of momentum conservation, while the second one expresses the law of mass conservation.

The uniqueness of the KdV equation is manifested, among others, in that it has an infinite number of conservation laws! Let us show this. For our purposes, it is convenient to present the KdV equation in the following form

$$P(u) = u_t - 6\, u\, u_x + u_{xxx} = 0, \tag{8.52}$$

which is connected with (8.41) by means of the scaling transformation together with the reflection $u \to -u$. We introduce the auxiliary transformation,

$$u = w + \varepsilon w_x + \varepsilon^2\, w^2, \tag{8.53}$$

treating for a while ε as a small parameter.

Theorem 8.1. *The ansatz (8.53) links the equation (8.52) with the Gardner equation*

$$Q(w) = w_t - 6\left(w + \varepsilon^2\, w^2\right)\, w_x + w_{xxx} = 0. \tag{8.54}$$

Proof. Inserting the formulas

$$\begin{aligned} u_t &= w_t + \varepsilon w_{xt} + 2\,\varepsilon^2\, w\, w_t, \\ u_x &= w_x + \varepsilon w_{xx} + 2\,\varepsilon^2\, w\, w_x, \\ u_{xx} &= w_{xx} + \varepsilon w_{xxx} + 2\,\varepsilon^2\,\left(w_x^2 + w\, w_{xx}\right), \\ u_{xxx} &= w_{xxx} + \varepsilon w_{xxxx} + 2\,\varepsilon^2\,\left(3\, w_x\, w_{xx} + w\, w_{xxx}\right) \end{aligned}$$

into the equation (8.52), we obtain:

$$\begin{aligned}
P(u) &= w_t + \varepsilon w_{xt} + 2\,\varepsilon^2\, w\, w_t - 6\big(w + \varepsilon w_x + \varepsilon^2\, w^2\big)\big(w_x + \varepsilon w_{xx} + 2\,\varepsilon^2\, w\, w_x\big) \\
&\quad + w_{xxx} + \varepsilon w_{xxxx} + 2\,\varepsilon^2\, \left(3\, w_x\, w_{xx} + w\, w_{xxx}\right) \\
&= w_t - 6\left(w + \varepsilon^2\, w^2\right)\, w_x + w_{xxx} \\
&\quad + \varepsilon \left\{ w_{tx} - 6\left[\left(w + \varepsilon^2\, w^2\right) w_{xx} + w_x \left(w_x + 2\,\varepsilon^2\, w\, w_x\right)\right] + w_{4x} \right\} \\
&\quad + 2\,\varepsilon^2\, w \left[w_t - 6\, w_x \left(w + \varepsilon^2\, w^2\right) + w_{xxx}\right] \\
&= \left[1 + \varepsilon\, \frac{\partial}{\partial x} + 2\,\varepsilon^2\, w\right] Q(w).
\end{aligned}$$

The equation (8.54) can be presented in the following equivalent form:

$$Q(w) = w_t + \left(w_{xx} - 3\, w^2 - 2\,\varepsilon^2\, w^3\right)_x = 0. \tag{8.55}$$

Next we present the function w as a series expansion

$$w = \sum_{j=0}^{\infty} \varepsilon^j\, w_j. \tag{8.56}$$

Substituting (8.56) into (8.53) and equating to zero the coefficients at different powers of ε, we obtain the equation

$$\begin{aligned}
u &= w + \varepsilon w_x + \varepsilon^2\, w^2 \\
&= \sum_{j=0}^{\infty} \varepsilon^j\, w_j + \varepsilon \sum_{j=0}^{\infty} \varepsilon^j\, w_{j,x} + \varepsilon^2 \left(\sum_{j=0}^{\infty} \varepsilon^j\, w_j\right)^2.
\end{aligned}$$

Now, collecting the coefficients of different powers of ε and equating them to zero, we obtain an infinite system of equations: (we present the first four equations here):

$$\begin{aligned}
\varepsilon^0 &: \quad u = w_0, \\
\varepsilon^1 &: \quad 0 = w_1 + w_{0,x}, \\
\varepsilon^2 &: \quad 0 = w_2 + w_{1,x} + w_0^2, \\
\varepsilon^3 &: \quad w_3 + w_{2x} + 2\, w_0\, w_1, \\
&\ldots\ldots\ldots\ldots\ldots\ldots
\end{aligned}$$

Solving the presented equations step by step, we get:

$$\begin{aligned}
w_1 &= -u_x, \\
w_2 &= -u^2 + u_{xx}, \\
w_3 &= 4\, u\, u_x - u_{xxx} \\
&\ldots\ldots\ldots
\end{aligned}$$

The countable number of conservation laws can be found by substituting (8.56) to the formula (8.55) and equating to zero the coefficients for the same powers of ε. The first three of them are as follows:

$$\begin{array}{ll} \varepsilon^0: & w_{0,t} + \left[w_{0,xx} - 3\,w_0^2\right]_x = 0, \\ \varepsilon^1: & w_{1,t} + \left[w_{1,xx} - 6\,w_0\,w_1\right]_x = 0, \\ \varepsilon^2: & w_{2,t} + \left[w_{2,xx} - 3\,\left(w_1^2 + 2\,w_0\,w_1\right)\right]_x = 0. \end{array}$$

Corollary 8.2. *The KdV equation possesses an infinite number of conservation laws.*

Nonlinear differential equations having an infinite number of conservation laws are generally in some sense equivalent to linear equations. In the following section, such relationship will be demonstrated for the KdV equation.

8.5.4 *Complete integrability of the KdV equation*

Lemma 8.4. *The equation (8.52) is invariant with respect to the Galilei transformations*

$$u' = u + \frac{1}{6}\,c, \quad t' = t, \quad x' = x - c\,t.$$

Proof. In new variables the partial derivatives are expressed as follows:

$$\frac{\partial}{\partial\,t} = \frac{\partial}{\partial\,t'} - c\,\frac{\partial}{\partial\,x'}, \qquad \frac{\partial}{\partial\,x} = \frac{\partial}{\partial\,x'}.$$

From this we have:

$$\begin{aligned} P(u) &= \left(\tfrac{\partial}{\partial\,t'} - c\,\tfrac{\partial}{\partial\,x'}\right)\,u' - 6\,\left(u' - \tfrac{1}{6}\,c\right)\,u'_{x'} + u'_{x'x'x'} \\ &= \tfrac{\partial\,u'}{\partial\,t'} - 6\,u'\,u'_{x'} + u'_{x'x'x'} = P(u') = 0. \end{aligned}$$

Let us consider the mKdV equation

$$K(v) = v_t - 6\,v^2\,v_x + v_{xxx} = 0. \tag{8.57}$$

Theorem 8.2. *Solutions of the equations (8.52) and (8.57) are connected by the transformation*

$$u = v_x + v^2. \tag{8.58}$$

Proof. Inserting into (8.52) the differential consequences of (8.58)

$$\begin{aligned} u_t &= v_{xt} + 2\, v\, v_t, \\ u_x &= v_{xx} + 2\, v\, v_x, \\ u_{xx} &= v_{xxx} + 2\, \left(v_x^2 + v\, v_{xx}\right), \\ u_{xxx} &= v_{xxxx} + 2\, \left(3\, v_x\, v_{xx} + v\, v_{xxx}\right), \end{aligned}$$

we get:

$$\begin{aligned} P(u) &= v_{xt} + 2\, v\, v_t - 6\,(v_x + v^2)(v_{xx} + 2\, v\, v_x) + v_{xxxx} + 2(3\, v_x\, v_{xx} + v\, v_{xxx}) \\ &= 2\, v\, \left(v_t - 6\, v^2\, v_x + v_{xxx}\right) \\ &= \left(2\, v + \frac{\partial}{\partial\, x}\right) K(v). \end{aligned}$$

The equation

$$\left(2\, v + \frac{\partial}{\partial\, x}\right) K(v) = 0 \tag{8.59}$$

can be linearized with the help of the Cole-Hopf–type ansatz

$$v = \frac{\Psi_x}{\Psi}.$$

Inserting this ansatz into the equation (8.58), we obtain:

$$u = v_x + v^2 = \frac{\Psi_{xx}}{\Psi},$$

or

$$-\, \Psi_{xx} + u(t,\, x)\, \Psi = 0. \tag{8.60}$$

This equation is connected, in turn, with the Schrödinger equation

$$\frac{\partial\, \tilde{\Psi}(\tau,\, x; t)}{\partial\, \tau} = -\frac{\partial}{\partial\, x}\, \tilde{\Psi}(\tau,\, x; t) + u(t,\, x)\, \tilde{\Psi}(\tau,\, x; t). \tag{8.61}$$

Let us note that in the above equation the variable t plays the role of the parameter.

We are looking for the solution of the equation (8.61) in the form

$$\tilde{\Psi}(\tau,\, x; t) = e^{-i\, \lambda\, \tau}\, \Psi(x; t).$$

and this way get the Sturm-Liouville problem

$$-\, \Psi_{xx} + u(t,\, x)\, \Psi = \lambda\, \Psi, \tag{8.62}$$

which coincides with Eq. (8.60) as $\lambda = 0$.

It turns out that the parameter λ can be inserted into the equation (8.60) using the particular form of the Galilean transformation, which leaves both the KdV equation (as it was shown earlier) and the Schrödinger equation invariant:

$$t' = t, \quad u' = u + \lambda, \quad \lambda = \frac{c}{6}, \quad x' = x - 6\,\lambda, \quad \Psi' = \Psi.$$

Performing such transformation, we get the equation

$$-\Psi'_{x'x'} + u'(t',\, x')\,\Psi' = \lambda\,\Psi'.$$

So, up to the change of variables, equation (8.60) coincides with (8.62). Therefore, the solution of the nonlinear KdV equation comes down to solving the so-called *inverse scattering problem* [Dodd *et al.* (1984)] for the equation (8.62). The only problem is that $\lambda = \lambda(t)$. The dependence on the parameter t, however, in this particular case is not significant, as it turns out that this problem is *isospectral.* In other words, the spectral parameter λ does not depend on t, so instead of $u(t,\, x)$ we can use in the equation (8.62) the initial condition $u(0,\, x) = \varphi(x)$. Thus, for a wide class of initial data by solving the Sturm-Liouville problem

$$-\,\Psi_{xx} + \varphi(x)\,\Psi = \lambda\,\Psi, \tag{8.63}$$

we get the solution of the Cauchy problem

$$u_t - 6\,u\,u_x + u_{xxx} = 0, \qquad u(0, x) = \varphi(x). \tag{8.64}$$

Let us show that $\lambda_t = 0$. Differentiating the expression

$$u = \frac{\Psi_{xx}}{\Psi} + \lambda(t),$$

which is equivalent to the equation (8.62), over x we get:

$$\Psi^4\,\dot{\lambda}(t) = \Psi^2\,\frac{\partial}{\partial\,x}\,[\Psi_x\,M - M_x\,\Psi] + \Psi^4\,P(u), \tag{8.65}$$

where

$$M = \Psi_t - 2(u + 2\,\lambda)\,\Psi_x + u_x\,\Psi.$$

Derivation of the equality (8.65) requires a lot of computational skill. It is easier to check the validity of this formula using the *Mathematica* commands as shown below.

Cell 8.1.

```
In[1]:
    u = D[Ψ[x, t], x, x]/Ψ[x, t] + λ[t];
In[2]:
    ut = Simplify[D[u, t]];
In[3]:
    ux = Simplify[D[u, x]];
In[4]:
    u3x = Simplify[D[u, x, x, x]];
In[5]:
    Vyr1 = Simplify[ut - 6uux + u3x];
In[6]:
    Vyr2 = Simplify[ Vyr1 Ψ[x, t]^4];
    M = Simplify[D[Ψ[x, t], t] - 2(u + 2λ[t])D[Ψ[x, t], x] + D[u, x]Ψ[x, t]];
    Vyr3 = Simplify[D[D[Ψ[x, t], x]M - D[M, x]Ψ[x, t], x]];
    Vyr4 = Simplify[Vyr2 - Ψ[x, t]^4λ'[t] + Vyr3Ψ[x, t]^2]
Out[9]:
    0
```

Assuming that Ψ belongs to the space of square integrable functions $L^2(\mathbb{R})$ and integrating the equality (8.65), we obtain:

$$\dot{\lambda}(t) \int_{-\infty}^{\infty} \Psi^2 \, dx = [\Psi_x \, M - M_x \, \Psi] \, |_{-\infty}^{\infty} = 0.$$

8.5.5 *Numerical scheme for the KdV equation and its implementation for solving the Cauchy problem*

We will now construct a numerical scheme for the equation (8.41), using the three-point template. Let us use the notation $u(t_k, x_j) = U_{k,j}$, $t_k = \tau(k-1)$, $x_j = a + (j-1)h$, $\tau > 0$, $h = (b-a)/K$, $x_1 = a$, $x_{K+1} = b$, $k, j \in N$ and consider the scheme [Zabusky and Kruskal (1965)]

$$\begin{aligned} \frac{U_{k+1,j} - U_{k-1,j}}{2\tau} &+ 12\frac{U_{k,j-1} + U_{k,j} + U_{k,j+1}}{3} \cdot \frac{U_{k,j+1} - U_{k,j-1}}{2h} \\ &+ \frac{U_{k,j+2} - 2U_{k,j+1} + 2U_{k,j-1} - U_{k,j-2}}{2h^3} = 0. \end{aligned} \tag{8.66}$$

In the scheme used, the values of $U_{k+1,j}$, corresponding to the highest time layer, can be calculated from the formulas

$$\begin{aligned} U_{k+1,j} = U_{k-1,j} &- \frac{4\tau}{h} \left(U_{k,j-1} + U_{k,j} + U_{k,j+1}\right) \left(U_{k,j+1} - U_{k,j-1}\right) \\ &- \frac{\tau}{h^3} \left[U_{k,j+2} - 2U_{k,j+1} + 2U_{k,j-1} - U_{k,j-2}\right], \quad 3 \le j \le K-2. \end{aligned}$$

For soliton solutions, the boundary conditions can be defined as follows:

$$U_{k+1,j} = U_{k,j}, \qquad j = 1,\, 2,\, K-1,\, K.$$

As the initial data, we use the one-soliton solution

$$U_{1,j} = u(x_j) \qquad U_{2,j} = u(x_j - s\tau).$$

Inserting to the linearized version of the numerical scheme the function of the form

$$U_{k,j} = q^k e^{i\theta j}$$

we get the equation

$$q = \frac{1}{q} - \frac{4\tau}{h} U_0 (e^{i\theta} - e^{-i\theta}) - \frac{\tau}{h^3}(e^{2i\theta} - 2e^{i\theta} + 2e^{-i\theta} - e^{-2i\theta}),$$

which can also be presented in more convenient form

$$q^2 + q\left(\frac{8i\tau}{h} U_0 \sin\theta - \frac{8i\tau}{h^3} \sin\theta \sin^2\frac{\theta}{2}\right) - 1 = 0. \qquad (8.67)$$

If the roots of the equation (8.67) were real, then, according to the Vieta formulas, their product would be equal to one, which would imply the instability of the scheme. So the roots of the equation cannot be real, which means that the discriminant of the equation (8.67) should be a positive number. Hence

$$\Delta = -\left(\frac{8\tau}{h} U_0 \sin\theta - \frac{8\tau}{h^3} \sin\theta \sin^2\frac{\theta}{2}\right)^2 + 4 \geq 0 \Rightarrow$$

$$\left|\frac{4\tau}{h} U_0 \sin\theta - \frac{4\tau}{h^3} \sin\theta \sin^2\frac{\theta}{2}\right| \leq 1.$$

The extrema of the function $\sin\theta \sin^2\theta/2$ attain the values $x_{\max} = 2\pi/3 + 2\pi n$, $x_{\min} = 4\pi/3 + 2\pi m$, and this implies the inequality

$$\left|\frac{4\tau}{h} U_0 \frac{\sqrt{3}}{2} - \frac{4\tau}{h^3}\left(\frac{\sqrt{3}}{2}\right)^3\right| \leq 1.$$

For $U_0 = 0$ a necessary condition of stability takes the form

$$\frac{\tau}{h^3} \leq \frac{2}{3\sqrt{3}}.$$

This condition is taken into account in the numerical simulations. Their implementation in the *Mathematica* package is presented below.

Cell 8.2.

In[1]:
```
uex[t_, x_, a_] = (a/4) Sech[Sqrt[a] (x - a t)/2]^2;
```
In[2]:
```
D[uex[t, x, a], t] + 12 uex[t, x, a] D[uex[t, x, a], x] + D[uex[t, x, a], x, 3] //
Simplify
```
Out[2]:
```
0
```
In[3]:
```
iniV[x_] = If[x <= 0, uex[0, x + 20, 0.5], uex[0, x - 30, 0.25] + uex[0, 0 + 20,
0.5] - uex[0, 0 - 30, 0.25]];
Lb = -60; Le = 100;
```
In[5]:
```
Plot[iniV[x], {x, Lb, Le}, PlotRange → All, BaseStyle → {Black, FontSize →
14}, AxesStyle → Directive[Black], AxesLabel → {x, u}] ;
```
Out[5]:

In[6]:
```
evolut = NDSolve[ {V_t[t, x] + 12 V[t, x] V_x [t, x] + V_{x,x,x} [t, x] == 0, V[0, x]
== u[x], V[t, a] == V[t, b]}, {V}, {t, 0, 400}, {x, a, b}, MaxStepSize → 0.25];
```
In[7]:
```
Plot3D[ Evaluate[First[{V[t, x]} /. evolut]], {x, Lb, Le}, {t, 0, 400}, PlotRange
→ All, PlotPoints → 150, AxesLabel → {x, t, u}, BaseStyle → {Black, Bold, FontSize
→ 14}, BoxStyle → Directive[Black], Mesh → None, ImageSize → Medium];
```
Out[6]:

In[6]:
```
ContourPlot[Evaluate[V[t, x] /. evolut], {x, Lb, Le}, {t, 0, 400}, PlotPoints →
150, ContourShading → False, PlotRange → All, Contours → Function[{min, max},
Range[0.01, 0.13, 0.01]], FrameLabel → {x, t}, BaseStyle→ {14, Black}, FrameStyle
→ Directive[Black]];
```
Out[6]:

Note that in the algorithm the periodic boundary conditions are used. Figure 9.1 shows the results of the numerical simulation of the initial value problem (2.26)–(8.41). Numerical simulations shows that the soliton evolves without changing its shape. This circumstance serves as an argument in favor of the stability of the solitary wave solution.

If to solve numerically the equation (8.41) using as initial data any smooth disturbance $\varphi(x) \in \mathcal{C}^k(\mathbb{R})$ with a compact support, then a finite number of solitons moving to the right are formed in the course of evolution, and in the long range they are also arranged one after another in the order of decreasing amplitude, see Fig. 8.12 (solution to this problem is presented in the file PM2_12.nb).

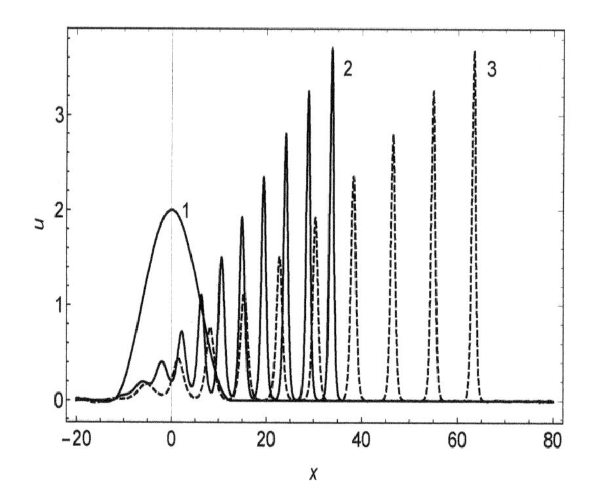

Fig. 8.12 Numerical solution of the initial value problem for the equation (8.41). The function $u(0,\, x) = \left[1 + \cos \frac{x\,\pi}{10}\right] * H(x + 10) * H(10 - x)$ is used as Cauchy data: 1 — $t = 0$; 2 — $t = 30$; 3 — $t = 150$.

In conclusion, we would like to note, that solitons are a special type of nonlinear waves that evolve without changing their shape. They behave like elastic balls, restoring their shape after the mutual interaction. The "elasticity" of the solitons' interaction is often attributed to the facts that the KdV equation possesses an infinite number of conservation laws and is equivalent in some sense to the linear Schrödinger equation.

Chapter 9

Techniques and methods for obtaining exact solutions of nonlinear evolutionary equations

9.1 Hirota method and multi-soliton solutions to the Korteweg-de Vries equation

It turns out that apart from the bell-shaped soliton solution (8.46), the KdV equation support the solutions with two, three, $\ldots N$ humps, called *multisolitons*. Such solutions can be constructed analytically, using the *Hirota method* described below.

Let us start from deriving the *bilinear equation*

$$\hat{F}(D_X, D_T) f \cdot f = 0, \tag{9.1}$$

where the Hirota operators D_X, D_T and their superpositions are defined as follows:

$$D_T^n D_X^m f \cdot g = \left(\frac{\partial}{\partial t} - \frac{\partial}{\partial t'}\right)^n \left(\frac{\partial}{\partial x} - \frac{\partial}{\partial x'}\right)^m f(x,t) \cdot g(x',t') \mid_{x=x',t=t'} . \tag{9.2}$$

Let us note that

$$D_T^n D_X^m f \cdot 1 = \frac{\partial^{n+m}}{\partial t^n \partial x^m} f(t,x),$$

that is, the operators act on the product $f \cdot 1$ like the usual differentiations of f. It appears directly from (9.2) the validity of the following formulas:

$$\begin{aligned}
&D_T^n D_X^m 1 \cdot f = (-1)^{n+m} \tfrac{\partial^{n+m}}{\partial t^n \partial x^m} f(t,x), \\
&D_T f \cdot g = f_t g - f g_t, \\
&D_X f \cdot g = f_x g - f g_x, \\
&D_T D_X f \cdot f = 2 f f_{xt} - 2 f_x f_t, \\
&D_X^4 f \cdot f = 2 f f_{xxxx} - 8 f_{xxx} f_x + 6 (f_{xx})^2,
\end{aligned} \tag{9.3}$$

$$D_T^n D_X^m e^{\omega_1 t + k_1 x} \cdot e^{\omega_2 t + k_2 x} = (\omega_1 - \omega_2)^n (k_1 - k_2)^m e^{(\omega_1+\omega_2)t + (k_1+k_2)x}. \tag{9.4}$$

The above properties enable to state that in case of the polynomial operator $\hat{F} = \sum_{\mu,\,\nu \in \mathbb{N}} C_{\mu\,\nu} D_T^{\mu} D_X^{\nu}$, the following formulas hold true:

$$\hat{F}e^{\theta_1} \cdot e^{\theta_2} = F(\omega_1 - \omega_2,\, k_1 - k_2)\, e^{\theta_1 + \theta_2}, \quad \theta_i = \omega_i t + k_i x + c_i, \quad i = 1, 2, \tag{9.5}$$

$$F(\omega_1 - \omega_2,\, k_1 - k_2) = \sum_{\mu,\,\nu \in \mathbb{N}} C_{\mu\,\nu}(\omega_1 - \omega_2)^{\mu}\,(k_1 - k_2)^{\nu}. \tag{9.6}$$

Let's apply the Hirota method to the equation (8.41). If we insert the substitution $u = w_x$ into the KdV equation, integrate the obtained equation with respect to x and equate the integration constant to zero, then we obtain the equation of the form

$$w_t + 6(w_x)^2 + w_{xxx} = 0.$$

The Cole-Hopf ansatz $w = \partial_x \log f = f_x/f$ reduces the above equation to the bilinear form

$$f f_{xt} - f_x f_t + 3(f_{xx})^2 - 4 f_x f_{xxx} + f f_{xxxx} = 0. \tag{9.7}$$

Using the formulas (9.2), (9.3), we can rewrite the equation (9.7) in the form

$$\hat{F} f \cdot f \equiv (D_X D_T + D_X^4) f \cdot f = 0. \tag{9.8}$$

We present the solution of (9.8) as formal series with respect to the parameter ε:

$$f = 1 + \varepsilon f_1 + \varepsilon^2 f_2 + \varepsilon^3 f_3 + \ldots \tag{9.9}$$

(we call this series formal, because at the end of the computations without loss of generality we can put the parameter ε to be equal to one). Inserting the ansatz (9.9) into the equation (9.8), and next equating to zero the coefficients at different powers of ε, we get the following system:

$$\varepsilon: \quad \hat{F} f_1 \cdot 1 = \hat{L} f_1 = 0, \quad \hat{L} = \frac{\partial^2}{\partial x \partial t} + \frac{\partial^4}{\partial x^4}; \tag{9.10}$$

$$\varepsilon^2: \quad 2\hat{L} f_2 = -\hat{F} f_1 \cdot f_1; \tag{9.11}$$

$$\varepsilon^3: \quad 2\hat{L} f_3 = -\hat{F}\,(f_2 \cdot f_1 + f_1 \cdot f_2); \tag{9.12}$$

$$\varepsilon^4: \quad 2\hat{L} f_4 = -\hat{F}\,(f_3 \cdot f_1 + f_2 \cdot f_2 + f_1 \cdot f_3); \tag{9.13}$$

....

$$\varepsilon^n: \quad 2\hat{L} f_n = -\hat{F}(f_{n-1} \cdot f_1 + f_{n-2} \cdot f_2 + \cdots + f_2 \cdot f_{n-2} + f_1 \cdot f_{n-1}); \tag{9.14}$$

....

The linear equation (9.10) will be satisfied by the exponential function $f_1 = e^{\theta_1}$, $\theta_1 = \omega_1 t + k_1 x + c_1$, if $\omega_1 = -k_1^3$. From (9.5), we immediately conclude that the equation (9.11) attains the form $\hat{L}\, f_2 = 0$. If we assume that $f_{2+\nu} = 0$, $\nu = 0, 1, 2, \ldots$, then this equation and all the subsequent equations will be satisfied. Thus, the series (9.9) can be truncated and the solution attains the form

$$f = 1 + e^{\theta_1} = 1 + e^{\omega_1 t + k_1 x + c_1}, \qquad \omega_1 = -k_1^3.$$

Note that we get rid of the dependence on the parameter ε using the fact that the argument of the exponential function contains arbitrary constant. If $c_1 = 0$, then the solution takes the form

$$u = \partial_x \left[\frac{f_x}{f}\right] = \frac{k^2}{4} \cosh^{-2}\left[\frac{k}{2}\left(x - k^2\, t\right)\right].$$

Making the substitution $s = k^2$, we get the formula (2.26).

Now let us note that, in accordance with the superposition principle, the function

$$\begin{aligned} f_1 &= \exp(\theta_1) + \exp(\theta_2), \\ \theta_i &= \omega_i t + k_i x + c_i = k_i(x - k_i^2\, t) + c_i, \quad i = 1,\, 2 \end{aligned} \tag{9.15}$$

will also satisfy the equation (9.10). Inserting (9.15) into the equation (9.11), we get:

$$\hat{L} f_2 = -\frac{1}{2}\hat{F}\left(e^{\theta_1}\cdot e^{\theta_1} + 2e^{\theta_1}\cdot e^{\theta_2} + e^{\theta_2}\cdot e^{\theta_2}\right) = -F(\omega_1 - \omega_2,\, k_1 - k_2)\, e^{\theta_1}\cdot e^{\theta_2}.$$

Making the substitution $f_2 = A\, e^{\theta_1}\cdot e^{\theta_2}$, we obtain the formula

$$A\, F(\omega_1 + \omega_2,\, k_1 + k_2) = -F(\omega_1 - \omega_2,\, k_1 - k_2), \quad \omega_i = -k_i^2, \quad i = 1,\, 2,$$

from which appears that

$$A = \left(\frac{k_1 - k_2}{k_1 + k_2}\right)^2.$$

Using the above formulas, we can rewrite the equation (9.12) as

$$\hat{L} f_3 = -A\hat{F}\left(e^{\theta_1 + \theta_2}\cdot e^{\theta_1 + \theta_2}\right) \equiv 0.$$

So the substitution $f_3 = 0$ turns this equation into identity. It also seen that $f_{3+k} = 0$, $k = 1, 2, \ldots$ resets all subsequent equations, and hence the function

$$f = 1 + \exp(\theta_1) + \exp(\theta_2) + A\exp(\theta_1 + \theta_2) \tag{9.16}$$

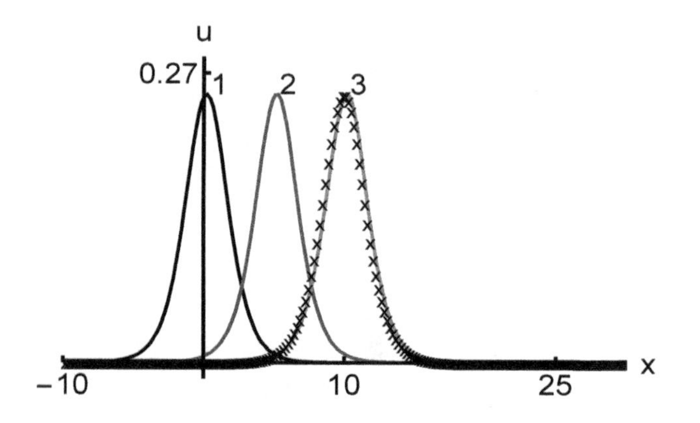

Fig. 9.1 Temporal evolution of the soliton pulse obtained with the help of the formula (2.26) for $s = 1$; 1: $t = 0$; 2: $t = 20$; 3: $t = 40$. The crosses in the graph corresponding to $t = 40$ denote the result of the numerical simulation carried out by means of the scheme (8.66). The function (2.26) was used as initial data. The comparison of the numerical solution with the analytical solution shows a very high accuracy of the numerical scheme.

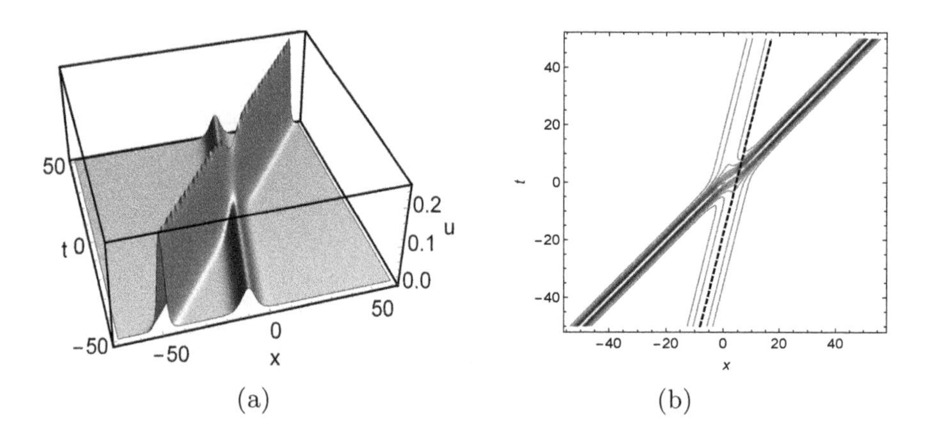

Fig. 9.2 Panel (a): bisoliton solution $u = \partial_{xx} \log f$, where f is described by (9.16). Following values of the parameters are used: $k_1 = 1$, $k_2 = 0.5$; panel (b): projection of the solution on the (t, x) plane.

is a solution to the equation (9.8). The function (9.16) corresponds to the *bi-soliton solution* of the equation (8.41). Its evolution is shown in Fig. 9.2.

Interpretation of the solution $u(t, x) = \partial_{xx} \log f(t, x)$ in *Mathematica* package for $f(t, x)$ described by the formula (9.16) is presented in the file PM2_12.nb. In this file, the analytical solution of the problem is compared with the results of the numerical simulation in which a pair of one-soliton solutions separated from each other by a sufficiently large distance are used as initial data.

Using the Hirota method, it is possible to construct analytical formula describing the n-soliton solution of the KdV equation. The solution behaves in such a way that, regardless of the initial configuration, in the process of evolution, the soliton with the largest amplitude is located on the right at the head of the train, followed by the next largest, etc.

Problem.

(1) Show that the function

$$u(t,\, x) = \frac{\partial^2}{\partial\, x^2} \log f(t,\, x), \tag{9.17}$$

where

$$\begin{aligned} f(t,\, x) = 1 + \varepsilon \sum_{k=1}^{3} \exp\theta_k + \varepsilon^2 \sum_{i<j} A_{i\,j} \exp[\theta_i + \theta_j] \\ + \varepsilon^3 A_{1\,2} A_{2\,3} A_{1\,3} \exp[\theta_1 + \theta_2 + \theta_3], \end{aligned} \tag{9.18}$$

$$\theta_i = k_i\, x - k_i^3\, t + c_i, \qquad k_i \neq k_j \quad \text{when} \quad i \neq j \tag{9.19}$$

with an appropriate choice of the coefficients $A_{i\,j}$ satisfies the KdV equation (8.41).

Hint.

- Unknown coefficients $A_{i\,j}$ are easiest to determine by solving the system (9.10)–(9.14). It is preferable to do this in the *Mathematica* package.
- Using the fact that the functions θ_i depend additively on arbitrary constants c_i, it is possible to set $\varepsilon = 1$ in the final formula.

Selecting appropriately the parameters k_i and c_i, represent graphically the process of three-soliton interaction.

(2) In Chapter 3 of the book [Scott (2003)], the following recipe for constructing N-soliton solution of the equation

$$u_t - 6\, u\, u_x + u_{xxx} = 0 \tag{9.20}$$

is given.

- The author uses the ansatz

$$u = -2\, \frac{\partial^2}{\partial\, x^2} f(t,\, x) \tag{9.21}$$

and, taking into account the fact that the function $w = -2f_x/f$ satisfies the equation

$$w_t - 3(w_x)^2 + w_{xxx} = 0,$$

falls into the standard bilinear equation coinciding with (9.7). Note that the ansatz $\tilde{u} = -u/2$ reduces (9.20) to the canonical equation (8.41).

- In Appendix B1 of the Scott's book, the following formula is given that determines the N-soliton solution of the equation (9.20): u is determined by means of the formula (9.21) where $f = \det M$, $M = [m_{i\,j}]$ is $N \times N$ matrix, and
$$m_{i\,j} = \frac{2\sqrt{\kappa_i\,\kappa_j}}{\kappa_i + \kappa_j}\exp\left[\frac{1}{2}(\kappa_i + \kappa_j)x - \frac{1}{2}(\kappa_i^3 + \kappa_j^3)t + c_{i\,j}\right] + d_{i\,j}, \tag{9.22}$$
where $\kappa_i,\ \kappa_j,\ c_{i\,j},\ d_{i\,j}$ are constant values and $\kappa_i \neq \kappa_j$ when $i \neq j$.

Using the formulas (9.21), (9.22) find the four-soliton solution. Selecting appropriately the parameters κ_i, $c_{i\,j}$, $d_{i\,j}$, $i,\ j = 1,\ 2,\ 3,\ 4$, represent graphically the process of four-solitons' interaction.

9.2 Application of the Hirota method and its modifications for solving the nonlinear evolutionary equations

9.2.1 *Traveling wave solution to the Burgers-like equation describing active media*

Let us consider the following equation:

$$u_t + u\,u_x + B\,u_x = \kappa\,u_{xx} + \lambda\,u\,(u - S)\,(u + Q). \tag{9.23}$$

Using the Hirota ansatz

$$u(t,\,x) = \frac{f_x}{f}, \quad f = f(t,\,x) = 1 + \epsilon e^z, \quad z = x - c\,t, \tag{9.24}$$

it is possible to find solutions describing localized wave patterns.

We insert the ansatz (9.24) into the equation (9.23) and, after reducing all the terms to the common denominator, obtain a very cumbersome algebraic expression, whose numerator is interpreted as a polynomial with respect to the powers of ϵ. Equating to zero the expressions for various powers of ϵ, we obtain the following system of algebraic equations:

$$(1 + Q)\,\lambda\,(1 - S) = 0, \tag{9.25}$$

$$-B + c + \kappa - Q\,S\,\lambda = 0, \tag{9.26}$$

$$-c + 1 + \kappa + 2\,Q\,S\,\lambda + B + (S - Q)\lambda = 0. \tag{9.27}$$

Let us consider the case $S = 1$ for which the equation (9.25) turns into identity. From (9.26), we express c:

$$c = \lambda Q + B - \kappa. \tag{9.28}$$

Inserting (9.28) into (9.27), we get the following expression for κ:

$$\kappa = -\frac{1+\lambda}{2}. \tag{9.29}$$

Substituting this formula into (9.28), we obtain

$$c = \frac{2(\lambda Q + B) - (1+\lambda)}{2}. \tag{9.30}$$

It can be shown that with such constraints (i.e., when $S = 1$ and the conditions (9.29), (9.30) are met), formula (9.24) does determine the solution of the initial equation. The easiest way to check it is to use the *Mathematica* package.

The case of $Q = -1$ is analyzed in the similar way. Instead of doing it "by hand", let us show how this problem is solved in the *Mathematica* package.

Cell 9.1.

In[1]:
```
f=1+ε Exp[z];
```
In[2]:
```
u=D[f,z]/f;
```
In[3]:
```
eq1=Together[-c1 D[u, z]+B u D[u, z]-κ D[u, {z, 2}]-λu(u-S)(u+Q) ];
```
In[4]:
```
eq2=Numerator[eq1//.z → 0];
```
In[5]:
```
CoefficientList[eq2,ε]
```
Out[5]:
$\{0, -c1 - \kappa + Q\,S\,\lambda, B - c1 + \kappa - Q\,\lambda + S\,\lambda + 2 + Q\,S\,\lambda, -\lambda - Q\lambda + S\lambda + QS\lambda\}$

In[6]:
```
aux1=Table[eq3[[i]],{i, 2, 4}];
```
In[7]:
```
Solve[aux1,{Q, c1, κ}]
```
Out[7]:
$\{\{c1 \to \frac{1}{2}(B + \lambda - 2S\lambda), \kappa \to \frac{1}{2}(-B - \lambda), Q \to -1\}\}$

9.2.2 *Traveling wave solutions supported by the hyperbolic modification of the Burgers equation*

Formally the Burgers equation can be derived from the conservation law

$$\frac{\partial}{\partial t}\int_{\Omega} u(t,\,x)\,d\,x=-\int_{\partial\,\Omega} G\,d\,\sigma,$$

where $G=u^2/2-\nu\,u_x$, under the condition that the domain Ω area can be shrinked to a point. However, if the modeled substance is multi-component, or, in general, has an internal structure different from that of the atomic structure (as is the case with the description of multi-phase mixtures, sands, soils, etc.), then this cannot be done and the structure must be taken into account. The simplest way of taking into account the presence of structure is to enter an inflow delay:

$$\frac{\partial}{\partial t}\,u(t+\tau,\,x)=-\frac{\partial}{\partial x}G(t,\,x).$$

Expanding the function $u(t+\tau,\,x)$ into the Maclaurin series over the parameter τ, and keeping only two terms of the decomposition, we get the following equation:

$$\tau\,u_{tt}+u_t+u\,u_x=\nu\,u_{xx}+O(\tau^2) \tag{9.31}$$

(we assume that $\tau\ll 1$). Note that in the above formula we can use the function G, which takes into account the nonlinearity of the transport coefficient ν, as well as the presence of mass sources and the inhomogeneity. So in general we are dealing with the equation

$$\tau\,u_{tt}+u_t+u\,u_x=\kappa\left[u^n\,u_x\right]_x+f(x,\,u,\,u_x). \tag{9.32}$$

Equation (9.32) with some specific functions $f(x,\,u,\,u_x)$ has a number of physically meaningful solutions of the traveling wave type (among others, the kinks, soliton-like and periodic solutions). Below we consider the equation with a polynomial source term:

$$\tau u_{tt}+Auu_x+Bu_t+H\,u_x-\kappa u_{xx}=f(u)=\lambda\,(u-m_1)(u-m_2)(u-m_3). \tag{9.33}$$

We are looking for the TW solutions of the form

$$u\;=\;v(t,\,x)\;=\;U(\xi),\qquad \xi=x+\mu\,t. \tag{9.34}$$

Inserting the ansatz (9.34) into the equation (9.33), we get the following ODE:

$$(\tau\,\mu^2-\kappa)\,U''+(H+B\,\mu)\,U'+A\,U\,U'=\lambda\,(U-m_1)(U-m_2)(U-m_3). \tag{9.35}$$

This equation, in general (i.e., with arbitrary values of the parameters), is not integrable. In order to extract a family of solutions which can be described in terms of analytical functions, we will present the solution in the Cole-Hopf form:

$$U(\xi) = \frac{\Psi'(\xi)}{\Psi(\xi)}. \tag{9.36}$$

This leads us to the equation

$$\begin{aligned}
&\Psi(\xi)^2 \left[\lambda\, m_1 m_2 m_3 \Psi(\xi) - \lambda\,(m_2 m_3 + m_1 m_2 + m_1 m_3)\Psi'(\xi)\right.\\
&\left.+(H + B\mu)\Psi''(\xi) + (\mu^2\tau - \kappa)\Psi'''(\xi)\right] + \Psi(\xi)\,\Psi'(\xi)\left[\Psi''(\xi)\times\right.\\
&\left.\left(A + 3\kappa - 3\mu^2\tau\right) - \Psi'(\xi)\,(H\ -\lambda\,(m_1 + m_2 + m_3) + B\mu)\right]\\
&+ \left[\Psi'(\xi)\right]^3 \left(A - \lambda - 2\kappa + 2\mu^2\tau\right) = 0.
\end{aligned} \tag{9.37}$$

It is easy to check that the following statement holds true:

Lemma 9.1. *The ansatz (9.36) leads to the linear equation*

$$\Psi''' - \sum_{i=1}^{3} m_i \Psi'' + \sum_{i \neq j} m_i\, m_j \Psi' - m_1\, m_2\, m_3\, \Psi = 0, \tag{9.38}$$

provided that $\lambda \neq 0$*, and the following conditions are fulfilled:*

$$A = -3(\kappa - \mu^2\tau), \tag{9.39}$$

$$H + B\,\mu^2 = \lambda \sum_{i=1}^{3} m_i, \tag{9.40}$$

$$\kappa - \mu^2\tau = \lambda. \tag{9.41}$$

Inserting the ansatz $\Psi = e^{\sigma\, z}$ into the equation (9.38), we get the characteristic equation

$$\sigma^3 - (m_1 + m_2 + m_3)\,\sigma^2 + (m_2\, m_3 + m_1\, m_2 + m_1\, m_3)\,\sigma - m_1\, m_2\, m_3 = 0. \tag{9.42}$$

The roots of the above algebraic equation occur to coincide with the numbers m_1, m_2 and m_3. Below we analyze the possible cases.

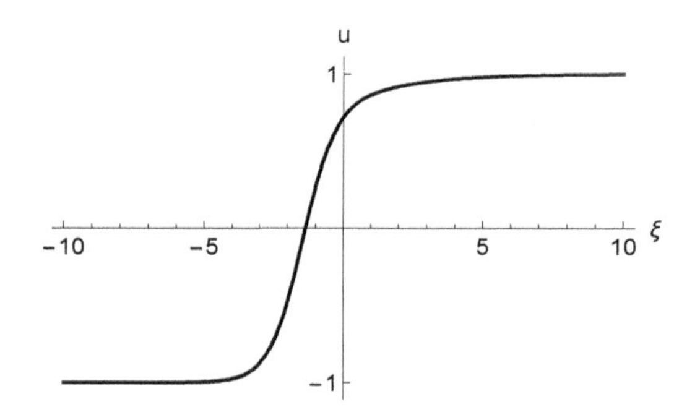

Fig. 9.3 Kink-like solution described by the formula (9.43).

(1) If $m_1 \neq m_2 \neq m_3$, then the solution is as follows

$$\Psi(\xi) = e^{m_1\,\xi} c_1 + e^{m_2\,\xi} c_2 + e^{m_3\,\xi} c_3.$$

If in addition $c_1 \neq 0$, then, dividing the numerator and denominator of (9.36) by this parameter, we get the solution

$$u(t,\,x) = \frac{m_1 e^{m_1 \xi} + c_2 m_2 e^{m_2 \xi} + c_3 m_3 e^{m_3 \xi}}{e^{m_1 \xi} + c_2 e^{m_2 \xi} + c_3 e^{m_3 \xi}}, \qquad \xi = x + \mu t. \tag{9.43}$$

If c_2 and c_3 are both positive, then the formula (9.43) describes a kink-like solution depicted in Fig. 9.3.

(2) If $m_1 \neq m_2 = m_3$, then

$$\Psi(\xi) = e^{m_1 \xi} + e^{m_2 \xi}(c_2 + c_3 \xi).$$

Inserting this function into (9.36), we get the following solution:

$$u(t,\,x) = \frac{m_1 e^{m_1 \xi} + c_3 e^{m_2 \xi}(c_2\, m_2 + c_3 + m_2\, c_3 \xi)}{e^{m_1 \xi} + e^{m_2 \xi}(c_2 + c_3 \xi)}. \tag{9.44}$$

The solution (9.44) tends to either m_1 or m_2 as $\xi \to \pm\infty$. In order that the solution be non-singular, the following condition should be met

$$e^{(m_1 - m_2)\xi} > -(c_2 + c_3 \xi). \tag{9.45}$$

It is possible when $m_1 - m_2 > 0$ and simultaneously $c_3 < 0$. There also exists another option when $m_2 - m_1 > 0$, and $c_3 > 0$. Figure 9.4 illustrates the first case. From the analysis of Fig. 9.4 appears that the condition (9.45) is fulfilled when $-c_2 < c_{cr}$, where

$$c_{cr} = -A\left[c_3 + e^{m_1 - m_2}\right], \qquad A = \frac{1}{m_1 - m_2} \log \frac{c_3}{m_2 - m_1}.$$

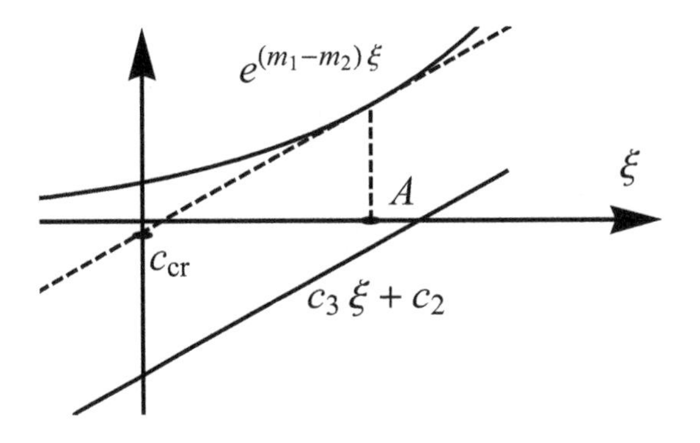

Fig. 9.4 Case 2: graph of the nonsingular solution. Continuous line corresponds to the graphs of the function $f_1(\xi) = e^{(m_1-m_2)\xi}$ and $f_2(\xi) = -(c_2 + c_3\xi)$. Dashed line defines the straight line tangent to $f_1(\xi)$, parallel to $f_2(\xi)$.

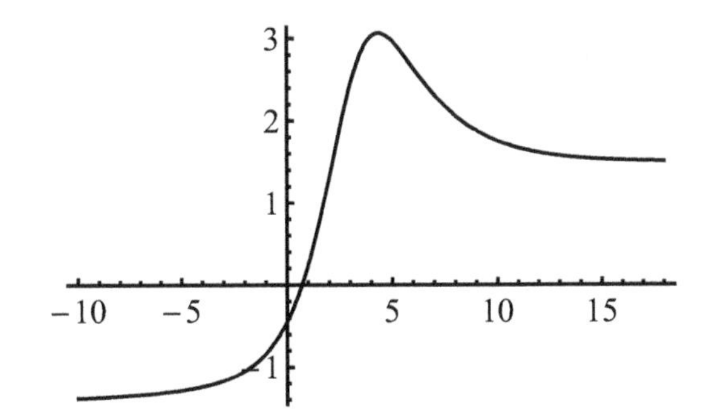

Fig. 9.5 Example of the traveling wave solution described by the formula (9.44).

(3) At $m_1 = m_2 = m_3 = m$

$$\Psi(\xi) = e^{m\xi}\left[c_3 + \xi\,(c_2 + \xi)\right].$$

Employing (9.36), we get the solution

$$u(t,\,x) = m + \frac{c_2 + 2\xi}{c_3 + \xi(c_2 + \xi)}. \tag{9.46}$$

The formula (9.46) defines the soliton-like solution with "heavy" tail (Fig. 9.6), under the condition that the following inequality takes place $c_2^2 - 4\,c_3 < 0$.

(4) When $m_1,\ m_2 = \alpha \pm \beta\,i$, while m_3, is real

$$u(\xi) = \frac{c_3 m_3 e^{m_3\xi} + 2e^{\alpha\xi}\left[\alpha\cos(\beta\xi) - \beta\sin(\beta\xi)\right]}{c_3 e^{m_3\xi} + 2\,e^{\alpha\xi}\cos(\beta\xi)}. \tag{9.47}$$

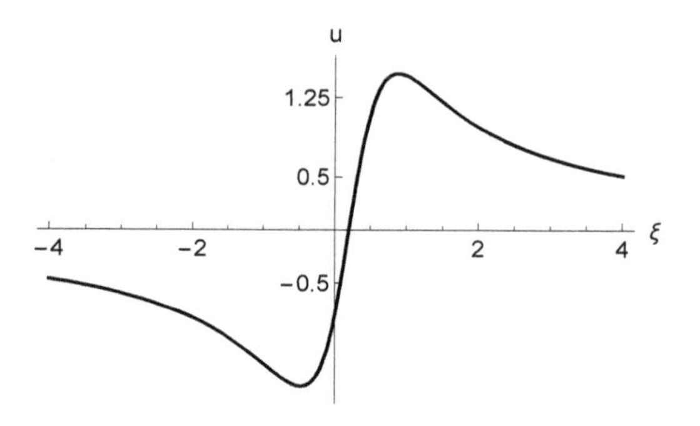

Fig. 9.6 Example of the solution described by the formula (9.46).

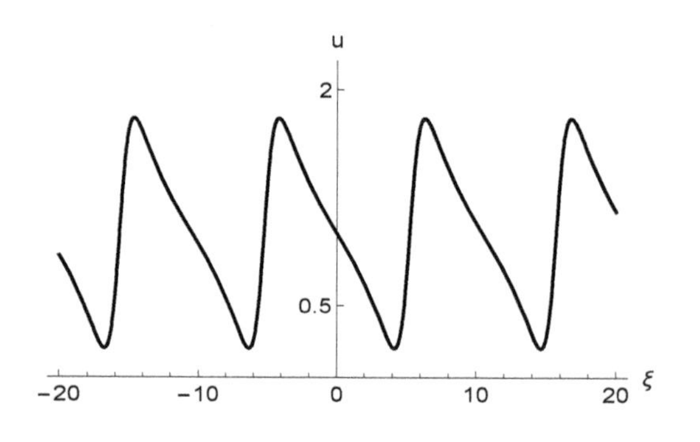

Fig. 9.7 Periodic solution of the equation (9.33) described by the formula (9.47).

This solution is not singular provided that $m_3 = \alpha$ and $|c_3| > 2$. In this case we deal with the periodic solution (Fig. 9.7).

When $f(u)$ is different, then, using the following modification of the formula (9.36)

$$u(\xi) = \frac{\sum_{\mu=0}^{m} a_\mu exp(\mu\alpha\xi)}{\sum_{\nu=0}^{n} b_\nu exp(\nu\alpha\xi)}, \tag{9.48}$$

it is still possible to obtain various localized solution without posing extra conditions leading to the linearization. Such representation is effective e.g. in the case of the equation

$$\tau u_{tt} + u_t - \kappa u_{xx} = \sum_{\nu=0}^{4} \lambda_{\nu+1} u^{\nu/2}. \tag{9.49}$$

Using the *Mathematica* package, we can show that the function

$$u(t,x) = \left[\frac{1 - e^{(x+vt)}}{1 + e^{(x+vt)}}\right]^2 \tag{9.50}$$

satisfies the equation (9.49) provided that

$$\lambda_1 = \tfrac{1}{2}(\tau\, v^2 - 1), \qquad \lambda_2 = -v,$$

$$\lambda_3 = -2(\tau\, v^2 - 1), \quad \lambda_4 = v, \quad \lambda_5 = \tfrac{3}{2}(\tau\, v^2 - 1).$$

The solution (9.50) describes the evolution of the localized solitary wave of rarefaction moving with the velocity v from right to the left. Solution of this sort is presented in Fig. 9.8.

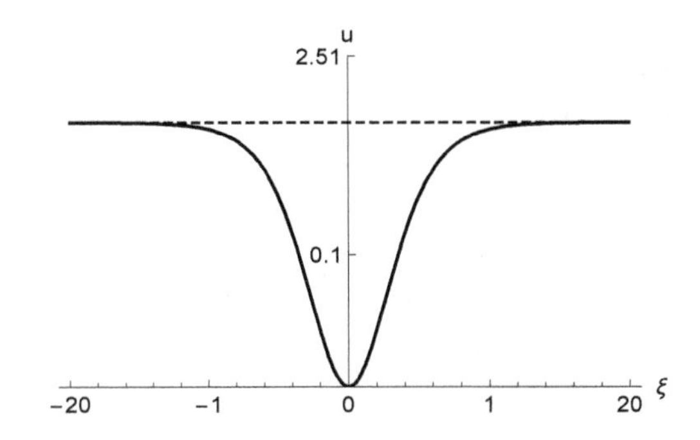

Fig. 9.8 "Dark" soliton described by the formula (9.50) at $t = 0$.

The code used in the *Mathematica* package to obtain conditions for the coefficients $\{\lambda_i\}_{i=1}^5$ is presented in cell 9.2.

Cell 9.2.

In[1]:=
Clear$[\varepsilon, v, \tau, \lambda_1, \lambda_2, \lambda_3, \lambda_4, \lambda_5]$;
In[2]:=
$f = \frac{1-\varepsilon Exp[z]}{1+\varepsilon Exp[z]}$; $u = f^2$;
In[4]:=
$eq1$ =Together$[(\tau v^2-1)D[u,\{z,2\}]+vD[u,\{z,1\}]-(\lambda_1+\lambda_2\, f+\lambda_3\, f^2+\lambda_4\, f^3+\lambda_5\, f^4)]$;
In[5]:=
$eq2$ =Numerator$[eq1//.z\to 0]$;
In[6]:=
$eq3$ =CoefficientList$[eq2,\varepsilon]$;
In[7]:=
Aux1=Table$[eq3[[i]] == 0, \{i = 1, 5\}]$;
In[8]:=
Aux2=Flatten[Solve[Aux1,$\{\lambda_1, \lambda_2, \lambda_3, \lambda_4, \lambda_5\}$]
Out[8]=
$\{\lambda_1 \to \frac{1}{2}(\tau\, v^2-1), \quad \lambda_2 \to -v, \quad \lambda_3 \to -2(\tau\, v^2-1), \quad \lambda_4 \to v, \quad \lambda_5 \to \frac{3}{2}(\tau\, v^2-1)\}$

```
(* VERIFICATION *)
In[9]:=
  v = 2.; τ = 0.5; ε = 1.; z = x − vt;
In[13]:=
  Verif=Simplify[eq2//.Aux2]
Out[8]=
  0.
```

Above, we've presented one of the simplest solutions to the equation (9.49) that can be found using the commands of the *Mathematica* package. More complex multiparameter families of solutions are presented in paper [Vladimirov *et al.* (2006)].

Problem.

(1) Show that the equation

$$\tau u_{tt} + u_t + u\,u_x = \frac{1}{\theta}(1 - \theta^2 u^p), \quad p > 0, \tag{9.51}$$

possesses a solution of the form

$$u(t,\,x) = U(z) = A\frac{e^z}{(1+e^z)^{s+1}}, \quad s > 0, \quad z = x + v\,t. \tag{9.52}$$

Express "free" parameters A and v as functions of the parameters included in the equation.

Hint. Parameters s and p can be determined by balancing the term with the second derivative and the non-linear term. The parameters should be chosen so that the system of algebraic equations that will be obtained as a result of substituting (9.52) into (9.51), reducing the expression to a common denominator and equating to zero the coefficients at various powers of the function e^z in the numerator should be minimal.

(2) Show that the equation

$$u_t = u_{xx} + u(1-u)$$

possesses the solution

$$u(t,\,x) = \frac{1}{(1+e^{\theta})^{s+1}}, \quad \theta = \omega\,t + k\,x + c, \quad s > 0$$

for certain values of the parameters ω and k. Regarding the definition of the parameter s, see the note to the previous problem.

(3) Consider the equation

$$q_t + q_x = q_{xx} + q(q-1).$$

Show that it possesses the solution of the form

$$q(t,\,x) = \frac{A}{\cosh^s \theta}, \quad \theta = \omega\,t + k\,x + c$$

for certain values of the parameters.

9.3 Bi-kink solutions of the convection-reaction-diffusion equation and its hyperbolic modification

9.3.1 *Introduction*

Hirota method delivers very effective tools for analytical description of traveling wave solutions within the nonlinear evolutionary PDEs. But most of the papers concerned with the model which are not completely integrable are concentrated on finding simple solutions such as single solitary wave or kink. Prior to late 80th of the XX century only few authors looked for "bi-soliton" solutions to non-integrable models. Being informal, we understand by this term solutions which describe the interaction of traveling waves. Perhaps the first advanced search of the bi-soliton solutions to PDEs that are not completely integrable was undertaken by H. Cornille and A. Gervois [Cornille and Gervois (1982, 1983)]. They describe a broad class of non-linear PDEs, possessing bi-soliton solutions. Yet, the methodology put forward by these authors applies merely to the so called "factorized" PDEs. A few years later T. Kawahara and M. Tanaka [Kawahara and Tanaka (1983)] considered the Fisher equation that does not fit the classification scheme from [Cornille and Gervois (1982, 1983)]. They succeeded in analytical description of interacting travelling fronts, using the classical Hirota method. In the following subsection we present the analytical description of interacting wave fronts supported by the convection-reaction-diffusion equation.

9.3.2 *Solutions of the convection-reaction-diffusion equation describing interacting wave fronts*

Let us consider the equation

$$u_t + u\, u_x - \kappa u_{xx} = \lambda u(u^2 - 1), \tag{9.53}$$

where κ, λ are constant parameters, $\kappa > 0$. Our goal is to obtain the analytical descriptions of the bi-kink solutions, basing on the Hirota method. Employment of this method to the models which are not completely integrable leads to the systems of non-linear algebraic equations that rather cannot be solved directly without any additional information about parameters being involved into the scheme. Processing the system of algebraic equations appearing within the Hirota method becomes more simple if we restrict ourselves to the class of functions possessing some given asymptotic features. In what follows, we will assume that bi-kink solutions tend on infinity to corresponding TW solutions. For this reason we begin with the analytical description of kink-like TW solutions, playing the role of asymptotics.

To obtain them, we use the ansatz

$$u = \frac{f_x}{f}, \tag{9.54}$$

assuming in addition that $f = 1 + \epsilon \exp(z)$, $z = a\,x - v\,t + c$, where a, v, c are the parameters to be determined, while ϵ is a formal decomposition parameter introduced for technical reasons.

Cell 9.3.

```
In[1]: Clear[ϵ, a, v, κ, λ];
In[2]:
    u = a ϵExp[z]/(1+ϵExp[z]);
    eq1 = Together[-vD[u,{z,1}]+a u D[u,{z,1}]-a^2κ D[u,{z,2}]-λ u(u^2 -1)];
In[4]: eq2 = Numerator[eq1//. z → 0];
In[5]: eq3 = CoefficientList[eq2, ϵ];
Out[5]: {0,-a v -a^3κ+ aλ, a^3 - a v+a^3κ +2 a λ, aλ -a^3λ}
In[6]: Aux1 = Table[eq3[[i]] ==0,{i,2,4}]
Out[6]: {-av-a^3κ + aλ == 0, a^3 - a v+a^3κ +2 a λ == 0, aλ -a^3λ == 0}
In[7]: Aux2 = Solve[Aux1, {a, λ, v}]
Out[7]:
    {{λ → −1−2κ, v→ −1−3κ, a→ 1},{λ → −1−2κ, v→ −1−3κ,a→ 1},{a→ 0}}
In[8]: Clear[ϵ, a, v, κ, λ];
In[9]:
    κ =1./6.;
    ϵ =1;
    z=ax-vt//.{Aux2[[1]]};
In[12]: Verifl = Simplify[eq2 //. Aux2[[1]]]
Out[12]: 0
In[13]: Clear[ϵ, a, v, κ, λ];
In[14]:
    κ =1./6.;
    ϵ =1;
    z=ax-vt//.{Aux2[[1]]};
In[17]: Verifl = Simplify[eq2 //. Aux2[[2]]]
Out[17]:
    0
```

Employing the *Mathematica* code shown in the cell 9.3, we obtain that the formula (9.54) satisfies the equation (9.53) if $\lambda = -(2\,\kappa + 1)$, $v = -(3\kappa + 1)$, c is arbitrary, and $a = \pm 1$. Note that the constant c can be chosen in such a way that $\epsilon e^c = 1$. Therefore, without loss of generality, we can put in the final formula $\epsilon = 1$.

And now we are looking for the solution of the form

$$f = 1 + \epsilon_1 \exp(z_1) + \epsilon_2 \exp(z_2), \tag{9.55}$$

where $z_1 = x - vt + c_1$, $z_2 = -x - vt + c_2$, c_1 and c_2 are arbitrary, $\epsilon_{1,2}$ are formal decomposition parameters (which in the final formula will be put equal to unity). Using the *Mathematica* code presented in the cell 9.4 we get the result that can be formulated in the form of the following statement.

Statement 9.1. The convection-reaction-diffusion equation (9.53) possess bi-kink solution

$$u(t,x) = \frac{\exp(z_1) - \exp(z_2)}{1 + \exp(z_1) + \exp(z_2)},$$

where $z_1 = +x - vt + c_1$, $z_2 = -x - vt + c_2$, if $\lambda = -(1+2\kappa)$, $v = -(1+3\kappa)$, $0 < \kappa$, c_1 and c_2 are arbitrary.

Cell 9.4.

```
In[1]: Clear[ϵ, a, v, κ, λ, f, u];
In[2]: f=1+ϵ1e^z1 + ϵ2e^z2;
    u = (D[f,z1]-D[f,z2])/f;
In[4]:
    u1 = D[u, {z1, 1}]- D[u, {z2, 1}];
    u2 = D[u1, {z1, 1}]- D[u1, {z2, 1}];
In[6]: eq1 = Together[-v (D[u, z1] + D[u, z2]) + u u1 - κ u2 - λ u (u^2 - 1)];
In[7]: eq2 = Numerator[eq1//. {z1 → 0, z2 → 0}];
In[8]: eq2 = Simplify[eq2//. {Log[e]→ 1, Log[e]^2 → 1, Log[e]^3 →1}];
In[9]: eq3 = CoefficientList[eq2, {ϵ1, ϵ2}]
Out[9]: {{0, v + κ - λ, -1 + v - κ - 2 λ}, {-v - κ + λ, 0, -4 (1 + 2 κ + λ)}, {1 -
v +κ + 2 λ, 4(1+2κ+λ),0}}
In[10]: Aux11 = Table[eq3[[1, i]] == 0, {i, 2, 3}]
Out[10]: {v + κ - λ == 0, -1 + v - κ - 2 λ == 0}
In[11]: Aux11A = Solve[Aux11, {λ, v}]
Out[11]: {{λ → -1 - 2 κ, v → -1 - 3 κ}}
In[12]: Aux12 = Table[eq3[[2, i]]==0, {i, 1, 3}]
Out[12]: {-v - κ + λ==0, True, -4 (1 + 2 κ + λ)==0}
In[13]: Aux12A = Solve[Aux12, {λ, v}]
Out[13]: {{λ → -1 - 2 κ, v → -1 - 3 κ}}
In[14]: Aux13 = Table[eq3[[3, i]]==0, {i, 1, 2}]
Out[14]: {1 - v + κ + 2 λ==0, 4 (1 + 2 κ+ λ)==0}
In[15]: Aux13A = Flatten[Solve[Aux13, {λ, v}]]
Out[15]: {λ → -1 - 2 κ, v → -1 - 3 κ}
In[16]: (*VERIFICATION*)
In[17]: Clear[f, u, v, κ, λ, c1, c2];
In[18]:
```

```
  z1 = x - v t + c1 //. Flatten[{Aux13A, c1 → 1, c2 → 2}];
  z2 = -x - v t + c2 //. Flatten[{Aux13A, c1 → 1, c2 → 2}];
In[20]:
  f=1+Exp[z1]+Exp[z2];
  u=D[f,x]/f;
In[22]:
  Verif1 = Simplify[D[u, t] + u D[u, x] - κ D[u, {x, 2}] - λ u (u^2 - 1)//. Aux13A]
Out[22]: 0
```

9.3.3 *Bi-kink solutions of the modified convection-reaction-diffusion equation*

Now, we consider the hyperbolic modification of the convection-reaction-diffusion equation:

$$\tau\, u_{tt} + u_t + u\, u_x + B\, u_x - \kappa u_{xx} = \lambda u(u - S)(u + Q), \tag{9.56}$$

where τ, κ, B, S, Q, λ are constant parameters, $\tau \geq 0$, $\kappa > 0$.

It is shown in the paper [Vladimirov and Maczka (2007)] that obtaining bi-soliton or bi-kink solutions of the equation (9.56) using the Hirota method is not possible, but this goal can be achieved with the help of more general substitution

$$u(t, x) = \frac{g}{f}, \tag{9.57}$$

where

$$g = \alpha\varepsilon_1 \exp(z_1) + \beta\varepsilon_2 \exp(z_2) + A\,\varepsilon_1\varepsilon_2 \exp(z_1 + z_2), \tag{9.58}$$

$$f = 1 + \varepsilon_1 \exp(z_1) + \varepsilon_2 \exp(z_2) + \varepsilon_2\varepsilon_2\, R\, \exp(z_1 + z_2). \tag{9.59}$$

Here $z_1 = hx - vt + c_1$, $z_2 = kx - wt + c_2$, h, k, v, w, S, Q, R, A, c_1 and c_2 are constant parameters, ε_1, ε_2 play the role of small parameters. So, the expression (9.57) is the quotient of the algebraic combinations of the exponential functions that are not strictly related to each other.

The search for bi-soliton solutions is carried out according to the same scheme as in the previous subsection: substituting (9.57) into (9.56), reducing the equation to the common denominator, we treat the numerator of the expression obtained as a polynomial equation with respect to the parameters ε_1 and ε_2. Equating to zero the coefficients at various degrees of $\varepsilon_1^m\, \varepsilon_2^n$, we obtain a system of nonlinear algebraic equations for unknown parameters. The solution to this system, is a very difficult technical problem, so we omit it here, giving only the final result.

Statement 9.2. If

$$B = \frac{2Q - S}{4}, \lambda = -\frac{2\,QS + S^2 - 4\,v}{4\,Q\,S + +2S^2}, \tau = -\frac{4\,S}{(2\,Q + S)(2\,Q\,S + S^2 - 4\,v)},$$

$$w = -\frac{1}{4}Q\,(2\,Q + S),$$

and

$$\kappa = -\frac{S(2\,Q + S)}{4(2\,Q\,S + S^2 - 4\,v)},$$

then equation (9.56) possesses the following solution

$$u(t, x) = \frac{S\exp(z_1)[1 + R\exp(z_2)]}{1 + \exp(z_1) + \exp(z_2) + R\exp(z_1 + z_2)}, \tag{9.60}$$

where

$$z_1 = S\,x - v\,t + c_1, \quad z_2 = -Q\,x - w\,t + c_2,$$

S, Q, R, v, c_1, c_2 are arbitrary constants.

Pattern of the interacting wave fronts described by the above formula is shown in Fig. 9.9.

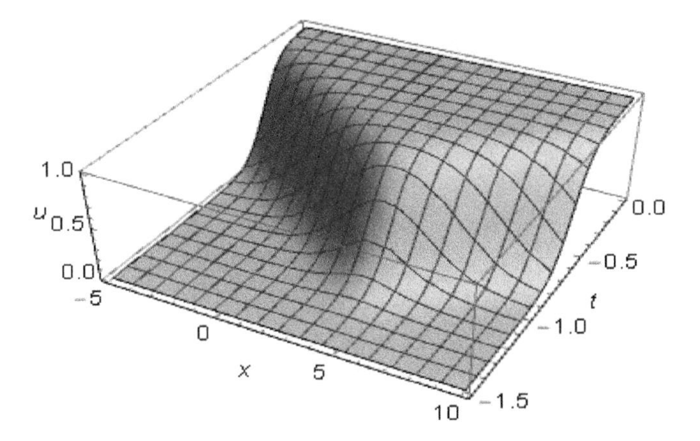

Fig. 9.9 Solution of the equation (9.56) described by the formula (9.60). The following values of the parameters are used: $S = 1$, $Q = -1.9$, $c_1 = 10$, $c_2 = -6$, $R = 0$, $v = -13.3$.

Remark. An algorithm for obtaining a family of bi-kink solutions of the equation (9.56) is presented in the file PM2_13.nb. To facilitate the calculations, we assume at the very beginning that $h = S$ and $k = -Q$. From the analysis of the above algorithm it follows that the problem has, apparently, another nontrivial solutions. Note that the formulation of the statement 9.2 contains an arbitrary parameter v, which is fixed in the work [Vladimirov and Maczka (2007)].

Chapter 10

Nonlinear wave patterns described by some non-integrable models

10.1 Introduction

As noted earlier, many researchers believe that the unique particle-like properties of solutions to the Korteweg-de Vries equation are related to the integrability of this equation and the presence of an infinite set of conservation laws. However, the experience of recent decades shows that, at least in part, the properties of "true" solitons are possessed by solutions of non-integrable equations, including dissipative-type equations, which, as a rule, have a very modest set of conservation laws. This chapter contains examples of models that support localized wave structures with soliton properties. Some of these models are derived from very realistic considerations.

10.2 Rosenau-Hyman equation: evolution of compactons

At the beginning of this chapter, let us analyze the properties of the solutions of the Rosenau-Hyman equation (2.28), which is a KdV-type equation with nonlinear dispersion term. The qualitative analysis of the system (2.30) describing the traveling wave solutions of the Rosenau-Hyman equation, performed in Section 2.3, shows that among the solutions the homoclinic trajectory exists, bi asymptotic to a topological saddle. Due to the degeneracy of the stationary point, it takes finite "time" to bypass this trajectory. As a result of this degeneracy, the homoclinic loop corresponds to a soliton with a compact support called *compacton*.

Unfortunately, we can say nothing about the existence of the analogs of the Hirota's method for (2.28), enabling to state the nonlinear superposition principle for compactons, except perhaps that the existence of such an ana-

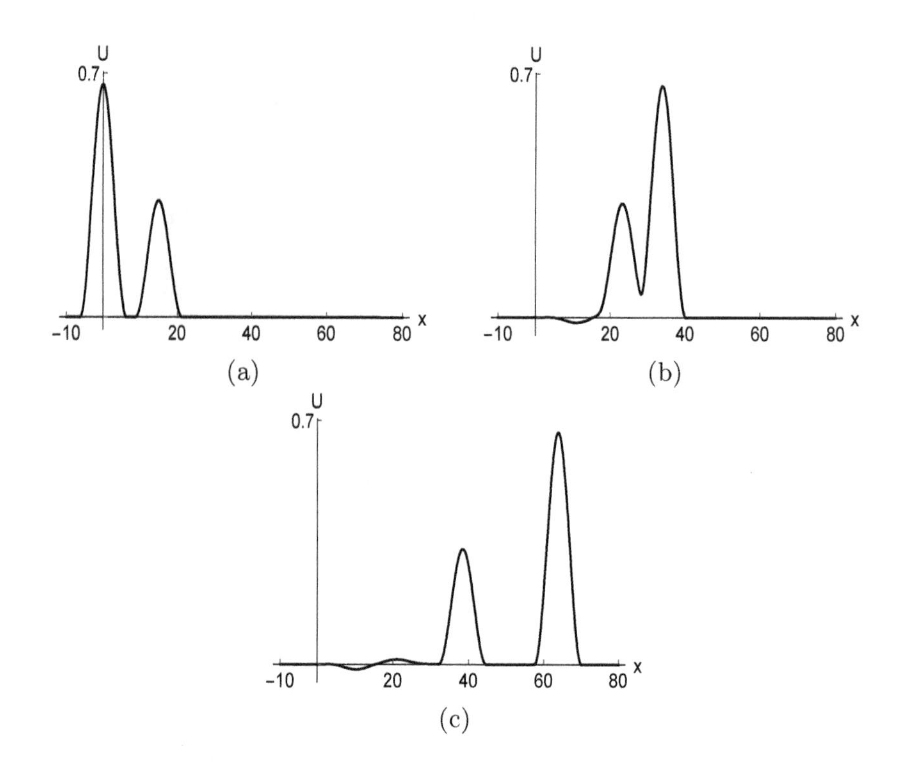

Fig. 10.1 Interaction of two compactons: (a) configuration at time $t = 0$; (b) configuration at time $t = 60$; (c) configuration at time $t = 120$. As initial data at the moment $t = 0$, compact solutions with the speeds $c_1 = 0.5$ (left compacton) and $c_1 = 0.25$ (right compacton) are used.

log seems highly unlikely. However, solving numerically the initial value problem (2.28), (2.38) one can see that, like the soliton, the compacton evolves without changing its shape and that compactons interact almost elastically. Consider a situation where two compactons whose support do not intersect, are selected as the Cauchy's data (see PM2_14.nb). By choosing the output configuration in such a way that the compacton with the larger amplitude is to the left of the compacton with the smaller amplitude (Fig. 10.1(a)), we can observe the interaction of the compactons over time, as both compactons move from left to the right, and the velocities of their movement are proportional to the amplitudes. Figure 10.1(b) shows the result of the numerical simulation at the moment when the compactons interact with each other. After the end of the impact, the compactons restore their shape, while the compacton with a larger amplitude advances to the front. Unlike the interaction of solitons, a pair consisting of "bright" and "dark" compactons is formed at the place of collision of compactons (this pair is clearly visible in Fig. 10.1(c)). The amplitudes of these newly created wave patterns are an order of magnitude smaller than the amplitudes of the interacting compactons.

Finally, we present the results of the numerical experiment performed in the PM2_14.nb file, in which as the initial data the function

$$u(0,\, x) = \varphi(x) = \begin{cases} 0.3\,\left[1 + \cos\frac{\pi}{30}(x+120)\right], & \text{as} \quad \left|\frac{x+120}{30}\right| < 1, \\ 0, & \text{elsewhere} \end{cases} \tag{10.1}$$

is taken, which is not the exact solution to (2.28). The result of numerical solving the initial value problem (2.28), (10.1) is presented in Fig. 10.2. Similarly to the behavior of an analogous problem in the case of the KdV equation, the impulse splits in the course of evolution into a series of compactons moving to the right and arranged in the order of increasing amplitude.

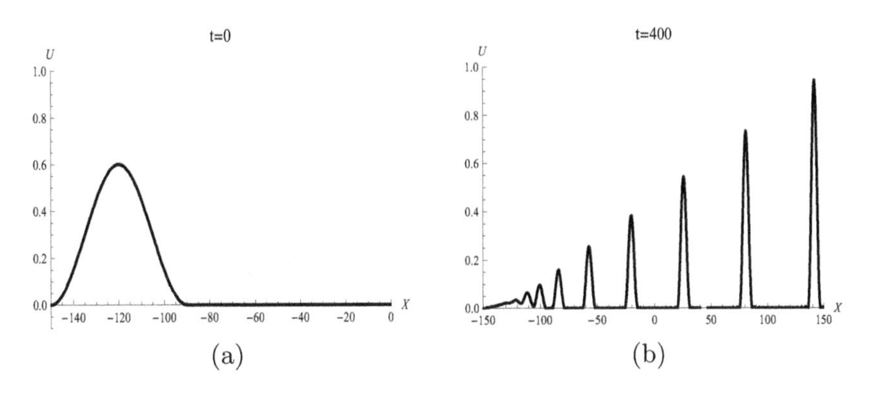

Fig. 10.2 Result of numerical solution of the Cauchy problem (2.28), (10.1): (a) initial configuration at time $t = 0$; (b) configuration in the moment of time $t = 400$.

Note that it has been rigorously proven in paper [Vodova (2013)] that the equation (2.28) is not completely integrable, that is, it is not related to a linear equation in any way and possesses only finite number of conservation laws. Nevertheless, the solutions to this equation retain some properties of the Korteweg-de Vries model. This is evidenced by the fact that the compactons restore their original shape after the interaction, although due to the presence of small compacton-anticompacton pair in the "tail", the interaction cannot be characterized as completely elastic.

Problem.

Consider the equation

$$u_t + \left(u^3\right)_x + \left(u^3\right)_{xxx} = 0. \tag{10.2}$$

(1) Construct the traveling wave solution having the form

$$u(t,\, x) = A\cos^{\gamma}(B\,z), \qquad z = x - v\,t. \tag{10.3}$$

(2) Construct on this basis the (generalized) compacton-like solution whose support lies in the set $-\frac{\pi}{2\,B},\ \frac{\pi}{2\,B}$. Basing on the numerical scheme presented in the file PM2_14.nb, construct the numerical scheme for the equation (10.2) and trace numerically the evolution of the compacton.

10.3 Evolutionary PDEs associated with a chain of pre-stressed granules

10.3.1 *Introduction*

In this section we deal with nonlinear evolutionary PDEs associated with dynamics of a one-dimensional chain of pre-stressed granules which arises in quite a number of applications. Since Nesterenko's pioneering works [Nesterenko (1983, 1994)] propagation of pulses in such media has been a subject of a great number of experimental studies and numerical simulations, see [Lazaridi and Nesterenko (1985); Coste *et al.* (1997); Nesterenko (2001); Daraio *et al.* (2006); Herbold and Nesterenko (2007); Ahnert and Pikovsky (2009); Yang *et al.* (2011); Vladimirov and Skurativskyi (2015, 2020)] and references therein. We consider a nonlinear evolutionary PDE which is derived from the infinite system of ODEs describing the dynamics of one-dimensional chain of elastic bodies interacting with each other by means of a nonlinear force. The PDE in question is obtained through the passage to continuum limit followed by the formal multi-scale decomposition. It occurs to possess generalized traveling wave solutions with compact support manifesting some features of solitons. We also perform the numerical study of the compacton solutions. Numerical simulations show that the compacton solutions recover their shapes after the collisions. In this connection note that compactons [Nesterenko (1983); Lazaridi and Nesterenko (1985); Rosenau and Hyman (1993)] exist for a number of physically relevant models and possess several interesting features making them a subject of intense research, cf. e.g. [Rosenau (1996); Destrade and Saccomandi (2006); Rosenau (2006); Vodova (2013); Cirillo *et al.* (2016); Zilburg and Rosenau (2017)] and references therein.

10.3.2 *Evolutionary PDEs associated with the granular pre-stressed chains*

As we already know, there exist non-integrable equations possessing localized TW solutions with the behavior quite similar to the KdV solitons. These are not only compact solutions obeying the $K(2,\,2)$ equation, but also similar solutions of the entire $K(m,\,n)$ hierarchy

$$K(m,n): u_t + (u^m)_x + (u^n)_{xxx} = 0, \qquad m \geq 2, \qquad n \geq 2. \tag{10.4}$$

The members of this hierarchy are not completely integrable at least for generic values of the parameters m, n, see [Rosenau (1996); Vodova (2013)] and references therein, and yet possess compactly-supported TW solutions exhibiting solitonic features [Rosenau and Hyman (1993); De Frutos *et al.* (1995)].

The $K(m,n)$ family was introduced in the 1990s as a formal generalization of the KdV hierarchy without referring to its physical context. Earlier V.F. Nesterenko [Nesterenko (1983)] considered the dynamics of a chain of pre-stressed granules described by the following ODE system:

$$\ddot{Q}_k(t) = F(Q_{k-1} - Q_k) - F(Q_k - Q_{k+1}), \qquad k \in \{0, \pm 1, \pm 2 \dots\}, \tag{10.5}$$

where $Q_k(t)$ is the displacement of the kth granule center-of-mass from its equilibrium position,

$$F(z) = Az^n, \qquad n > 1. \tag{10.6}$$

He has described for the first time within this model the formation and evolution of the compactly supported wave patterns [Nesterenko (1983, 1994); Lazaridi and Nesterenko (1985)]. In [Nesterenko (1983, 1994, 2001)] the nonlinear evolutionary PDEs are presented, being the quasi-continual limits of the discrete models; in this connection cf. also [Nesterenko (2018)].

As in the case of the FPU problem, the transition to the continual model is achieved via the substitution

$$Q_k(t) = u(t, k \cdot a) \approx u(t, x), \tag{10.7}$$

where a is the average distance between granules. We shall also employ the substitutions

$$\begin{aligned} Q_{k\pm 1} &= u(t, x \pm a) = \exp(\pm a D_x) u(t,x) \\ &= \sum_{j=0}^{4} \frac{(\pm a)^j}{j!} \frac{\partial^j}{\partial x^j} u(t,x) + O\left(a^5\right). \end{aligned} \tag{10.8}$$

Following Nesterenko [Nesterenko (2001)], we introduce the notation

$$N = -a\, u_x,$$
$$\varphi = \tfrac{a^2}{2}\, u_{xx} - \tfrac{a^3}{6}\, u_{xxx} + \tfrac{a^4}{24}\, u_{xxxx} + \ldots,$$
$$\psi = -\left(\tfrac{a^2}{2}\, u_{xx} + \tfrac{a^3}{6}\, u_{xxx} + \tfrac{a^4}{24}\, u_{xxxx} + \ldots\right).$$

Using this notation we can rewrite the quasi-continuous approximation corresponding to the system (10.5) in the following form:

$$\begin{aligned} u_{tt} &= A\left\{(N+\varphi)^n - (N+\psi)^n\right\} \\ &= A\left[nN^{n-1}(\varphi-\psi) + \frac{n(n-1)}{2}N^{n-2}(\varphi^2-\psi^2)\right. \\ &\quad \left. + \frac{n(n-1)(n-2)}{6}N^{n-3}(\varphi^2-\psi^2) + \ldots\right]. \end{aligned}$$

Substituting into this formula the easily verifiable equalities

$$\varphi - \psi = 2\sum_{m=1}^{3} \tfrac{a^{2\,m}}{(2\,m)!}\partial_x^{2\,m} Q + O(|a|^7),$$
$$\varphi^2 - \psi^2 = -\tfrac{a^5}{3} Q_{xx} Q_{xxx} + O(|a|^7),$$
$$\varphi^3 - \psi^3 = \tfrac{a^6}{4} Q_{xx}^3 + O(|a|^7),$$
$$\varphi^{3+\nu} - \psi^{3+\nu} = O(|a|^7), \quad \nu > 1,$$

we get the equation

$$u_{tt} = A\, a^{n+1}\left\{n(-Q_x)^{n-1}\left[Q_{xx} + \tfrac{a^2}{12} Q_{4\,x}\right] - \tfrac{n(n-1)\,a^2}{4}(-Q_x)^{n-2}\, Q_{2\,x}\, Q_{3\,x} \right.$$
$$\left. + \tfrac{n(n-1)(n-2)a^2}{24}(-Q_x)^{n-3}\, Q_{x\,x}^3\right\} + O(|a|^{n+4}).$$

Differentiating the left and right sides with respect to the variable x and introducing the notation $u_x = -S$, we can represent this expression in the following form:

$$S_{tt} = C\left\{S^n + \beta S^{\frac{n-1}{2}}\left[S^{\frac{n+1}{2}}\right]_{xx}\right\}_{xx}, \tag{10.9}$$

where $C = Aa^{n+1}$, $\beta = \frac{na^2}{6(n+1)}$ (from now on we'll omit the terms of the order $O(a^{n+4})$).

Equation (10.9) occurs to possess a one-parameter family of compacton TW solutions describing the propagation of the waves of compression.

Lemma 10.1. *Equation (10.9) possesses the following generalized solution:*

$$S(t,\, x) = A\begin{cases} \cos^{\frac{2}{n-1}}[B\,z] & \text{when } |B\,z| < \pi/2, \\ 0 & \text{elsewhere}, \end{cases} \tag{10.10}$$

where $z = x - Vt$,

$$A = \left[\frac{\lambda\,(n+1)}{2}\right]^{\frac{1}{n-1}}, \qquad B = \frac{1}{\sqrt{\beta}}\frac{n-1}{n+1}, \qquad \lambda = \frac{V^2}{C}.$$

The presence of the above solution is established directly by substituting a solution of the form $S(z) = A\cos^{\gamma}[B\,z]$, $z = x - V\,t$ into the factorized and twice integrated equation

$$\lambda\,S = S^n + \beta S^{\frac{n-1}{2}}\left[S^{\frac{n+1}{2}}\right]'',$$

which can be obtained from (10.9) using the substitution $S = S(z)$, integrating the resulting equation twice within the interval $(-\infty, z)$ and using the boundary conditions $S^{(k)}(\pm\infty) = 0$, $k = 0, 1, \ldots$. Balancing the powers of the function $\cos(B\,z)$ in the resulting equation, we come to the conclusion that a nonzero solution can be obtained only if $\gamma = 2/(n-1)$. Further, equating to zero the coefficients of the functions $\cos^{n\gamma}(B\,z)$ and $\cos^{\gamma}(B\,z)$, we obtain a pair of algebraic equations, from which the above values of the coefficients A and B can be easily obtained.

Unfortunately, the compacton solutions (10.10) are unstable. A similar situation occurs in the case of the Boussinesq equation, obtained as a continuum limit of the Fermi–Pasta–Ulam system of coupled oscillators. The Boussinesq equation (8.49) possesses unstable soliton-like solutions, and the KdV equation, supporting the stable uni-directional solitons, is extracted from the Boussinesq equation.

Our approach to finding a "proper" compacton-supporting equation is as follows. We start from the discrete system (10.5) in which the interaction force has the form

$$F(z) = Az^n + Bz. \tag{10.11}$$

In addition, we assume that $B = \gamma a^{n+1}$, where $|\gamma| = O(|1|)$.

Inserting (10.7), (10.8) into the formula (10.5) and assuming that the interaction is described by (10.11), we obtain, up to the terms of the order $O(a^{n+4})$ and higher in the expansion of the right-hand side of (10.5), cf. the discussion after (10.8), the equation

$$u_{tt} = -C\left\{(-u_x)^n + \beta\,(-u_x)^{\frac{n-1}{2}}\left[(-u_x)^{\frac{n+1}{2}}\right]_{xx}\right\}_x - \gamma a^{n+3}\,(-u_x)_x\,.$$

Differentiating the above equation with respect to x and introducing the new variable $S = -u_x$, we obtain the following equation:

$$S_{tt} = C\left\{S^n + \beta S^{\frac{n-1}{2}}\left[S^{\frac{n+1}{2}}\right]_{xx}\right\}_{xx} + \gamma a^{n+3} S_{xx}. \tag{10.12}$$

Now we use a series of scaling transformations. Employing the scaling $\tau = \sqrt{\gamma a^{n+3}}t$ enables us to rewrite the above equation in the form

$$S_{\tau\tau} = \frac{C}{\gamma a^{n+3}}\left\{S^n + \beta S^{\frac{n-1}{2}}\left[S^{\frac{n+1}{2}}\right]_{xx}\right\}_{xx} + S_{xx}.$$

Next, the transformation $\bar{T} = \frac{1}{2}a^q\tau$, $\xi = a^p(x-\tau)$, $S = a^r W$ is used. If we assign the following values to the parameters $q = 1$, $p = -1$, $r = 2/(n-1)$, then the higher-order coefficient $O(a^2)$ will be that of the second derivative with respect to $\bar{T}$. So, dropping the terms proportional to $O(a^2)$, we obtain, after the integration with respect to ξ, the equation:

$$W_{\bar{T}} + \frac{A}{\gamma}\left\{W^n + \frac{n}{6(n+1)}W^{\frac{n-1}{2}}\left[W^{\frac{n+1}{2}}\right]_{\xi\xi}\right\}_{\xi} = 0.$$

Performing the rescaling

$$T = \frac{A}{\gamma}L\bar{T}, \qquad X = L\xi,$$

where $L = \sqrt{\frac{6(n+1)}{n}}$, we finally obtain the sought-for equation

$$W_T + \left\{W^n + W^{\frac{n-1}{2}}\left[W^{\frac{n+1}{2}}\right]_{XX}\right\}_X = 0, \tag{10.13}$$

to which we shall hereinafter refer as to the *Nesterenko equation.*

The description of waves of rarefaction in the case $n = 2k$ requires the following modification of the interaction force:

$$F(z) = -Az^{2k} + Bz \tag{10.14}$$

(for $n = 2k+1$ the formula (10.11) describes automatically both waves of compression and of rarefaction). Applying the above machinery to (10.5) with the interaction (10.14), we obtain, in the same notation, the equation

$$W_T - \left\{W^n + W^{\frac{n-1}{2}}\left[W^{\frac{n+1}{2}}\right]_{XX}\right\}_X = 0, \quad n = 2k. \tag{10.15}$$

Thus, the universal equation describing waves of compression and rarefaction for arbitrary $n \in \mathbb{N}$ can be written in the form

$$W_T + [\mathrm{sgn}(W)]^{n+1}\left\{W^n + W^{\frac{n-1}{2}}\left[W^{\frac{n+1}{2}}\right]_{XX}\right\}_X = 0. \tag{10.16}$$

In closing, note that equations (10.13), (10.15) and (10.16) are obtained by formal application of the multiscale decomposition method which cannot be substantiated in our case because of negativity of the index p (cf. with [Rosenau (2003)] where this problem is discussed in a more general fashion). Further study of these equations is justified by the fact that they possess a set of generalized compacton solutions

$$W_c^{\epsilon}(t,x) = \begin{cases} \epsilon\, M\cos^{\gamma}(K\,z), & \text{as} \quad |K\,z| > \frac{\pi}{2}, \\ 0, & \text{otherwise}, \end{cases} \tag{10.17}$$

where

$$M = \left[\frac{c(n+1)}{2}\right]^{\frac{1}{n-1}}, \quad K = \frac{n-1}{n+1}, \quad \gamma = \frac{2}{n-1}, \quad \epsilon = \pm 1.$$

Depending on sign of ϵ, formula (10.17) describes bright or dark compactons possessing interesting dynamical features. As will be shown next, these solutions describe well enough propagation of short impulses in the chain of pre-stressed blocks.

10.3.3 Numerical simulations for dynamics of compactons

The dynamics of solitary waves is studied by means of direct numerical simulation based on the finite-difference scheme.

To derive a finite-difference scheme, say, for the model equation (10.13), we modify the scheme presented in [De Frutos *et al.* (1995)]. In agreement with the methodology proposed in that paper we introduce the artificial viscosity by adding the term εW_{4x}, where ε is a small parameter. Thus, instead of (10.13) we have for the case of $n = 3$ the following equation:

$$W_t + \left\{W^3\right\}_x + \left\{W\left[W^2\right]_{xx}\right\}_x + \varepsilon W_{4x} = 0. \tag{10.18}$$

Let us approximate the spatial derivatives as follows:

$$\begin{aligned}&\frac{1}{120}(\dot{W}_{j-2} + 26\dot{W}_{j-1} + 66\dot{W}_j + 26\dot{W}_{j+1} + \dot{W}_{j+2})\\&\quad+\frac{1}{24h}(-W_{j-2}^3 - 10W_{j-1}^3 + 10W_{j+1}^3 + W_{j+2}^3)\\&\quad+\frac{1}{24h}(-L_{j-2} - 10L_{j-1} + 10L_{j+1} + L_{j+2})\\&+\varepsilon\frac{1}{h^4}(W_{j-2} - 4W_{j-1} + 6W_j - 4W_{j+1} + W_{j+2}) = 0,\end{aligned} \tag{10.19}$$

where $L_j = W_j \frac{W_{j-2}^2 - 2W_j^2 + W_{j+2}^2}{h^2}$.

To integrate the system (10.19) in time, we use the midpoint method. Then the quantities W_j and $\dot{W}_j$ are represented in the form

$$W_j \to \frac{W_j^{n+1} + W_j^n}{2}, \qquad \dot{W}_j \to \frac{W_j^{n+1} - W_j^n}{\tau}.$$

The resulting nonlinear algebraic system with respect to W_j^{n+1} can be solved by iterative methods.

We test the scheme (10.19) by considering the movement of a single compacton. Assume that the model parameters $c = 1$ and the scheme parameters $N = 600$, $h = 30/N$, $\tau = 0.01$, $\varepsilon = 10^{-3}$ are fixed. The application of the scheme (10.19) gives us Fig. 10.3(a). The starting profile providing the initial condition for the numerical scheme is chosen according to (10.17) where $n = 3$, $c_1 = 1$ and $c_2 = 1/4$, namely,

$$W_{1,2} = \begin{cases}\epsilon\sqrt{2c_{1,2}}\cos\left((z - z_{1,2})/2\right), & \text{if } |(z - z_{1,2})/2| < \pi/2.\\ 0, & \text{otherwise.}\end{cases}$$

We put in the above formula $z_1 = 5$, $z_2 = 13$, $\epsilon = +1$ for Fig. 10.3 while $\epsilon = -1$ for Fig. 10.4 (note that W_i corresponds to i-th figure).

To study the interaction of two bright compactons, we combine the compacton having the velocity $c = 1$ with the slow one characterized by

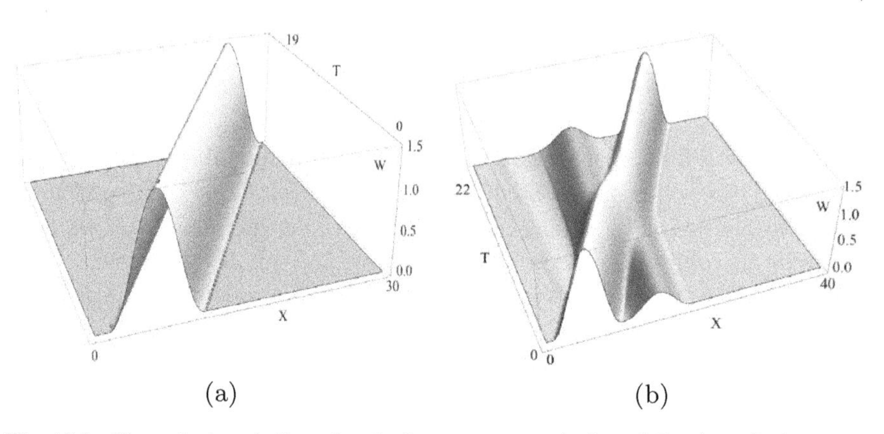

Fig. 10.3 Numerical evolution of a single compacton solution of Eq. (10.18) characterized by the velocity $c = 1$ (a) and a pair of compacton solutions characterized by the velocities $c = 1$ and $c = 1/4$ (b), respectively.

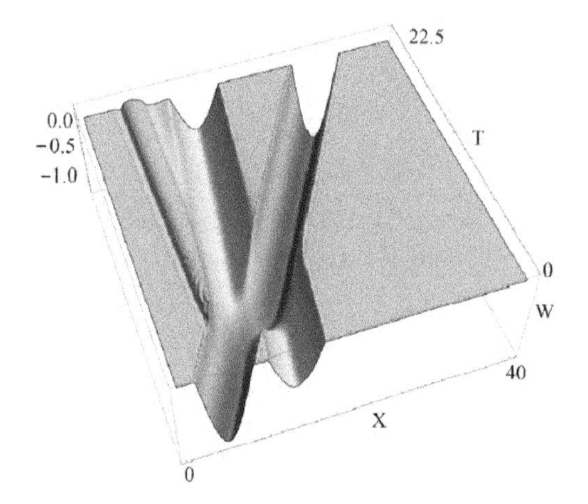

Fig. 10.4 Numerical evolution of a pair of dark compactons characterized by the velocities $c = 1$ and $c = 1/4$, respectively.

the velocity $c = 1/4$ and being shifted to the right at the initial moment of time. The result of modeling is presented in Fig. 10.3(b). The interaction of two dark compactons has similar properties and is depicted in Fig. 10.4.

As we have already mentioned at the end of Section 10.3.2, there is no way of selecting the scales in the model equations (10.13), (10.15) and (10.16), so the scaling decomposition employed there is rather formal. Nevertheless, it leads to interesting equations possessing localized solutions with solitonic features.

Now we are going to compare the evolution of the compacton solutions with corresponding solutions of the finite (but long enough) discrete system of blocks. Since the average distance a between adjacent blocks does not play the role of a small parameter anymore, we assume from now on that it is equal to one. With this assumption in mind, we can write equation (10.16) in the initial variables t, x as follows:

$$W_t + Q\left[\operatorname{sgn}(W)\right]^{n+1}\left\{W^n + \hat{\beta} W^{\frac{n-1}{2}}\left[W^{\frac{n+1}{2}}\right]_{xx}\right\}_x = 0, \tag{10.20}$$

where

$$Q = \frac{A}{\gamma}, \qquad \hat{\beta} = \frac{n}{6(n+1)}.$$

It is easy to verify that equation (10.20) possesses the following compacton solutions:

$$W_c^{\epsilon}(z) = \epsilon W_c(z) = \begin{cases} \epsilon \tilde{M} \cos^{\gamma}\left(\tilde{B} z\right), & \text{if } |\tilde{B} z| < \frac{\pi}{2}, \\ 0 & \text{otherwise,} \end{cases} \tag{10.21}$$

where $\epsilon = \pm 1$, $z = x - ct$,

$$\tilde{M} = \left[\frac{c(n+1)}{2Q}\right]^{\frac{1}{n-1}}, \qquad \tilde{B} = \frac{n-1}{(n+1)\sqrt{\beta}}, \qquad \gamma = \frac{2}{n-1}.$$

We introduce the functions $R_k = Q_{k-1} - Q_k$ being the discrete analogs to the strain field $W(t, x)$. In accordance with (10.5), (10.11), these functions satisfy the system

$$\begin{aligned}
&\ddot{R}_1(t) = 0, \\
&\ddot{R}_k(t) = A\left[R_{k-1}|R_{k-1}|^{n-1} - 2R_k|R_k|^{n-1} + R_{k+1}|R_{k+1}|^{n-1}\right] \\
&\qquad\qquad + \gamma\left[R_{k-1}|R_{k-1}|^{n-1} - 2R_k|R_k|^{n-1} + R_{k+1}|R_{k+1}|^{n-1}\right], \\
&\qquad k = 2, \ldots m-1, \\
&\ddot{R}_m(t) = 0.
\end{aligned} \tag{10.22}$$

We solve this system with the following initial conditions induced by the compacton solution (10.21) in the respective nodes:

$$R_k(0) = \begin{cases} \epsilon \tilde{M} \cos^{\gamma}[\tilde{B} k - I] & \text{if } |\tilde{B} k - I| < \pi/2 \\ 0 & \text{otherwise,} \end{cases} \tag{10.23}$$

$$\dot{R}_k(0) = \begin{cases} \epsilon \tilde{M} c \gamma \tilde{B} \cos^{\gamma-1}[\tilde{B} k - I] \sin[\tilde{B} k - I] & \text{if } |\tilde{B} k - I| < \pi/2 \\ 0, & \text{otherwise,} \end{cases} \tag{10.24}$$

$$R_1(0) = \dot{R}_1(0) = R_m(0) = \dot{R}_m(0) = 0, \tag{10.25}$$

where I is a constant phase, $k = 2, 3, \ldots, m-1$. Note that A and γ appear in equation (10.20) in the form of the ratio $Q = A/\gamma$, whereas in the system (10.22) they appear as independent parameters. Therefore, one should not expect a one-to-one correspondence between the solutions of the discrete and continuous problems for arbitrary values of the parameters. The numerical experiments confirm this hypothesis by showing that synchronous evolution of the same compacton perturbation within two models can be observed for a unique value of the velocity $c = c_0$. This value depends strongly on the parameter γ and depends also on the parameter A, but in a much weaker fashion. It has been observed that at $c < c_0$ the discrete compacton moves quicker than its continuous analogue while at $c > c_0$ the opposite effect occurs. The result of comparison for a single compacton is shown in Fig. 10.5. One can see that at the chosen values of the parameters the main perturbations move synchronously and do not change their form. However, in the tail part of the discrete analogue small non-vanishing oscillations appear after a while.

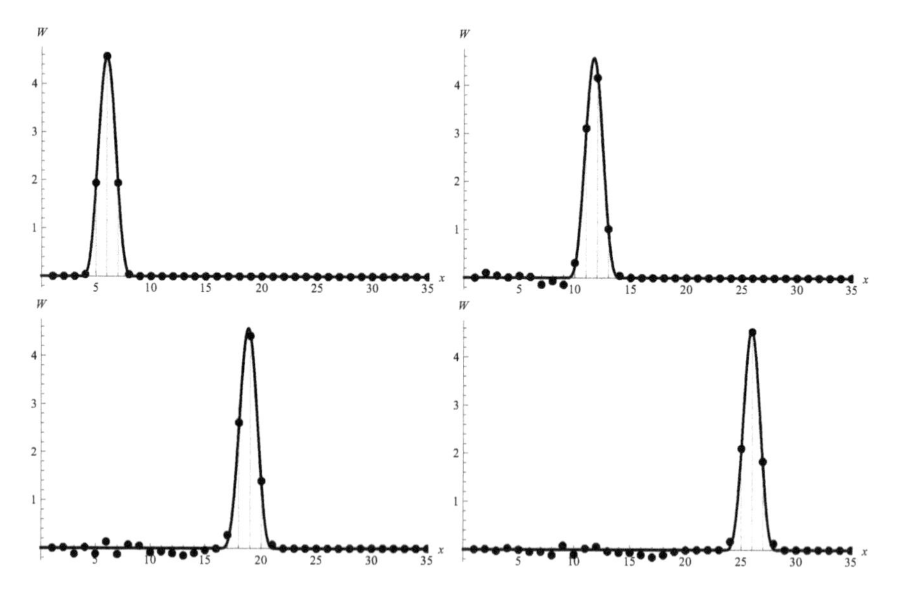

Fig. 10.5 Evolution of the initial perturbation in the granular media (marked with dots) on the background of the corresponding evolution of the compacton (marked with solid lines) obtained at the following values of the parameters: $n = 3/2$, $c = 1.425$, $A = 0.25$, $B = 0.3$. Upper row: left: $t = 0$; right: $t = 4$; lower row: left: $t = 9$; right: $t = 14$.

Since for every value of the parameter γ there is a unique value of the wave pack velocity for which the discrete and continuous compacton perturbations move synchronously, one should not expect that the collision of compactons within these two models will proceed in the same way for any set of values of parameters. However, collision processes display not much of qualitative differences for the discrete pulses which interact elastically like their continuous analogues. This is illustrated in Fig. 10.6 showing the evolution of two initially separated discrete compactons. For convenience, the continuous compactons which coincide with the right-hand side of the initial data (10.23) at $t = 0$ (the leftmost graph in the first row) are also shown in this figure. Continuous curves shown on the following graphs are obtained by appropriate translations. They are presented in order to emphasize the quasi-elastic nature of interaction of the discrete pulses.

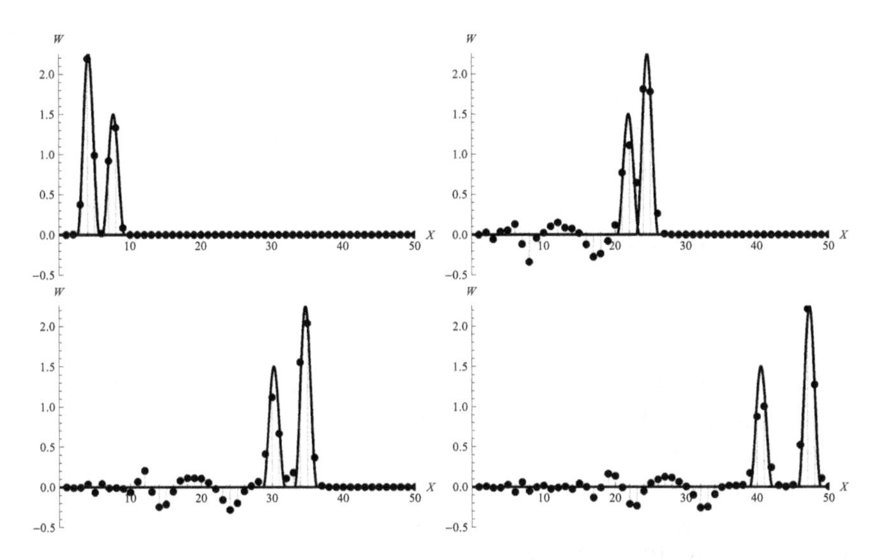

Fig. 10.6 Evolution of two initially separated compacton perturbation in the granular media (marked with dots) on the background of the corresponding compacton solutions of the continual model (marked with solid lines), obtained at the following values of the parameters: $n = 2$, $c_1 = 1.5$, $c_2 = 1.0$, $A = B = 1$. Upper row: left: $t = 0$; right: $t = 12$; lower row: left: $t = 18.25$; right: $t = 26$.

10.3.4 *Conclusions and discussion*

In this section we have studied compacton solutions supported by the nonlinear evolutionary PDEs. The equations we considered, (10.13), (10.15), and (10.16), are obtained from the dynamical system (10.5) describing one-dimensional chain of pre-stressed elastic bodies.

In contrast with the above, equations (10.13) (resp. (10.15)), which are obtained using formal multiscale decomposition, possess families of bright (resp. dark) compacton solutions which appear to be stable. This is backed by the results of the numerical simulations. As was shown in [Sergyeyev *et al.* (2019)], for generic values of the parameter n equation (10.13) does not possess an infinite set of higher symmetries or other signs of complete integrability such as infinite hierarchies of conservation laws. Nevertheless the compacton solutions to this equation possess some features which are characteristic for "genuine" soliton solutions.

A characteristic feature of equations (10.13), (10.15) related to the decomposition we used is that they describe processes with "long" temporal and "short" spatial scales. Hence it is rather questionable whether these equations can adequately describe a localized pulse propagation in discrete media in the situation when the distance between the adjacent particles is comparable to the compacton width Δx. In fact, making the "reverse" transformations $X \to \xi \to x$ we get the following formula for the width of the compacton solution (10.17) in the initial coordinate system:

$$\Delta x = \pi a \sqrt{\frac{n(n+1)}{6(n-1)^2}}.$$

This is nothing but equation (1.130) from [Nesterenko (2001)]. For $n = 3/2$, corresponding to the Hertzian force between spherical particles, we get $\Delta x \approx 4.96a$. It is then interesting to notice that the same results for the particles with the spherical geometry were obtained in the course of numerical simulations, and experimental studies [Nesterenko (1983); Lazaridi and Nesterenko (1985); Nesterenko (1994); Nesterenko *et al.* (1995)]. We wish to stress that results of our analysis as well as the main conclusions are in agreement with the earlier publications by other authors. In particular, Ph. Rosenau notes, when considering the general models of dense chains [Rosenau (2006)], that the natural separation of scales leading to an unidirectional PDE of first order in time does not exist.

10.4 Relaxing hydrodynamic-type model and its qualitative investigations

We consider the travelling wave solution of the system (5.16). Factorization of the system (5.16) is performed by means of the following ansatz:

$$u = U(\omega), \quad p = (x_0 - x)\Pi(\omega), \quad V = \frac{R(\omega)}{x_0 - x}, \quad \omega = \xi t + \log \frac{x_0}{x_0 - x},$$

where $0 < x_0$ is a point "at infinity", as will be shown below. In fact, inserting this ansatz into the second equation of system (5.16), we get the quadrature

$$U = \xi R + \text{const}$$

and the following dynamical system:

$$\begin{aligned} \xi \Delta R' &= R\,[\sigma R \Pi - \kappa + \tau \xi R \gamma]\,, \\ \xi \Delta \Pi' &= \xi\,\{\xi R\,(\kappa - R\Pi) - \chi\,(\Pi + \gamma)\}\,, \end{aligned} \tag{10.26}$$

where $(\cdot)' = d\,(\cdot)\,/d\omega$, $\Delta = \tau(\xi R)^2 - \chi$, $\sigma = 1 + \tau\xi$.

In case $\gamma < 0$ the system (10.26) has three stationary points in the right half-plane. One of them, having the coordinates $R_0 = 0$, $\Pi_0 = -\gamma$, lies in the vertical coordinate axis. Another one, having the coordinates

$$R_1 = -\kappa/\gamma, \quad \Pi_1 = -\gamma,$$

is the only stationary point lying in the physical parameters range. Besides, there is a pair of stationary points with the coordinates

$$R_{2,3} = \pm\sqrt{\frac{\chi}{\tau\xi^2}}, \quad \Pi_{2,3} = \frac{\kappa - \tau\xi\gamma R_{2,3}}{\sigma R_{2,3}},$$

lying on the line of singular points $\Delta = \tau(\xi R)^2 - \chi$.

Stationary point $A_1(-\kappa/\gamma, -\gamma)$ can be associated with a stationary invariant solution of the form

$$u_0 = 0, \quad p_0 = \gamma(x_l - x_0), \quad V_0 = \kappa/[\gamma(x_l - x_0)] \tag{10.27}$$

where we identify the coordinate x with x_l in order to stress the difference between the physical Eulerian coordinate x_e and the mass Lagrange coordinate used in this model. Note that the Lagrange coordinate x_l and the corresponding Eulerian coordinate are connected as follows:

$$x_e = (\kappa/\gamma)\,\ln(x_0 - x_l), \quad x_0 - x_l > 0. \tag{10.28}$$

If we introduce the parameter $D = \kappa\xi/\gamma$, then the stationary solution (10.27) can be rewritten in the following form:

$$u_0 = 0, \qquad p_0 = \kappa\,\rho_0, \qquad \rho_0 = (\Pi_1/\kappa)\exp(\xi x_e\,/D),$$

where $\rho_0 = V_0^{-1}$. Thus, a one-to-one correspondence is established between the stationary invariant solutions in the Euler and Lagrange representations. Note that according to (10.28), the Lagrangian coordinate $x_l = x_0$ corresponds in the Euler representation to the point at $+\infty$.

Our next goal is to study the conditions assuring the appearance of the nonlinear periodic solution in proximity of the stationary point (R_1, Π_1). To do this, we move the origin to the stationary point. Performing the change of variables $x = R - R_1$, $y = \Pi - \Pi_1$, we get the system

$$\xi\Delta \begin{pmatrix} x \\ y \end{pmatrix}' = \begin{bmatrix} \kappa, & R_1^2\sigma \\ -\kappa\xi^2, & -\left[(\xi R_1)^2 + \chi\xi\right] \end{bmatrix} \begin{pmatrix} x \\ y \end{pmatrix} + \begin{pmatrix} H_1 \\ H_2 \end{pmatrix}, \tag{10.29}$$

where

$$H_1 = x\left(\Pi_1 x + 2\sigma R_1 y + \sigma x y\right),$$
$$H_2 = -\xi^2\left(\Pi_1 x^2 + 2R_1 x y + x^2 y\right).$$

Conditions that guarantee the appearance of periodic solutions in the system (10.29) are formulated as follows:

$$(\xi R_1)^2 + \chi\xi = \kappa, \tag{10.30}$$

$$\Omega^2 = \kappa\xi\Delta > 0. \tag{10.31}$$

The inequality (10.31) will be fulfilled if $\xi < 0$ and the coordinate R_1 belongs to the segment $(0, \sqrt{\chi/(\tau\xi^2)})$, at which $\Delta < 0$. So the critical value of the parameter $\xi = \xi_{cr}$ is expressed by the formula

$$\xi_{cr} = -\frac{\chi + \sqrt{\chi^2 + 4\kappa R_1^2}}{2R_1^2}. \tag{10.32}$$

We are going to calculate the real part of the Floquet index $Re\, C_1$. For this purpose, we pass to the canonical coordinates, in which the linearization matrix has the form $M_{ij} = \Omega(\delta_{2\,i}\,\delta_{1\,j} - \delta_{1\,i}\,\delta_{2\,j})$. The transition to canonical coordinates, which uses the standard procedure described in Section 3.5.5 and further on, is given in the file PM2_15.nb. The same file describes the procedure for estimating the sign of the real part of the first Floquet index. This procedure is very cumbersome, so we do not present it here. The main conclusion from the analysis carried out with the significant use of the tools of the *Mathematica* package is that the limit cycle is stable if the condition (10.31) is satisfied, $\xi < 0$, and R_1 belongs to the segment $(0, \sqrt{\chi/(\tau\xi^2)})$.

Numerical studies of the behavior of the system (10.26) reveal the following sequence of regime changes: the stationary point $A(R_1, \Pi_1)$ is a stable

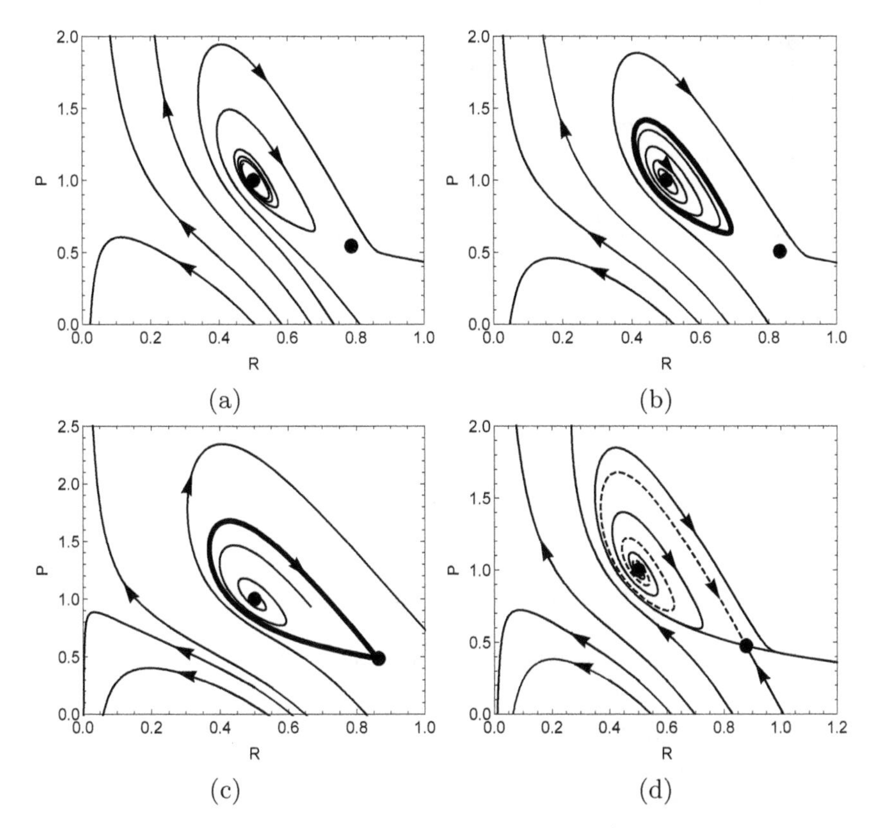

Fig. 10.7 Change of regimes described by the system (10.26) in vicinity of the stationary point $A(R_1, \Pi_1)$: (a) stable focus; (b) stable limit cycle; (c) the homoclinic loop; (d) unstable focus.

focus as long as $\xi < \xi_{cr}$ (Fig. 10.7(a)). Above the critical value, a stable limit cycle is softly generated (Fig. 10.7(b)). The radius of periodic solution increases with increase of the parameter ξ until this parameter reaches the second critical value, at which a homoclinic loop is formed (Fig. 10.7(c)). Above this value, the stationary point $A(R_1, \Pi_1)$ becomes an unstable focus (Fig. 10.7(d)).

10.4.1 *Numerical study of the relaxing hydrodynamic-type model*

For numerical studies of system (5.16) we construct the numerical scheme basing on the Godunov method [Godunov (1959); Rozhdestvenskij and Yanenko (1983)]. Let us consider the calculating cell $a\,b\,c\,d$ (see Fig. 10.8)

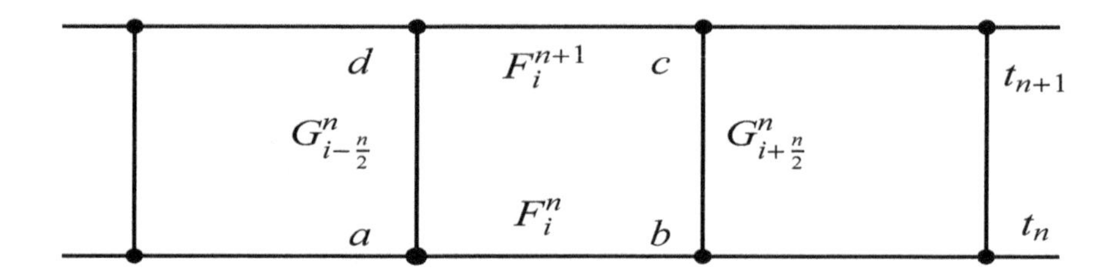

Fig. 10.8 Scheme of calculating cell for the system (10.33).

lying between nth and $(n+1)$th temporal layers of the uniform rectangular mesh. It is easy to see that system (5.16) can be presented in the following vector form:

$$\frac{\partial F}{\partial t} + \frac{\partial G}{\partial x} = H, \tag{10.33}$$

where

$$F = \left(u,\, V,\, p - \frac{\chi}{\tau V}\right)^{tr}, \quad G = (p,\, -u,\, 0)^{tr}, \quad H = \left[\gamma,\, 0,\, \frac{1}{\tau}\left(\frac{\kappa}{V} - p\right)\right]^{tr}.$$

From (10.33) it arises the equality of integrals

$$\iint_\Omega \left(\frac{\partial F}{\partial t} + \frac{\partial G}{\partial x}\right) dx\, dt = \iint_\Omega H\, dx\, dt,$$

where Ω coincides with the rectangle $a\,b\,c\,d$. Due to the Green formula, integral in the LHS of the above equation can be presented in the form

$$\iint_\Omega \left(\frac{\partial F}{\partial t} + \frac{\partial G}{\partial x}\right) dx\, dt = \oint_{\partial\Omega} G\, dt - F\, dx. \tag{10.34}$$

Let us denote the distance between the ith and $(i+1)$th nodes of the OX axis by Δx while the corresponding distance between the temporal layers by Δt. Then, up to $O\left(|\Delta x|^2, |\Delta t|^2\right)$, we get from (10.33)–(10.34) the following difference scheme:

$$\left(F_i^{n+1} - F_i^n\right)\Delta x + \left(G^{n+1}_{i+\frac{1}{2}} - G^n_{i-\frac{1}{2}}\right)\Delta t = H_i^n\, \Delta x\, \Delta t, \tag{10.35}$$

where $G^{n+1}_{i+\frac{1}{2}}$, $G^n_{i-\frac{1}{2}}$ are the values of the vector-function G on the segments $b\,c$ and $a\,d$, correspondingly. In the Godunov method these values are defined by solving the Riemann problem for the homogeneous system

$$\begin{aligned} u_t + p_x &= 0, \\ V_t - u_x &= 0, \\ p_t + \frac{\chi}{\tau V^2} V_t &= 0. \end{aligned} \tag{10.36}$$

For the system (10.36) the Riemann problem is formulated as follows. It is assumed that there are constant values (V_1, u_1, p_1) when $x < 0$ and generally speaking different constant values (V_2, u_2, p_2) when $x > 0$. The problem is posed to describe the evolution of the jump of the indicated quantities. The solution to this problem needs a certain kind of explanation. First of all, we note that the system (10.36) is equivalent in the acoustic approximation to the following system written in the characteristic form:

$$\frac{d}{dt}\left(p - \frac{\chi}{\tau V}\right) = 0, \qquad \frac{dx}{dt} = 0,$$

$$\frac{d}{dt}(p + \varepsilon C u) = \frac{\partial}{\partial t}(p + \varepsilon C u) + \varepsilon C \frac{\partial}{\partial x}(p + \varepsilon C u) = 0, \qquad \frac{dx}{dt} = \varepsilon C, \tag{10.37}$$

where $C = \sqrt{\frac{\chi}{\tau V_0}}$, $\varepsilon = \pm 1$. The functions

$$r_0 = p - \frac{\chi}{\tau V}, \qquad r_\pm = p \pm C u$$

are called *the Riemann invariants*. It follows from the formula (10.37) that the Riemann invariants retain constant values along the corresponding lines, called *characteristics*, namely: the invariant r_0 is constant along the line $x = 0$, the invariants $r_\pm$ retain the constant values on the characteristics $x = \pm C t$, respectively.

After this short introduction, we can start solving the Riemann problem. Let us immediately note that for arbitrarily chosen values of the parameters to the left and to the right of the point $x = 0$, when passing through the jump, the conditions for the constancy of the Riemann invariants along the corresponding characteristics will not be satisfied. These conditions can be satisfied if we assume that between the primary regions of constant values of the parameters in the half-plane $t > 0$, there appear new areas with different values of the parameters, separated by the characteristics.

Characteristics $x = \pm C t$ and $x = 0$ divide the half-plane $t > 0$ into four sectors (Fig. 10.9) and the problem is to find the values of the parameters in sectors II and III, basing at the values (u_1, V_1, p_1), (u_2, V_2, p_2) and assuming that $V_0 = \frac{1}{2}(V_1 + V_2)$. The scheme of calculating the values U_{II}, P_{II} is based on the property of the Riemann invariants to retain their values along the corresponding characteristics. From this we get the system of algebraic equations (cf. with Fig. 10.9):

$$p_1 + C u_1 = P_{II} + C U_{II},$$
$$p_2 + C u_2 = P_{II} - C U_{II}.$$

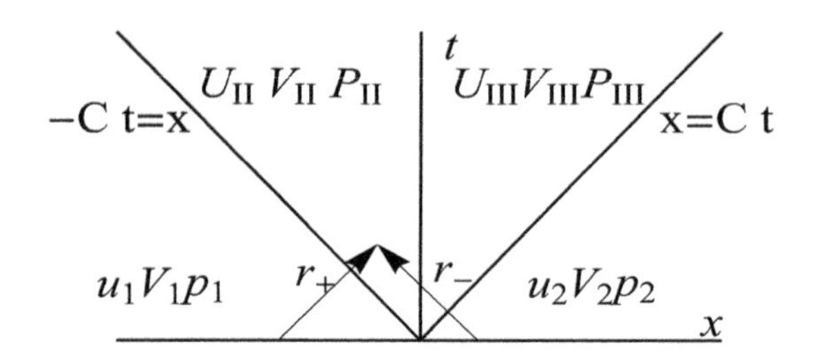

Fig. 10.9 Scheme of solving the Riemann problem for the system (10.36).

The system of determining equations for U_{III}, P_{III} occurs to be the same, so the values of the parameters U, P in the sector $-C\,t < x < C\,t$, $C = \sqrt{\chi/(\tau\, V_0^2)}$ are given by the formula

$$
\begin{aligned}
U &= \tfrac{u_1+u_2}{2} + \tfrac{p_1-p_2}{2\,C},\\
P &= C\,\tfrac{u_1-u_2}{2} + \tfrac{p_1+p_2}{2}.
\end{aligned}
\tag{10.38}
$$

Expression for the function V is omitted since it does not take part in the construction of the scheme on this step.

Thus, the difference scheme for the system (5.16) has the form

$$
\begin{gathered}
(u_i^n - u_i^{n+1})\Delta x - (p_{i+1/2}^n - p_{i-1/2}^n)\Delta t = -\gamma \Delta t \Delta x,\\
(V_i^n - V_i^{n+1})\Delta x + (u_{i+1/2}^n - u_{i-1/2}^n)\Delta t = 0,\\
\left(p_i^n - \frac{\chi}{\tau V_i^n}\right)\Delta x - \left(p_i^{n+1} - \frac{\chi}{\tau V_i^{n+1}}\right)\Delta x = -\frac{f}{\tau}\Delta t \Delta x,
\end{gathered}
\tag{10.39}
$$

where $(u_{i-1/2}^n, p_{i-1/2}^n)$ and $(u_{i+1/2}^n, p_{i+1/2}^n)$ are the solutions to Riemann problems $(V_{i-1}^n, u_{i-1}^n, p_{i-1}^n)$, (V_i^n, u_i^n, p_i^n) and (V_i^n, u_i^n, p_i^n), $(V_{i+1}^n, u_{i+1}^n, p_{i+1}^n)$, correspondingly,

$$
f = f(p_i^k, V_i^k) = \frac{\kappa}{V_i^k} - p_i^k,
$$

$k = n$ or $n+1$. The choice $k = n$ leads to the explicit Godunov scheme

$$
\left\{
\begin{aligned}
u_i^{n+1} &= u_i^n + \tfrac{\Delta t}{\Delta x}\left(p_{i-1/2}^n - p_{i+1/2}^n\right) + \gamma \Delta t,\\
V_i^{n+1} &= V_i^n + \tfrac{\Delta t}{\Delta x}(u_{i+1/2}^n - u_{i-1/2}^n),\\
p_i^{n+1} &= p_i^n + \tfrac{\chi}{\tau}\left(\tfrac{1}{V_i^{n+1}} - \tfrac{1}{V_i^n}\right) + f/\tau(p_i^n, V_i^n)\Delta t.
\end{aligned}
\right.
\tag{10.40}
$$

The numerical scheme was tested on the invariant solutions

$$
u = U(\omega), \quad p = P(\omega), \quad V = V(\omega), \quad \omega = x - Dt. \tag{10.41}
$$

Inserting (10.41) into the first two equations of the system (5.16), one obtains the following expressions for U and P:

$$\begin{cases} U = u_1 + D(V_1 - V), \\ P = p_1 + D^2(V_1 - V), \end{cases} \tag{10.42}$$

where $V_1 = \lim_{\omega \to \infty} V(\omega)$. We assume in addition that $u_1 = 0$ and $p_1 = \kappa/V_1$. Under these assumptions the critical point (u_1, p_1, V_1) represents the stationary solution of the initial system.

Inserting (10.42) into the third equation of the system (5.16), we get the equation

$$\frac{dV}{d\omega} = -V \frac{[D^2V^2 - SV + \kappa]}{\tau D[C_{T\infty}^2 - (DV)^2]} = F(V), \tag{10.43}$$

where $C_{T\infty} = \sqrt{\chi/\tau}$, $S = p_1 + D^2V_1$. Equation (10.43) has three stationary points:

$$V = V_0 = 0, \qquad V = V_1, \qquad V = V_2 = \kappa/(V_1D^2).$$

If $V_2 = \kappa/(V_1D^2) < V_1$, and the line $\chi/\tau - (DV)^2 = 0$ lies outside the segment (V_2, V_1), then the solution of the equation (10.43) describes a smooth wave of compression. The stationary solution at the $-\infty$, described by the formula

$$u_{-\infty} = u_2 = \frac{(DV_1)^2 - \kappa}{DV_1} > 0, \qquad p_{-\infty} = p_2 = \kappa/V_2, \qquad V_{-\infty} = V_2,$$

under the above conditions satisfies the initial system.

Since the frozen and the equilibrium acoustical speeds obey the inequality $\kappa < \chi/\tau$, the points V_1, V_2 and $V_3 = \sqrt{\chi/(\tau D^2)}$ satisfy the inequality

$$V_1V_2 < V_3^2.$$

Since we assume that $V_2 < V_1$, the stationary point V_3 can be located either to the right of point V_1, or inside the interval (V_2, V_1), yet in the last case a continuous solution connecting the stationary points V_2 and V_1 does not exist.

Effective width L of the TW solution described by the equation (10.43), defined as for the case of the viscous shock-wave solution of the Burgers equation, is expressed by the following formula:

$$L = 2D\tau \frac{4(C_{T\infty}V_1D)^2 - [(V_1D)^2 + \kappa]^2}{(V_1D)^4 - \kappa^2}.$$

Thus, at fixed values of $C_{T\infty}$ and κ effective width of the wave front is proportional to the time of relaxation τ. Let us note that L is positive at the arbitrary choice of $V_0 \in (V_2, V_1)$ as long as the inequalities $V_2 < V_1 < V_3$ hold.

The results of numerical simulation of the Cauchy problem in which the wave front played the role of Cauchy data is presented in Fig. 10.10. The numerical simulation shows that the viscous shock wave is stable and evolves in the self-similar mode.

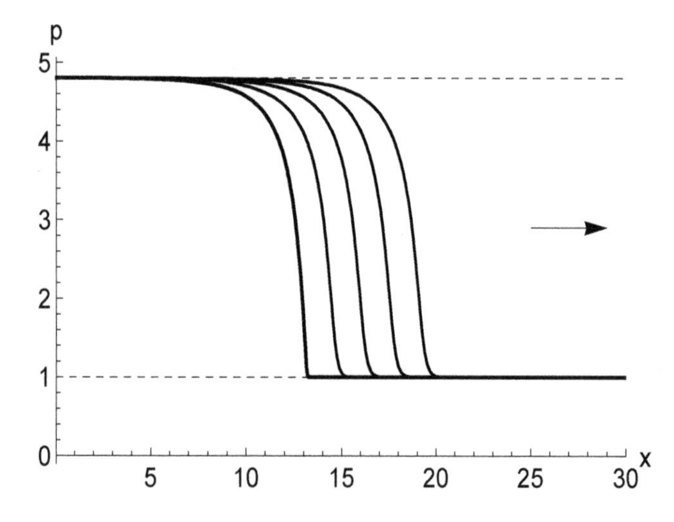

Fig. 10.10 Temporal evolution of the solution of the system (5.16) with the solution of (10.43) taken as a Cauchy data at the following values of the parameters: $\kappa = 0.5$, $\chi = 0.25$, $\tau = 0.1$, $V_1 = 0.5$, $D = 3.1$. Numerical calculations were displayed on a graph with an interval of $\Delta t = 0.5$.

When

$$V_2 < V_3 < V_1, \tag{3.7.18}$$

then the system (5.16) possesses shock wave solutions. Assuming that the medium ahead of the wave front is unmoved, we get, using (10.42), following expressions for the parameters V_3^*, p_3^*, u_3^* behind the shock front:

$$V_3^* = \frac{\chi}{\tau V_1 D^2}, \quad p_3^* = \frac{\kappa}{V_1} + \frac{\tau (V_1 D)^2 - \chi}{\tau V_1}, \quad u_3^* = \frac{\tau (V_1 D)^2 - \chi}{\tau V_1 D}.$$

Figure 10.11 represents the result solution of the Cauchy problem for the system (5.16) performed with the help of the Godunov scheme in which the shock wave solution to the equation (10.43) was employed as Cauchy data. The numerical experiment shows that the shock wave evolves in a stable self-similar mode.

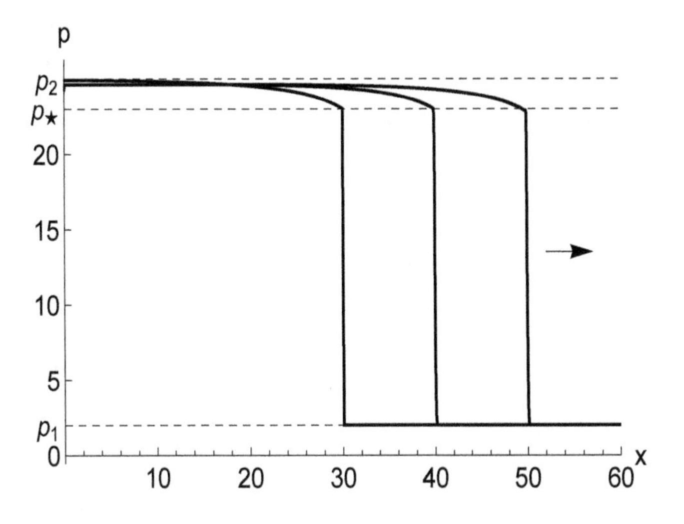

Fig. 10.11 Temporal evolution of the shock wave solutions to (10.43) taken as a Cauchy data in numerical solution of the system (5.16) by means of the Godunov method at the following values of the parameters: $\kappa = 2$, $\chi = 4$, $\tau = 1$, $V_1 = 1$, $D = 5$. Numerical calculations were displayed on a graph with an interval of $\Delta t = 2$.

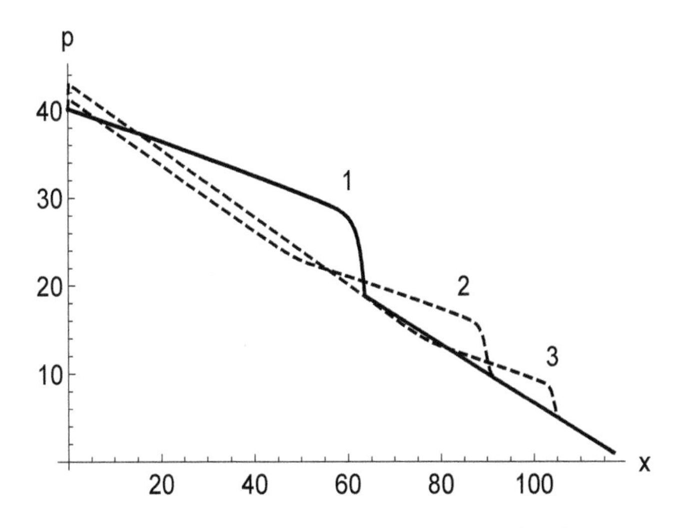

Fig. 10.12 Temporal evolution of the solution of the system (5.16) with the compacton solution of (10.26) taken as a Cauchy data at the following values of the parameters: $\kappa = 10$, $\chi = 1.5$, $\tau = 0.07$, $\gamma = -0.04$.

In numerical experiments we also used the initial data obtained by solving the factorized system (10.26). Of particular interest is the case when the homoclinic trajectory doubly asymptotic to the topological saddle of the system (10.26) was taken as the Cauchy data. Results of the numerical simulation are shown in Fig. 10.12. It is seen that wave pack evolves for a long time in a stable self-similar mode. Attention is drawn to the fact that the wave disturbance sags somewhat in the tail section. Without a doubt, this is due to the fact that the saddle point of the system (10.26) does not correspond to any stationary solution of the initial system (5.16). Nevertheless, for the chosen values of the parameters, at which the stationary point B_+ lies in close proximity to the stationary point A, a stable evolution of the wave packet associated with compacton solution of the system (10.26) is observed. It should be noted that the wave perturbation associated with the compacton has another interesting property. As shown in paper [Vladimirov (2008)], under certain conditions, an arbitrary perturbation taken as the initial condition splits into two impulses, and the impulse that rushes to the right, in the process of evolution, approaches the indicated compacton solution. This effect weakly depends on the shape of the initial perturbation, and to a much greater extent depends on its total energy.

Appendix A

Elements of calculus of functions of complex variables

In this appendix we briefly discuss the basic facts about the differential and integral calculus of complex functions [Ditkin and Prudnikov (1975); Kelly (2006)]. The main goal is to present the applications of *theory of residues* for calculating integrals, so we will consistently omit material that is not directly related to calculating integrals by residua. The interested reader can find a detailed discussion of the issues and concepts covered in any standard textbook on single complex variable function theory and operation calculus.

A.1 Holomorphic functions and their properties

Let us consider a complex-valued function $u(x,\, y)$, $(x,\, y) \in \mathbb{R}^2$:

$$u(x,\, y) = u_1(x,\, y) + i\, u_2(x,\, y), \tag{A.1}$$

where u_1, u_2 are real. Introducing new variables

$$z = x + i\, y, \qquad \bar{z} = x - i\, y, \tag{A.2}$$

we can present (A.1) as a function of complex variables:

$$u(x,\, y) = f(z,\, \bar{z}) + i\, g(z,\, \bar{z}), \tag{A.3}$$

where

$$f(z,\, \bar{z}) = u_1\left(\frac{z+\bar{z}}{2},\, \frac{z-\bar{z}}{2\,i}\right), \quad g(z,\, \bar{z}) = u_2\left(\frac{z+\bar{z}}{2},\, \frac{z-\bar{z}}{2\,i}\right). \tag{A.4}$$

Assuming that the functions u_2 and u_2 belong at least to $C^1(R^2)$, we can define for them the differentiation in the complex sense. Using (A.2) and the inverse transformation

$$x = \frac{z+\bar{z}}{2}, \qquad y = \frac{z-\bar{z}}{2\,i},$$

we get the following chain of equalities:

$$du = \frac{\partial u}{\partial x} dx + \frac{\partial u}{\partial y} dy = \frac{1}{2}\left(\frac{\partial u}{\partial z} + \frac{\partial u}{\partial \bar z}\right)(dz + d\bar z)$$
$$+ \frac{1}{2}\left(\frac{\partial u}{\partial z} - \frac{\partial u}{\partial \bar z}\right)(dz - d\bar z) = \frac{\partial u}{\partial z} dz + \frac{\partial u}{\partial \bar z} d\bar z.$$

Let us note that

$$\frac{\partial u}{\partial z} = \frac{1}{2}\left(\frac{\partial u}{\partial x} + \frac{1}{i}\frac{\partial u}{\partial y}\right), \qquad \frac{\partial u}{\partial \bar z} = \frac{1}{2}\left(\frac{\partial u}{\partial x} - \frac{1}{i}\frac{\partial u}{\partial y}\right). \tag{A.5}$$

In the following we will use the notation

$$\partial u = \frac{\partial u}{\partial z} dz, \qquad \bar\partial u = \frac{\partial u}{\partial \bar z} d\bar z. \tag{A.6}$$

Definition A.1. We say that the C^1 function $u:\ \mathbb{R}^2 \supset \Omega \ni (x, y) \to \mathbb{C}$ is holomorphic on the set $\Omega \subset \mathbb{R}^2$, if $\bar\partial u = 0$.

The holomorphic function is also called a *complex differentiable* function.

We shall denote the set of functions holomorphic on $\Omega \subset \mathbb{R}^2$ as $\mathcal{A}(\Omega)$. It is easy to show the correctness of the following statement:

Corollary A.1. *If (A.1) is the holomorphic function, then the functions* $u_1(x, y),\ u_2(x, y)$ *satisfy the Cauchy-Riemann system*

$$\frac{\partial u_1}{\partial x} = \frac{\partial u_2}{\partial y}, \qquad \frac{\partial u_2}{\partial x} = -\frac{\partial u_1}{\partial y}. \tag{A.7}$$

Below we present the main properties of the holomorphic functions:

(1) if $f,\ g \in \mathcal{A}(\Omega)$, then $\alpha f + \beta g \in \mathcal{A}(\Omega)$ and $f \cdot g \in \mathcal{A}(\Omega)$, where α, β are arbitrary complex numbers;
(2) if $f \in \mathcal{A}(\Omega)$ i $f\big|_{\Omega} \neq 0$, then $1/f \in \mathcal{A}(\Omega)$;
(3) if $f \in \mathcal{A}(\Omega)$, Ω is a connected set and $Im\, f\big|_{\Omega} = 0$, then $f = \text{const}$;
(4) if $f \in \mathcal{A}(\Omega)$, Ω is a connected set and $|f| = \text{const}$, then $f = \text{const}$.

A.2 Integration of the complex-valued functions

The basic formula in the calculus of the complex functions is the *Cauchy's formula*, connecting the value of the function $u \in \mathcal{A}(\Omega)$ at the point $\zeta \in \Omega$ with some line integral. Note that the integral of a complex function is defined in a similar way as in the case of the real function, namely, let ℓ be a continuous curve contained in the set Ω, $\{\ell_k\}_{k=0}^{n}$ is a division of this curve, $\Delta \ell_j$ is the line between point the ℓ_j and the point ℓ_{j+1}, $\Delta z_j = (x_{j+1} - x_j) + i\,(y_{j+1} - y_j)$ is the increment of the argument of the function on this segment. Then the integral of the u function along the curve ℓ is defined as follows:

$$\int_{\ell} u(z)\,d\,z = \lim_{\substack{n\to\infty \\ \lambda\to 0}} \sum_{k=0}^{n} u(\xi_k)\,\Delta\,z_k,$$

where $\xi_k \in \Delta\,\ell_k$, $\lambda = \max_k |\Delta\,z_k|$. Note that the following formula takes place:

$$\int_{l} u(z)\,d\,z = \int_{l} u_1\,d\,x - u_2\,d\,y + i\int_{l} u_1\,d\,y + u_2\,d\,x. \qquad \text{(A.8)}$$

It appears from the above formula that the complex integral inherits some properties of the real-valued integral, in particular, the following ones:

(1)

$$\int_{l} (\alpha_1 f_1(z) + \alpha_2 f_2(z))\;d\,z = \alpha_1 \int_{l} f_1(z)\,d\,z + \alpha_2 \int_{l} f_2(z)\,d\,z,$$

(2)

$$\int_{AB} f(z)\,d\,z = -\int_{BA} f(z)\,d\,z,$$

(3)

$$\int_{AB} f(z)\,d\,z = \int_{AC} f(z)\,d\,z + \int_{CB} f(z)\,d\,z,$$

(4)

$$\int_{AB} |d\,z| = l_{AB},$$

where l_{AB} is the length of the arc of the curve AB,

(5)

$$\left|\int_{l} f(z)\,d\,z\right| \le \int_{l} |f(z)|\,d\,z.$$

Treating the real and imaginary parts of the integral over the complex domain as a convenient line integrals, we can also directly employ the formula (A.8) and this implies the validity of the following statement.

Lemma A.1. *If Ω is a bounded simply connected set, while $u(z) = u_1(z) + i\,u_2(z) \in \mathcal{A}(\Omega)$, then*

$$\int_{\partial\Omega} u(z)\,dz = 0. \tag{A.9}$$

Proof. Employing to (A.9) the Gauss-Ostrogradski theorem and using (A.7), we can make sure that the real and the imaginary parts of the integral are equal to zero.

Corollary A.2. *If $u(z) = \in \mathcal{A}(\Omega)$, Ω is a simply connected set, A, $B \in \Omega$, then the integral*

$$\int_{AB} u(z)\,dz$$

does not depend on the path of integration.

Thus, if $u(z) \in \mathcal{A}(\Omega)$, Ω is a simply-connected set and z_0, $z \in \Omega$, then the integral

$$\int_{z_0}^{z} u(z)\,dz = F(z)$$

defines a one-to-one function. It is easy to see by considering

$$\lim_{\Delta z \to 0} \frac{\Delta F}{\Delta z},$$

that the complex derivative of $F(z)$ is well-defined in any point $z \in \Omega$, hence F is the analytical function and, besides, the formula $F'(z) = f(z)$ takes place. Both the derivatives of the analytical function and the integral of the analytical functions are calculated exactly the same as in the real case, for example

$$\int e^z\,dz = e^z + \text{const}, \qquad \int \cos(z)\,dz = \sin(z) + \text{const},$$

etc.

Now let us formulate the Cauchy theorem.

Theorem A.1. *Let Ω be a bounded simply connected set, and $f \in \mathcal{A}(\Omega)$. Then for any $a \in \operatorname{int}\Omega$ the following formula takes place:*

$$f(a) = \frac{1}{2\pi i}\int_{\partial\Omega} \frac{f(z)}{z-a}\,dz. \tag{A.10}$$

Outline of the proof. Let us considered a circle $K(a,\,\epsilon)$ centered at $a \in \mathbb{R}^2$, and lying inside Ω. We denote by Ω_ϵ the set Ω from which the set $K(a,\,\epsilon)$ is cut. Since $f \in \mathcal{A}(\Omega_\epsilon)$, then the following formula takes place:

$$0 = \int\limits_{\partial\Omega_\epsilon} \frac{f(z)}{z-a}\,dz = \int\limits_{\partial\Omega} \frac{f(z)}{z-a}\,dz - i\int_0^{2\pi} f(a+\epsilon\,e^{i\varphi})\,d\varphi.$$

Taking the limit $\lim\limits_{\epsilon\to 0}$ in this formula we obtain what is needed.

Remark. The formula (A.10) is called the *Cauchy formula.*

A.3 Functions holomorphic in the ring. Laurent series expansion and classification of singular points

Definition A.2. The set

$$R(z_0,\,r_1,\,r_2) = \{z \in \mathbb{C} : r_1 < |z - z_0| < r_2\}$$

is called an open ring centered at $z_0 \in \mathbb{C}$, with the inner radius $r_1 \geq 0$ and the outer radius $r_2 > r_1$.

Theorem A.2. *If $f(z)$ is the function holomorphic in $R(z_0,\,r_1,\,r_2)$, then it can be presented in the form of the following series*

$$f(z) = \sum_{n=-\infty}^{+\infty} a_n (z - z_0)^n, \tag{A.11}$$

where

$$a_n = \int\limits_{|\zeta - z_0| = \rho} \frac{f(\zeta)}{(\zeta - z_0)^{n+1}}\,d\zeta, \quad r_1 < \rho < r_2. \tag{A.12}$$

The series (A.11) is absolutely convergent in $R(z_0,\,r_1,\,r_2)$.

Outline of the proof. Let $r_1 < \rho_1 < \rho_2 < r_2$. We use the Cauchy's formula to the closure of the set $R(z_0,\,\rho_1,\,\rho_2)$:

$$\begin{aligned} f(z) &= \frac{1}{2\,i\,\pi} \int\limits_{\partial R(z_0,\,\rho_1,\,\rho_2)} \frac{f(\zeta)}{\zeta - z}\,d\zeta \\ &= \frac{1}{2\,i\,\pi}\left\{ \int\limits_{|\zeta - z_0| = \rho_2} \frac{f(\zeta)}{\zeta - z}\,d\zeta - \int\limits_{|\zeta - z_0| = \rho_1} \frac{f(\zeta)}{\zeta - z}\,d\zeta \right\}, \end{aligned}$$

where $z \in R(z_0, \rho_1, \rho_2)$. The above formula can be presented in the form

$$f(z) = \frac{1}{2\,i\,\pi} \left\{ \int\limits_{|\zeta - z_0| = \rho_2} \frac{f(\zeta)}{(\zeta - z_0)\left(1 - \frac{z - z_0}{\zeta - z_0}\right)}\, d\,\zeta + \int\limits_{|\zeta - z_0| = \rho_1} \frac{f(\zeta)}{(z - z_0)\left(1 - \frac{\zeta - z_0}{z - z_0}\right)}\, d\,\zeta \right\}.$$

Since $z \in R(z_0, \rho_1, \rho_2)$, then the following inequalities hold:

$$\left| \frac{z - z_0}{\zeta - z_0} \right| < 1 \quad \text{as} \quad |\zeta - z_0| = \rho_2,$$

$$\left| \frac{\zeta - z_0}{z - z_0} \right| < 1 \quad \text{as} \quad |\zeta - z_0| = \rho_1,$$

hence, using the formula for the sum of geometric progression, we get:

$$f(z) = \frac{1}{2\,i\,\pi} \sum_{n=0}^{\infty} \left\{ (z - z_0)^n \int\limits_{|\zeta - z_0| = \rho_2} \frac{f(\zeta)}{(\zeta - z_0)^{n+1}}\, d\,\zeta + \frac{1}{(z - z_0)^{n+1}} \int\limits_{|\zeta - z_0| = \rho_1} f(\zeta)(\zeta - z_0)^n\, d\,\zeta \right\}.$$

Since the integrand is holomorphic on $R(z_0, r_1, r_2)$, then the integrals in the above formula do not depend on the radius of the circle ρ as long as the inequalities $r_1 < \rho < r_2$ are fulfilled. From this appears the correctness of the formula (A.11) (details of the proof can be seen, for example, in the book [Maurin (1973)], Part II, Chapter XV, # 3).

Let $f \in \mathcal{A}(R(z_0, 0, r))$, and therefore it can be expanded into the Laurent series (A.11) inside the ring $R(z_0, 0, r)$. The following cases can occur:

(1) $a_n = 0$ for $n < 0$; then, assuming that $f(z_0) = a_0$, we'll obtain the formula

$$f(z) = \sum_{n=0}^{+\infty} a_n (z - z_0)^n, \qquad z \in R(z_0, 0, r);$$

in this case $f(z)$ is holomorphic on $K(z_0, r)$.

(2) There exist a natural number k such that $a_n = 0$ as $n < -k$ and $a_{-k} \neq 0$. Then

$$f(z) = \sum_{m=-k}^{-1} a_m (z - z_0)^m + \sum_{n=0}^{+\infty} a_n (z - z_0)^n, \qquad z \in R(z_0, 0, r).$$

The point z_0 is called the *isolated pole of the order k*.

(3) For any natural k there exists such $k_1 > k$ that $a_{-k_1} \neq 0$. Then

$$f(z) = \sum_{n=-\infty}^{+\infty} a_n (z - z_0)^n, \qquad z \in R(z_0, 0, r).$$

The point z_0 is called the *essentially singular*.

Definition A.3. If z_0 is the isolated singular point, then the coefficient a_{-1} of the decomposition of $f(z)$, into the Laurent series around this point is called the *residuum* of this function in z_0, and is denoted as $\operatorname{Res} f(z_0)$.

So the residuum is defined by the following formula:

$$\operatorname{Res} f(z_0) = a_{-1} = \frac{1}{2\pi i} \int\limits_{|\zeta - z_0| = \rho < r} f(\zeta)\, d\zeta, \tag{A.13}$$

where $f \in \mathcal{A}(R(z_0, 0, r))$, $r > 0$. Following statement holds true:

Lemma A.2. *If the function f has the pole of $n-th$ order in z_0, then*

$$\operatorname{Res} f(z_0) = \frac{1}{(n-1)!} \lim_{z \to z_0} \frac{d^{n-1}}{d\, z^{n-1}} \{(z - z_0)^n f(z)\}. \tag{A.14}$$

The proof is elementary.

Definition A.4. Let $\Omega \subset \mathbb{C}$ be an open set, while P be the set consisting of isolated singular points $z_k \in \Omega$, while $\Omega - P$ be open in $\mathbb{C}$. The function f is called *meromorphic* on Ω if it is holomorphic on $\Omega - P$, and the points z_k are its poles. The set of all functions meromorphic on Ω is denoted as $\mathcal{M}(\Omega)$.

Theorem A.3. *Let $\Omega \in \mathbb{C}$ be an open set, $P = \bigcup_{k=1}^{n} \{z_k\}$, $z_k \in \Omega$ be the set of isolated points. If the set $D \in \Omega$ is such that $\partial D \cap P$ is an empty set, then for any function $f \in \mathcal{M}(\Omega - P)$ the following formula takes place:*

$$\oint\limits_{\partial D} f(z)\, d\, z = 2\pi i \sum_{k=1}^{m} \operatorname{Res} f(z_k), \qquad z_k \in D \cap P. \tag{A.15}$$

We omit the proof of this statement.

The above theorem, called the *residues theorem* is employed, e.g., when calculation integral

$$\int_{-\infty}^{+\infty} R(x)\, d\, x,$$

in which the integrand $R(\cdot)$:

(a) does not possess singular points on the real axis;
(b) has the finite number of poles, and no other singular points;
(c) $\lim_{|z|\to\infty} z\,R(z) = 0$.

We will show how the residues theorem works in this case. We choose the integration contour Γ, which consists of the segment $(-r,\ +r)$ of the real axis and the semicircle K_r with the radius r, lying in the upper half-plane. The equality

$$\oint_\Gamma R(z)\,d\,z = 2\pi\,i\sum_k \mathrm{Res}R(z_k),$$

follows from the residues theorem. The summation in the above formula extends to all singular points lying inside the contour Γ.

There remains to show the validity of the formula

$$\lim_{r\to\infty}\int_{K_r} R(z)e^{i\,a\,z}\,d\,z = 0.$$

It is a consequence of the following statement, called the *Jordan lemma*:

Theorem A.4. *Let $f(z)$ be a function defined in the upper half-plane. If $a>0$, $\lim_{|z|\to\infty} f(z) = 0$, then*

$$\lim_{r\to\infty}\int_{K_r} f(z)\,e^{i\,a\,z}\,d\,z = 0.$$

Proof. Assume that

$$J_r = \int_{K_r} f(z)\,e^{i\,a\,z}\,d\,z = i\int_0^\pi re^{i\,\varphi}f(r\,e^{i\,\varphi})e^{i\,a\,r(\cos\varphi+i\sin\varphi)}\,d\,\varphi$$

the following estimation takes place:

$$|J_r| \le \int_0^\pi \left|f(r\,e^{i\,\varphi})\right|\,r\,e^{-a\,r\,\sin\varphi}\,d\,\varphi \le \sup_{0\le\varphi\le\pi}\left|f(r\,e^{i\,\varphi})\right|\int_0^\pi r\,e^{-a\,r\,\sin\varphi}\,d\,\varphi.$$

From the inequality $2\,\varphi\,\pi \le \sin\varphi$, $0\le\varphi\le\pi/2$, being the consequence of the concavity of the function $\sin\varphi$ on the segment $[0,\ 2\,\pi]$, appears that

$$\int_0^\pi r\,e^{-a\,r\,\sin\varphi}\,d\,\varphi = 2\int_0^{\pi/2} r\,e^{-a\,r\,\sin\varphi}\,d\,\varphi \le 2\int_0^{\pi/2} r\,e^{-2\,a\,r\,\varphi/\pi}$$
$$= \frac{\pi}{a}(1-e^{-a\,r}) \le \frac{\pi}{a}.$$

Hence

$$|J_r| \le \frac{\pi}{a} \sup \left| f(r\,e^{i\,\varphi}) \right| \underset{r\to\infty}{\to} 0 \,.$$

Estimation of K_r on the half-circle is as follows:

$$\left| \int_{K_r} R(z)\,d\,z \right| = \left| \int_0^{\pi} i\,\rho\, R(\rho e^{i\,\varphi})\, e^{i\,\varphi}\, d\,\varphi \right| \le \pi \sup_{0\le\varphi\le\pi} \left| \rho\, R(\rho e^{i\,\varphi}) \right| \underset{\rho\to\infty}{\to} 0.$$

When ρ tends to infinity, we get the formula

$$\int_{-\infty}^{+\infty} R(x) d\,x = 2\pi\, i \sum_k \mathrm{Res} R(z_k),$$

in which the summation is carried out over all singular points of the function $R(z)$ lying in the upper half-plane.

Example A.1. Let us consider the integral

$$I = \int_{-\infty}^{+\infty} \frac{1}{(1+x^2)^3} d\,x.$$

The conditions (a) are (b) evidently fulfilled; in addition the following estimation holds:

$$\lim_{|z|\to\infty} \frac{z}{(1+z^2)^3} = 0.$$

The function $f(z) = \frac{1}{(1+z^2)^3}$ has the poles in the points $z_{\pm} = \pm i$. Thus merely the point $z_+ = i$ lies in the upper half-plane. Calculating the residuum in this point, we get:

$$\mathrm{Res} f(i) = \frac{1}{2!} \lim_{z\to i} \frac{d^2}{d\,z^2} \left[(z-i)^3\, f(z) \right] = \frac{1}{2} \lim_{z\to i} \frac{d^2}{d\,z^2} \left[\frac{1}{(z+i)^3} \right] = -\frac{3}{16}\, i.$$

Hence $I = \frac{3}{8}\,\pi$.

Appendix B

Certain statements justifying the use of integral transformations in solving differential equations

B.1 The space of tempered distributions

Let us consider the set of test functions $\mathcal{S}(\mathbb{R}^n) \in C^\infty(\mathbb{R}^n)$ tending to zero together with all their partial derivatives as $|x| \to \infty$ quicker than any power of $(1+|x|)^{-m}$ for arbitrary natural m. Topology in the set $\mathcal{S}(\mathbb{R}^n)$ is defined as follows: a sequence $\phi_1, \phi_2, \ldots$ belonging to $\mathcal{S}$ tends to $\phi \in \mathcal{S}$, if for any multi-indices $\alpha = (\alpha_1, \ldots, \alpha_n)$, $\beta = (\beta_1, \ldots, \beta_n)$

$$x^\beta D^\alpha \phi_k(x) \Longrightarrow x^\beta D^\alpha \phi(x) \text{ as } k \to \infty.$$

Definition B.1. The space of *tempered distribution* $\mathcal{S}'(\mathbb{R}^n)$ is a set of linear functionals which act on $\mathcal{S}(\mathbb{R}^n)$, and are continuous in the weak topology.

The value of tempered distribution $f \in \mathcal{S}'(\mathbb{R}^n)$ on the test function $\phi(x) \in \mathcal{S}(\mathbb{R}^n)$ is denoted as

$$(f, \phi) \in \mathbb{C}.$$

Linearity of $f \in \mathcal{S}'(\mathbb{R}^n)$ means that $\forall\ \alpha_1, \alpha_2 \in \mathbb{C}$ and $\forall\ \phi_1, \phi_2 \in \mathcal{S}(\mathbb{R}^n)$

$$(f, \alpha_1 \phi_1 + \alpha_2 \phi_2) = \alpha_1 (f, \phi_1) + \alpha_2 (f, \phi_2).$$

The weak continuity of $f \in \mathcal{S}'(\mathbb{R}^n)$ means that $\mathcal{S} \ni \phi_k \underset{k\to\infty}{\to} \phi \in \mathcal{S}$, then

$$(f, \phi_k) \underset{k\to\infty}{\to} (f, \phi).$$

The convergence is treated as convergence of the numerical sequence in $\mathbb{C}$. The set $\mathcal{S}'(\mathbb{R}^n)$ will form a linear space if we define the linear combination $\lambda f + \mu g$, $f, g \in \mathcal{S}'(\mathbb{R}^n)$, $\lambda, \mu \in \mathbb{C}$ as follows:

$$(\lambda f + \mu g, \phi) = \lambda (f, \phi) + \mu (g, \phi), \qquad \phi \in \mathcal{S}(\mathbb{R}^n).$$

The simplest tempered distribution can be determined by means of any locally integrable function $f(x)$, which grows at infinity no faster than some finite degree polynomial function, in other words, if there is a number $m \in \mathbb{N} \cup \{0\}$ such that

$$\int_{\mathbb{R}^n} |f(x)|\,(1+|x|)^{-m}\,dx < \infty,$$

then the functional

$$(f,\,\phi) = \int_{\mathbb{R}^n} f(x)\,\phi(x)dx, \qquad \phi \in \mathcal{S} \tag{B.1}$$

is well defined. Each functional that can be defined by means of the formula (B.1) is called *a regular distribution.* In addition to regular distributions, which form a dense subset of $\mathcal{S}'(\mathbb{R}^n)$, there are so-called *singular distributions.* An example of a singular distribution is the Dirac delta $\delta(x)$ which can be defined as a limit of the regular distribution

$$f_\epsilon(x) = \begin{cases} C_\epsilon & \text{as} \quad |x| < \epsilon, \\ 0 & \text{as} \quad |x| \geq \epsilon, \end{cases}$$

where C_ϵ^{-1} is the volume of the n-dimension ball having the radius ϵ. The action of the δ-function on the test function $\phi(x) \in \mathcal{S}$ is defined as follows:

$$(\delta,\,\phi) = \lim_{\epsilon \to 0+} \int_{\mathbb{R}^n} f_\epsilon(x)\,\phi(x)\,dx, \qquad \phi \in \mathcal{S}. \tag{B.2}$$

Let us show, that

$$(\delta,\,\phi) = \phi(0).$$

Indeed,

$$\lim_{\epsilon \to 0} \left| \int_{\mathbb{R}^n} f_\epsilon(x)\,\phi(x)\,dx - \phi(0) \right| = \lim_{\epsilon \to 0} \left| \int_{|x|<\epsilon} f_\epsilon(x)\,\phi(x)\,dx - \phi(0) \int_{|x|<\epsilon} f_\epsilon(x)\,dx \right|$$
$$= \lim_{\epsilon \to 0} \left| \int_{|x|<\epsilon} f_\epsilon(x)\,[\phi(x) - \phi(0)]\,dx \right| \leq \lim_{\epsilon \to 0} \sup_{|x|<\epsilon} |\phi(x) - \phi(0)| = 0$$

and hence the statement is correct.

Properties of the tempered distributions. Let $f(x) \in \mathcal{S}'(\mathbb{R}^n)$ be the regular distribution, and $x = Ay + b$, $\det A \neq 0$ be the non-singular linear transformation of $\mathbb{R}^n$ into itself. Then $\forall\,\phi \in \mathcal{S}(\mathbb{R}^n)$

$$(f(Ay+b),\,\phi) = \int_{\mathbb{R}^n} f(Ay+b)\,\phi(y)\,dy = \frac{1}{|\det A|} \int_{\mathbb{R}^n} f(x)\,\phi\left[A^{-1}(x-b)\right]\,dy$$
$$= \frac{1}{|\det A|}\left(f,\,\phi\left[A^{-1}(x-b)\right]\right).$$

This equality is taken as the definition of the distribution $f(A\,y+b)$ in general case: $\forall\, f\,\in\,\mathcal{S}'(\mathbb{R}^n)$

$$(f(A\,y+b),\,\phi) := \frac{1}{|\det\,A|}\,(f,\,\phi\left[A^{-1}(x-b)\right]), \qquad \phi\,\in\,\mathcal{S}(\mathbb{R}^n). \qquad \text{(B.3)}$$

In particular, the following formula holds true

$$(\delta(x-x_0),\,\phi) = (\delta,\,\phi(x+x_0)) = \phi(x+x_0)\,\Big|_{x=0} = \phi(x_0).$$

Another actions on the tempered distributions are defined in a similar way as for the regular ones. The rationale behind this is that the set of regular distributions is dense in $\mathcal{S}'(\mathbb{R}^n)$ and any such definition can be introduced by means of the procedure of taking the limit. Proofs of this kind make significant use of the fact that the spaces of tempered distributions is complete in the weak topology.

Let us introduce two actions essentially used in Chapter 6.

(1) Suppose that

$$\begin{aligned}\theta_M = \{&g(x)\,\in\,C^\infty(\mathbb{R}^n)\,|\forall\alpha\;\exists\;m_\alpha\,\in\,\mathbb{N}\text{ and }C_\alpha>0:\\ &|D^\alpha\,g(x)|\,\leq\,C_\alpha\,(1+|x|)^{m_\alpha}\},\end{aligned}$$

where

$$\alpha=(\alpha_1,...,\alpha_n),\;\alpha_i\,\in\,N\cup\{0\},\;i=1,\,2,...,n,\quad D^\alpha=\partial_{x_1^{\alpha_1}}\,\partial_{x_2^{\alpha_2}}...\partial_{x_n^{\alpha_n}}.$$

Multiplication of $f\,\in\,\mathcal{S}'(\mathbb{R}^n)$ by $a(x)\,\in\,\theta_M$ is defined by the formula

$$(a\cdot\,f,\,\phi) := (f,\,a\cdot\phi), \qquad \phi\,\in\,\mathcal{S}(\mathbb{R}^n). \qquad \text{(B.4)}$$

(2) Let $f(x)\,\in\,C^p(\mathbb{R}^n)$ and $D^\alpha\,f(x)$ for $|\alpha|\,=\,\alpha_1+...+\alpha_n\,\leq\,p$ be the functions growing at the infinity not quicker than a finite degree polynomial. Then the following equality holds:

$$\begin{aligned}(D^\alpha f,\,\phi) &= \int_{\mathbb{R}^n} D^\alpha f(x)\,\phi(x)\,d\,x = (-1)^{|\alpha|}\int_{\mathbb{R}^n} f(x)\,D^\alpha\phi(x)\,d\,x\\ &= (-1)^{|\alpha|}(f,\,D^\alpha\phi).\end{aligned}$$

The above equality is taken as the definition of the derivative of the tempered distribution:

$$(D^\alpha f,\,\phi) := (-1)^{|\alpha|}(f,\,D^\alpha\phi), \quad f\,\in\,\mathcal{S}'(\mathbb{R}^n), \quad \phi\,\in\,\mathcal{S}(\mathbb{R}^n). \qquad \text{(B.5)}$$

Let us show that $D^\alpha f \in \mathcal{S}'(\mathbb{R}^n)$ $\forall\,\alpha$ or, in other words, that every tempered distribution is infinitely differentiable.

Indeed, the functional defined this way is linear, because $\forall\mu,\,\nu \in \mathbb{C}$, $\forall\phi,\,\psi \in \mathcal{S}(\mathbb{R}^n)$

$$\begin{aligned}&(D^\alpha f,\, \mu\,\phi + \nu\,\psi) = (-1)^{|\alpha|}(f,\, D^\alpha(\mu\,\phi + \nu\,\psi))\\ &= (-1)^{|\alpha|}\mu\,(f,\, D^\alpha\,\phi) + (-1)^{|\alpha|}\nu\,(f,\, D^\alpha\,\psi)\\ &= \mu\,(D^\alpha\, f, \phi) + \nu\,(D^\alpha\, f, \psi).\end{aligned}$$

Next we show that the functional $D^\alpha f$ is continuous. Indeed, if $\phi_k \overset{\mathcal{S}}{\to} \phi$, then

$$(D^\alpha\, f,\, \phi_k) = (-1)^{|\alpha|}(f,\, D^\alpha\,\phi_k) \underset{k\to\infty}{\to} (-1)^{|\alpha|}(f,\, D^\alpha\,\phi) = (D^\alpha\, f,\, \phi_k).$$

It is shown in the similar way that the result of taking the mixed derivative does not depend on the order:

$$\frac{\partial}{\partial\,x_i}\left(\frac{\partial\,f}{\partial\,x_j}\right) = \frac{\partial}{\partial\,x_j}\left(\frac{\partial\,f}{\partial\,x_i}\right).$$

B.2 Fourier transform of the tempered distributions

Assume first that the function $|f(x)|$ is integrable $\mathbb{R}^n$. For such a function the Fourier transform is define as follows:

$$F[f](\xi) = \int\limits_{\mathbb{R}^n} f(x)\, e^{i(\xi,\, x)}\, d\,x.$$

It can be easily seen that

$$|F[f](\xi)| \le \int\limits_{\mathbb{R}^n} |f(x)|\, d\,x < \infty.$$

Thus, $F[f](\cdot)$ being the bounded function, correctly defines the regular distribution:

$$(F[f],\, \phi) = \int\limits_{\mathbb{R}^n} F[f](\xi)\, \phi(\xi)\, d\,\xi, \qquad \phi \in \mathcal{S}(\mathbb{R}^n).$$

Using the Fubini theorem on the conversion of the order of integration, this formula can be transformed as follows:

$$\begin{aligned}&\int\limits_{\mathbb{R}^n} F[f](\xi)\,\phi(\xi)\,d\,\xi = \int\limits_{\mathbb{R}^n}\left(\int\limits_{\mathbb{R}^n} f(x)\, e^{i(\xi,\,x)}\,d\,x\right)\phi(\xi)\,d\,\xi\\ &= \int\limits_{\mathbb{R}^n} f(x)\left(\int\limits_{\mathbb{R}^n}\phi(\xi)\,e^{i(\xi,\,x)}\,d\,\xi\right)d\,x = \int\limits_{\mathbb{R}^n} f(x)\,F[\phi](x)\,d\,x.\end{aligned}$$

Following this equation, we define the Fourier transform $F[f]$ of any distribution $f \in \mathcal{S}'(\mathbb{R}^n)$ by the formula

$$(F[f], \phi) = (f, F[\phi]), \quad f \in \mathcal{S}'(\mathbb{R}^n), \quad \phi \in \mathcal{S}(\mathbb{R}^n). \tag{B.6}$$

It can be shown that $F[f]$ is a linear functional, correctly defined on the set $\mathcal{S}(\mathbb{R}^n)$, or, in other words, that $F[f] \in \mathcal{S}'(\mathbb{R}^n)$.

Let us introduce in $\mathcal{S}'(\mathbb{R}^n)$ the following operation:

$$F^{-1}[f] = \frac{1}{(2\pi)^n} F[f(-x)], \qquad f \in \mathcal{S}'(\mathbb{R}^n) \tag{B.7}$$

and show that F^{-1} is the operator inverse to F, that is

$$F^{-1}[F(f)] = f, \qquad F[F^{-1}(f)] = f, \quad f \in \mathcal{S}'(\mathbb{R}^n). \tag{B.8}$$

So, on virtue of (6.6), (6.7), (B.6) and (B.7) the following equalities hold:

$$\begin{aligned}
&\left(F^{-1}[F[f]], \phi\right) = \tfrac{1}{(2\pi)^n}\left(F[F[f](-\xi)], \phi\right) \\
&= \tfrac{1}{(2\pi)^n}\left(F[f](-\xi), F[\phi]\right) = \tfrac{1}{(2\pi)^n}\left(F[f], F[\phi](-\xi)\right) \\
&= \left(F[f], F^{-1}[\phi]\right) = \left(f, F\left[F^{-1}[\phi]\right]\right) = (f, \phi) \\
&= \left(f, F^{-1}\left[F[\phi]\right]\right) = \left(F^{-1}[f], F[\phi]\right) = \left(F\left[F^{-1}[f]\right], \phi\right).
\end{aligned}$$

This way the correctness of the formula (B.8) is shown. From the formula (B.8), in turn, it follows that each distribution $f \in \mathcal{S}'$ is a Fourier transform of distribution $g = F^{-1}[f] \in \mathcal{S}'$, and $F[f] = 0$, if and only if $f = 0$. This means that F and F^{-1} map $\mathcal{S}'$ into itself and these mappings are one-to-one.

B.3 Conditions enabling to reduce the search for the inverse Laplace transform to calculation of residues

For the given function $F(p)$, $p \in \mathbb{C}$ the inverse Laplace transform is given by the Melline formula

$$\mathcal{L}^{-1}[F](t) = \frac{1}{2\pi i}\int_{\gamma - i\infty}^{\gamma + i\infty} F(p)\, e^{pt}\, dp, \quad \gamma > \gamma_0 > 0$$

provided that the integral in the r.h.s. exists. Below the conditions are formulated, under which the finding of the inverse Laplace transform comes down to computing the residua of the integrand [Ditkin and Prudnikov (1975); Trim (1996)].

Theorem B.1. *Suppose that*

(1) $F(p)$ is an analytic function of the complex variable everywhere except a set of isolated poles $p_1, p_2, ..., p_n, ...$ ($p_1 \leq p_2 \leq ... \leq p_n \leq ...$), and $\forall n \; Re\, p_n < \gamma_0$ for certain $\gamma_0 > 0$.

(2) There exist the limit

$$\lim_{R\to\infty} \frac{1}{2\,\pi\, i} \int_{\gamma - i\,R}^{\gamma + i\,R} F(p)\, e^{p\,t}\, d\,p = \frac{1}{2\,\pi\, i} \int_{\gamma - i\,\infty}^{\gamma + i\,\infty} F(p)\, e^{p\,t}\, d\,p, \quad \gamma > \gamma_0.$$

(3) There exists a sequence of piecewise smooth function $\mathcal{G}_n$ (contours) consisting of a segment of the straight line $Re\, p = \gamma$ and an arc adjacent to it at the points $\gamma - i\,\beta_n$, $\gamma + i\,\beta_n$ (see Fig. B.1), placed at the half-plane $Re\, p < \gamma$ and not containing function singularities; there is a relationship $\mathcal{G}_{n-1} \subset \mathcal{G}_n$, and each contour $\mathcal{G}_n$ contains the origin and the first m_n poles; the condition $\lim\limits_{n\to\infty} \beta_n = \infty$ is fulfilled.

(4) $\forall t > 0$

$$\lim_{n\to\infty} \oint\limits_{\mathcal{G}_n} F(p)\, e^{p\,t}\, d\,p = 0.$$

Then the following formula takes place:

$$\frac{1}{2\,\pi\, i} \int_{\gamma - i\,\infty}^{\gamma + i\,\infty} F(p)\, e^{p\,t}\, d\,p = \sum_{n\in\mathcal{M}} Res\left[F(p_n)\, e^{p_n\,t}\right].$$

Example B.1. Let us find

$$\mathcal{L}^{-1}\left[\frac{1}{p(1+e^p)}\right].$$

The function

$$F(p) = \frac{1}{p(1+e^p)}$$

has the simple pole at the point $p = 0$. Solving the equation

$$e^p = -1 = e^{(2\,n+1)\,i\,\pi},$$

we find the coordinates of the remaining singular points:

$$p_n = (2\,n+1)\,i\,\pi, \quad n = 0, \pm 1, \pm 2, \ldots.$$

Let us note that

$$(1+e^p)'\,\Big|_{p=p_n} = e^{(2\,n+1)\,i\,\pi} = -1 \neq 0$$

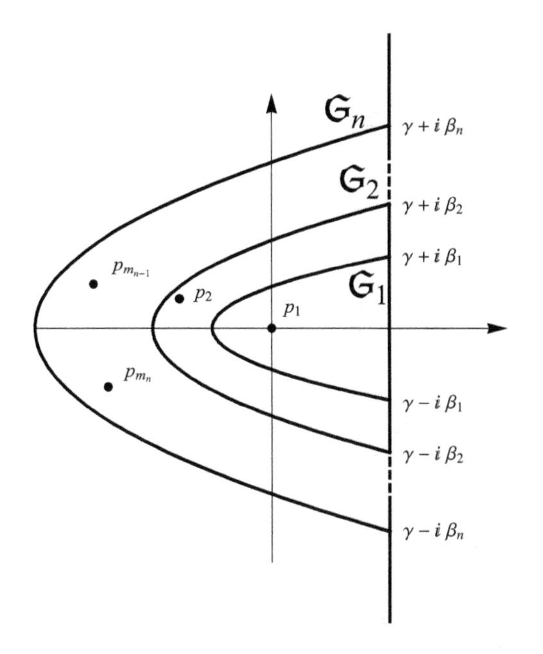

Fig. B.1 An example of a sequence of the piecewise smooth contours $\mathcal{G}_n$ based on the line $\mathrm{Re}\, p = \gamma > 0$.

and hence the poles p_n are simple. Calculating the residua in the corresponding points, we get:

$$\mathrm{Res}\,\left[e^{t\,p}\,F(p)\right]\,\Big|_{p=0} = \lim_{p\to 0} p\,e^{t\,p}\,F(p) = \tfrac{1}{2},$$

$$\mathrm{Res}\,\left[e^{p}\,F(p)\right]\,\Big|_{p=p_n} = \frac{e^{t\,p_n}}{[p(1+e^p)]'\,\big|_{p=p_n}} = \frac{e^{t\,p_n}}{p_n\,e^{p_n}} = -\frac{e^{t(2\,n+1)\,i\,\pi}}{(2\,n+1)\,i\,\pi}.$$

Hence the following formula takes place:

$$\text{the sum of residua} = \frac{1}{2} - \frac{2}{\pi}\sum_{n=0}^{\infty}\frac{1}{2\,n+1}\sin(2\,n+1)\pi\,t.$$

Now it remains to show that the assumptions of Theorem B.1 are satisfied.

At first, let us consider the integrals along the curve CD (see Fig. B.2). At this curve $p = 2\,\pi\,n\,e^{\varphi}$, and φ satisfies the inequalities $\pi/2 \le \varphi \le 3\pi/2$. Let us find the estimation for the modulus of the function $G(p) = 1 + e^p$:

$$|G(p)| = \left|1 + e^{2\,\pi\,n\,\cos\varphi}\left[\cos(2\,\pi\,n\,\sin\varphi) + i\sin(2\,\pi\,n\,\sin\varphi)\right]\right| \ge$$
$$1 + e^{2\,\pi\,n\,\cos\varphi}\,\cos(2\,\pi\,n\,\sin\varphi).$$

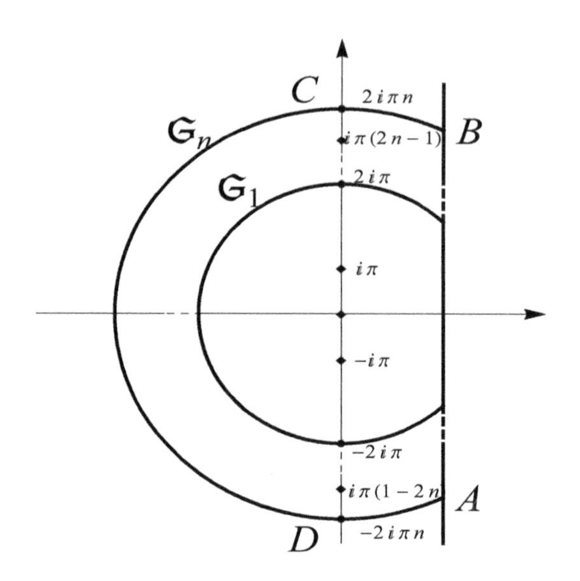

Fig. B.2 Piecewise smooth closed curves $\mathcal{G}_n$ used in Example B.1 to determine the inverse Laplace transform. The diamonds determine the coordinates of the poles of the function $F(p)$.

The function at the r.h.s. attains the maximal value on the ends of the segment $\varphi \in [\pi/2,\, 3\,\pi/2]$:

$$1 + e^{2\,\pi\,n\,\cos\varphi}\,\cos(2\,\pi\,n\,\sin\varphi)\Big|_{\varphi=\pi/2,3\pi/2} = 2.$$

This function attains its minimal value $b_n = 1 - e^{2\,\pi\,n}$ at $\varphi = \pi$. So

$$|F(p)| \le \frac{b_n^{-1}}{|p|},$$

and from this appears that

$$\lim_{n\to\infty} \int\limits_{C\,D} e^{t\,p} F(p)\,d\,p = 0.$$

It remains to show that the integrals along the arcs $B\,C$ and $D\,A$ also disappear when $n \to \infty$. Let's estimate, for example, the integral along the segment $B\,C$. Note that the argument in this segment changes in the range $\varphi \in \left[\frac{\pi}{2} - \delta_0,\, \frac{\pi}{2}\right]$, where $\delta_0 = \arcsin\frac{x_0}{2\,\pi\,n}$. For the modulus of function $G(p)$ following estimation takes place:

$$|G(p)| = \left|1 + e^{2\,\pi\,n}\left[\cos(\frac{\pi}{2} - \delta) + i\,\sin(\frac{\pi}{2} - \delta)\right]\right|$$

$$= \sqrt{1 + 2\,e^{2\,\pi\,n\,\sin\delta}\cos(2\,\pi\,n\,\cos\delta) + e^{4\,\pi\,n\,\sin\delta}}.$$

The absolute value of $G(p)$ will be greater than one as long as $\cos(2\,\pi\,n\,\cos\,\delta) > 0$. This inequality will be true for $\delta \in [0,\,\delta_0]$ if

$$\cos(2\,\pi\,n\,\cos\,\delta_0) > 0 \quad \text{or} \quad \sqrt{(2\,\pi\,n)^2 - x_0^2} > \frac{3\,\pi}{2}.$$

The last inequality is fulfilled for sufficiently large n. So the following estimation takes place

$$\left|\int\limits_{BC} e^{t\,p} F(p)\,d\,p\right| \le \int_{\pi/2-\delta_0}^{\pi/2} e^{t\,2\,\pi\,n\cos(\pi/2-\delta)}\,d\,\delta$$

$$= \int_{\pi/2-\delta_0}^{\pi/2} e^{t\,2\,\pi\,n\sin(\delta)}\,d\,\delta < \int_{\pi/2-\delta_0}^{\pi/2} e^{t\,x_0}\,d\,\delta = e^{t\,x_0}\,\frac{x_0}{2\,\pi\,n} \underset{n\to\infty}{\longrightarrow} 0.$$

The fact that the integral tends to zero on the segment $D\,A$ is shown analogously. So we have shown that the following formula takes place:

$$\mathcal{L}^{-1}\left[\frac{1}{p(1+e^p)}\right] = \frac{1}{2} - \frac{2}{\pi}\sum_{n=0}^{\infty}\frac{1}{2\,n+1}\sin(2\,n+1)\pi\,t.$$

Appendix C

An introduction into the theory of special functions

C.1 Introduction

The method of separating variables used in solving the initial-boundary problems of mathematical physics leads to the *Sturm-Liouville problem*, which is formulated as follows:

Find the values of the parameter λ for which the boundary value problem

$$\nabla\, \kappa(x) \nabla v(x) + \lambda\, v(x) = 0, \qquad \nabla = \left(\frac{\partial}{\partial x_1}, \frac{\partial}{\partial x_2}, \ldots, \frac{\partial}{\partial x_n} \right), \qquad v \Big|_{\partial \Omega} = 0,$$

where $\kappa(x)$ is known smooth function, has nontrivial solutions in the bounded domain $\Omega \subset \mathbb{R}^n$.

In case when $n = 1$, $0 < \kappa = \text{const}$, and $\Omega = (a,\, b)$, the eigenfunction $v_n(x)$ corresponding to the eigenvalue $\lambda_n > 0$ is the trigonometric function. In case when $n \geq 2$ while $\partial\,\Omega$ is not the boundary of an n-dimensional cube (or the rectangle, when $n = 2$) eigenfunctions $v_n(M)$, $M \in \mathbb{R}^n$ belong, as a rule, to the set of *special functions.* The general form of the equation having the special functions as its solution is as follows:

$$\hat{L} y + \lambda\, \rho(x)\, y = 0, \quad a < x < b, \quad \rho(x) \Big|_{[a,\, b]} > 0, \tag{C.1}$$

$$\hat{L}(y) = \frac{d}{d\,x} \left[\kappa(x)\, \frac{d\,y}{d\,x} \right] - q(x)\, y, \quad \kappa(x) \geq 0, \quad q(x) \geq 0.$$

Solution to the equation (C.1) can be found in the form of the power series, while the conditions that guarantee the existence of the analytical solution are formulated in the form of the following statement [Golubev (1950); Elsgolts (1977)]:

Theorem C.1. *If the functions $p_i(x)$, $i = 0, 1, 2$ are analytical in certain vicinity of the point $x_0 \in \mathbb{R}$, and $p_0(x_0) \neq 0$, then the solution of the equation*

$$p_0(x)\, y'' + p_1(x)\, y' + p_2(x)\, y = 0 \tag{C.2}$$

is also the analytical function defined in some vicinity of the point x_0, so that one can search for the solution to the equation (C.2) in the form

$$y = \sum_{k=0}^{\infty} a_k\, (x - x_0)^k.$$

The following statement formulates the conditions that guarantee the existence of solutions allowing the decomposition into the generalized series.

Theorem C.2. *If the function $p_i(x)$, $i = 0, 1, 2$ appearing in the equation (C.2) are analytical in some vicinity of the point x_0, while the function $p_0(x)$ has in the point x_0 kth order zero ($k < \infty$), while the function $p_1(x)$ has in this point $(k-1)$th order zero, or higher (if $k > 1$), and the function $p_2(x)$ has zero of at least the order $k - 2$ (if $k > 2$), then there exists at least one non-trivial solution of the equation (C.2) having the form*

$$y = \sum_{m=0}^{\infty} a_m\, (x - x_0)^{m+\sigma},$$

where σ is a real parameter.

Let us return to the equation (C.1), completing it with conditions

$$y(a) = A, \quad |A| < \infty, \qquad y(b) = B, \quad |B| < \infty. \tag{C.3}$$

The boundary value problem (C.1)–(C.3) has a number of general properties [Tikhonov and Samarskii (1990)], which are enumerated below.

(1) There exist a countable number of eigenvalues $\lambda_1 < \lambda_2 < \cdots < \lambda_n < \ldots$, and a countable number of eigenfunctions $y_1(x), y_2(x), \ldots, y_n(x), \ldots$ corresponding to them.
(2) If $q(x) \geq 0$, then all the eigenvalues are non-negative: $\lambda_k \geq 0$.
(3) The eigenfunctions $y_m(x)$, $y_n(x)$, corresponding to the different eigenvalues $\lambda_m \neq \lambda_n$ are orthogonal in the weighted scalar product:

$$(y_m, y_n) = \int_a^b y_m(x)\, y_n(x)\, \rho(x)\, d\,x = 0, \qquad m \neq n.$$

(4) A function $f(x)$ is decomposed into the series

$$f(x) = \sum_{n=1}^{\infty} f_n y_n(x), \qquad f_n = \frac{(f, y_n)}{(y_n, y_n)},$$

if

- $f(x)$ is continuously differentiable on the segment $a < x < b$ and has piecewise continuous second derivative;
- $f(x)$ satisfies boundary conditions compatible with the boundary conditions of the problem (C.1)–(C.3), namely,

$$|f(a)| < \infty \quad \text{as} \quad 0 \leq q(a) < \infty,$$

$$f(a) = 0 \quad \text{as} \quad q(x) \underset{x \to a}{\to} \infty.$$

C.2 Cylindrical functions

In many problems of the mathematical physics the following equation appears:

$$\frac{d^2 y}{d x^2} + \frac{1}{x}\frac{d y}{d x} + \left(1 - \frac{\nu^2}{x^2}\right) y = 0. \tag{C.4}$$

The above equation is called cylindrical equation of the order ν or the *Bessel equation.* Equation

$$x^2 y'' + x y' + (x^2 - \nu^2) y = 0, \tag{C.5}$$

which is equivalent to (C.4), satisfies the conditions of Theorem C.2, and has thus the solution of the form

$$y(x) = x^\sigma \left(a_0 + a_1 x + a_2 x^2 + \ldots\right). \tag{C.6}$$

Inserting the series (C.6) into the equation (C.5) and equation to zero the coefficients at the same powers of x, we get the conditions $\sigma = \pm\nu$, $a_1 = 0$ and recursion

$$(\sigma + k + \nu)(\sigma + k - \nu) a_k + a_{k-2} = 0, \qquad k > 1.$$

From this appears that all the odd coefficients a_{2k+1} nullify. Since the coefficient a_0 is undefined, we can choose it however we like. Putting the coefficient equal to

$$a_0 = \frac{1}{2^\nu \Gamma(\nu + 1)}, \qquad \Gamma(s) = \int_0^\infty e^{-x} x^{s-1}\, d x,$$

we can express the solution in the form

$$J_{\pm\nu}(x) = \sum_{k=0}^{\infty} \frac{1}{\Gamma(k+1)\,\Gamma(k\pm\nu+1)} \left(\frac{x}{2}\right)^{2\,k\pm\nu}. \tag{C.7}$$

The function $J_\nu(x)$ is called the *Bessel function of the first kind of the order* ν. Let us note, that the functions $J_\nu(x)$ and $J_{-\nu}(x)$ are linearly independent if only ν is not an integer. In case when $\nu = n \in \mathbb{N}$, the following relationship exists between the functions

$$J_{-n}(x) = (-1)^n\, J_n(x).$$

C.3 Legendre polynomials

Spherical functions appear in solutions of the three-dimensional Laplace equation when the boundary conditions are invariant with respect to rotations. They are closely related to the *Legendre polynomials* $P_n(x)$, which can be introduced as the coefficients of the decomposition of so called generating function

$$\Psi(\rho,\, x) = \frac{1}{\sqrt{1+\rho^2-2\,\rho\, x}}, \qquad 0<\rho<1, \quad -1\le x\le 1$$

into the series with respect to the variable ρ:

$$\Psi(\rho,\, x) = \sum_{n=0}^{\infty} P_n(x)\,\rho^n.$$

An explicit form of Legendre polynomials can be obtained using the following formula:

$$P_n(x) = \frac{1}{n!}\frac{\partial\,\Psi}{\partial\,\rho}\Big|_{\rho=0} = \frac{1}{2^n\, n!}\frac{d^n}{d\,x^n}(x^2-1)^n.$$

Legendre polynomials $P_n(x)$ are eigenfunctions corresponding to the eigenvalues $\lambda_n = n(n+1)$ in the spectral problem, which is formulated as follows:

Find all possible values λ for which there exist nontrivial solutions of the Legendre equation

$$\frac{d}{d\,x}\left[(1-x^2)\,\frac{d\,y}{d\,x}\right] + \lambda\, y = 0, \qquad -1<x<1. \tag{C.8}$$

Equation (C.8) is connected with so called *associated Legendre equation*

$$\frac{d}{d\,x}\left[(1-x^2)\,\frac{d\,y}{d\,x}\right] + \left(\lambda - \frac{m^2}{1-x^2}\right) y = 0, \qquad -1<x<1. \tag{C.9}$$

For solutions of the equation (C.9) that satisfy the boundedness conditions

$$|y(\pm 1)| < \infty \tag{C.10}$$

requirements of Theorem C.2 are met, for this equation is a special case of the equation (C.1) with $k(x) = 1 - x^2$, $q(x) = m^2/(1 - x^2)$, $a = -1$ and $b = 1$. Since $k(x)$ attains zero at the points $x = -1$ and $x = 1$, then the constraint conditions are posed at both ends of the segment $(-1, 1)$, where the solution should have zeros of the order $m/2$. Hence the solution of the problem (C.9)–(C.10)is naturally presented as

$$y(x) = (1 - x^2)^{m/2}\, v(x), \qquad v(\pm 1) \neq 0.$$

It can be easily seen that the function $v(x)$ satisfies the following equation:

$$(1 - x^2)\, v'' - 2\, x(m + 1)\, v' + [\lambda - m(m + 1)]\, v = 0. \tag{C.11}$$

The similar equation is satisfied by the function $d^m\, P_n(x)/d\, x^m$, where non-zero solutions to this equation are bounded if and only if $\lambda = n(n + 1)$, $n \in \mathbb{N}$ and, besides, the condition $m \leq n$ is satisfied. Thus, the functions

$$P_n^m(x) = (1 - x^2)^{m/2}\, \frac{d^m\, P_n(x)}{d\, x^m}, \qquad m \leq n$$

called the *associated Legendre functions*, satisfy the equation

$$\frac{d}{d\,x}\left[(1 - x^2)\,\frac{d\,P_n^m}{d\,x}\right] + \left[n(n + 1) - \frac{m^2}{1 - x^2}\right] P_n^m = 0, \qquad -1 < x < 1, \qquad m \leq n. \tag{C.12}$$

The associated Legendre functions are an essential part of the spherical harmonics.

Bibliography

Ahnert, K. and Pikovsky, A. (2009). Compactons and chaos in strongly nonlinear lattices, *Phys. Rev. E* **79**, p. 026209.

Aramanovich, I. G., Volkovyskii, L. I., and Lunts, G. L. (2004). *A Collection of Problems on Complex Analysis* (Doover Books on Mathematics, New York).

Barenblatt, G. (2012). *Scaling, Self-similarity and Intermediate Asymptotics* (Cambridge University Press).

Bellomo, N., Preziosi, L., and Romano, A. (2002). *Mechanics and Dynamical Systems with Mathematica* (Springer, New York).

Berezin, I. S. and Zhidkov, N. P. (1965). *Computing Methods, vol. 2* (Pergamon Press, UK).

Bhatnagar, P. L. and Prasad, P. (1970). *Nonlinear Waves in One-Dimensional Dispersive Systems* (Oxford University Press, UK).

Cattaneo, C. (1948). Sulla conduzione del calore, *Atti Semin. Mat. Fis. Univ. Modena* **3**, p. 3.

Christie, I., Griffiths, D. F., Mitchell, A. R., and Sanz-Serna, J. M. (1981). Product approximation for nonlinear problems in the finite element method, *IMA J. Numer. Anal.* **1**, pp. 253–266.

Cirillo, E., Ianiro, N., and Sciarra, G. (2016). Compacton formation under Allen-Cahn dynamics, *Proc. R. Soc. Lond. Ser. A Math. Phys. Eng. Sci.* **472**, 2188, p. 20150852.

Cornille, H. and Gervois, A. (1982). Bi-soliton solutions of a weakly nonlinear evolutionary PDEs with quadratic nonlinearity, *Physica D* **6**, pp. 1–50.

Cornille, H. and Gervois, A. (1983). Connection between the existence of bisolitons for quadratic nonlinearities and a factorization of the associate linear operator, *Journal of Mathematical Physics* **24**, pp. 2042–2055.

Coste, C., Falcon, E., and Fauve, S. (1997). Solitary waves in a chain of beads under Hertz contact, *Phys. Rev. E* **56**, pp. 6104–6117.

Daraio, C., Nesterenko, V. F., Herbold, E. B., and Jin, S. (2006). Energy trapping and shock disintegration in a composite granular medium, *Phys. Rev. Lett.* **96**, p. 058002.

De Frutos, J., Lopez-Marcos, M. A., and Sanz-Serna, J. M. (1995). A finite-difference scheme for the K(2, 2) compacton equation, *J. Comput. Phys.* **120**, pp. 248–252.

Destrade, M. and Saccomandi, G. (2006). Solitary and compactlike shear waves in the bulk of solids, *Phys. Rev. E* **73**, p. 065604.

Ditkin, V. A. and Prudnikov, A. P. (1975). *Operational Calculus* (Vysshaja Shkola, Moscow).

Dodd, R. K., Eilbeck, J. C., Gibbon, J. D., and Morris, H. C. (1984). *Solitons and Nonlinear Wave Equations* (Academic Press, London).

Dubin, D. (2003). *Numerical and Analytical Methods for Scientists and Engineers with Mathematica* (John Wiley & Sons, New Jersey).

Elsgolts, L. (1977). *Differential Equations and the Calculus of Variations* (Mir, Moscow).

Engelbrecht, J. (1992). Mathematical modelling of nerve pulse transmission, *Period. Polytech., Mech. Eng.* **36**, 2, pp. 153–161.

Enns, R. H. and Mc Gurie, G. C. (2001). *Nonlinear Physics with Mathematica for Scientists and Engineers* (Birkhauser, Boston).

Gelfand, I. M. (1971). *Wyklady z Algebry Liniowej* (PWN, Warszawa).

Godunov, S. K. (1959). Difference scheme for numerical solution of discontinuos solution of hydrodynamic equations (translated US Joint Publ. Res. Service, JPRS 7226, 1969), *Matematicheskii Sbornik* **47**, pp. 271–306.

Golubev, V. V. (1950). *Lectures on analytical Theory of Differential Equations* (Gostekhizdat, Moscow).

Guckenheimer, J. and Holmes, P. (1987). *Nonlinear Oscillations, Dynamical Systems and Bifurcations of Vector Fields* (Springer, NY).

Haken, H. (1983). *Synergetics. An Introduction* (Springer, Berlin).

Herbold, E. and Nesterenko, V. F. (2007). Shock wave structure in strongly nonlinear lattice with viscous dissipation, *Phys. Rev. E* **75**, p. 021304.

Joseph, D. D. and Preziozi, L. (1989). Heat waves, *Reviews of Modern Physics* **61**, 1, pp. 41–73.

Kapitza, P. L. (1951). Dynamic stability of the pendulum with vibrating suspension point, *Journal of Experimental and Theoretical Physics* **21**, 5, pp. 588–598.

Kawahara, T. and Tanaka, M. (1983). Interactions of traveling fronts: An exact solution of a nonlinear diffusion equation, *Physics Letters A* **97**, 8, pp. 311–314.

Kelly, J. J. (2006). *Graduate Mathematical Physics: With Mathematica supplements* (WILEY-VCH, Weinheim).

Korn, G. A. and Korn, T. M. (1968). *Mathematical Handbook for Scientists and Engineers* (McGraww-Hill, NY).

Koszalka, I. (2006). *Vibrating pendulum and stratified fluids* (Woods Hole Oceanography Institution. Technical Report, Geophysical Fluid Dynamics Proceedings Volumes, WHOI, 2005).

Kythe, P. K. (1997). *Partial differential equations and Mathematica* (CRC Press, Boca Raton, Florida).

Landau, L. D. and Lifszyc, J. M. (1965). *Mechanika* (PWN, Warszawa).

Lazaridi, A. N. and Nesterenko, V. F. (1985). Observation of a new type of solitary waves in a one-dimensional granular medium, *J. Appl. Mech. Tech. Phys.* **26**, pp. 405–408.

Li, Y. A., Olver, P. J., and Rosenau, P. (1998). Non-analytic solutions of nonlinear wave models, Tech. Rep. 1591, Institute of Mathematics and Applications, Minnesota, `https://hdl.handle.net/11299/3266`.

Marchuk, G. I. (1982). *Methods of Numerical Mathematics* (Springer-Verlag, New York).

Maugin, G. A. and Engelbrecht, J. (1994). A thermodynamical viewpoint on nerve pulse dynamics, *J. Non-Equilibrium Thermodyn.* **19**, 2, pp. 9–23.

Maurin, K. (1973). *Analiza, cz. 1,2* (PWN, Warszawa).

Metropolis, N., Stein, M. L., and Stein, P. R. (1973). On finite limit sets for transformations on the unit interval, *Journal of Combinatorial Theory, Series A* **15**, 1, pp. 25–44.

Nayfeh, A. H. (2011). *Introduction to Perturbation Techniques* (Wiley, New Jersey).

Nesterenko, V. F. (1983). Propagation of nonlinear compression pulses in granular media, *J. Appl. Mech. Tech. Phys.* **24**, pp. 733–743.

Nesterenko, V. F. (1994). Solitary waves in discrete media with anomalous compressibility and similar to "sonic vacuum", *Journal de Physique* **4**, pp. 729–734.

Nesterenko, V. F. (2001). *Dynamics of Heterogeneous Materials* (Springer-Verlag, New York).

Nesterenko, V. F. (2018). Waves in strongly nonlinear discrete systems, *Philos. Trans. R. Soc.* **376**, 2127, p. 20170130.

Nesterenko, V. F., Lazaridi, A. N., and Sibiryakov, E. B. (1995). The decay of soliton at the contact of two "acoustic vacuums", *J. Appl. Mech. Tech. Phys.* **36**, pp. 166–168.

Peitgen, H. O. and Richter, P. H. (1984). Harmonie in chaos und kosmos, *Annals of Phys.* **94**, pp. 226–274.

Perestyuk, M. O. and Marynets, V. V. (2001). *The Theory of Equations of Mathematical Physics* (Lybid', Kyiv).

Rocsoreanu, C., Georgescu, A., and Giurgiteanu, N. (2012). *The FitzHugh-Nagumo Model: Bifurcation and Dynamics* (Springer-Verlag, New York).

Rosenau, P. (1996). On solitons, compactons, and lagrange maps, *Phys. Lett. A* **211**, 5, pp. 265–275.

Rosenau, P. (1997). On nonanalytic solitary waves formed by a nonlinear dispersion, *Physics Letters A* **230**, pp. 305–318.

Rosenau, P. (2003). Hamilton dynamics of dense chains and lattices: Or how to correct the continuum, *Phys. Lett. A* **31**, pp. 39–52.

Rosenau, P. (2006). On a model equation of traveling and stationary compactons, *Phys. Lett. A* **356**, pp. 44–50.

Rosenau, P. and Hyman, J. (1993). Compactons: Solitons with finite wavelength, *Phys. Rev. Lett.* **70**, pp. 564–567.

Rozhdestvenskij, B. L. and Yanenko, N. N. (1983). *Systems of Quasilinear Equations and Their Applications to Gas Dynamics* (Translations of Mathematical Monographs, vol. 55, American Mathematical Society, Providence).

Schuster, H. G. (2005). *Deterministic Chaos: An Introduction* (Wiley-VCH, Weinheim).

Scott, A. (2003). *Nonlinear Science. Emergence and Dynamics of Coherent Structures* (Oxford University Press).

Sergyeyev, A., Skurativskyi, S., and Vladimirov, V. (2019). Compacton solutions and (non)integrability of nonlinear evolutionary PDEs associated with a chain of prestressed granules, *Nonlinear Analysis: Real World Applications* **47**, pp. 68–84.

Simon, B. (2005). Sturm oscillation and comparison theorems, in W. Amrein, A. Hinz, and D. Pearson (eds.), *Sturm-Liouville Theory: Past and Present* (Birkhauser Verlag, Basel), pp. 29–43.

Tikhonov, A. N. and Samarskii, A. A. (1990). *Equations of Mathematical Physics* (Dover Publications Inc., NY).

Trim, D. (1996). *Introduction to Complex Analysis and Its Applications* (PWS Publishing Company, USA).

Vladimirov, V. A. (2007). *Wstap do Teorii Dystrybucji (skrypt)* (AGH im S. Staszica, http://wms.mat.agh.edu.pl/ vladimir/frame.htm).

Vladimirov, V. A. (2008). Compacton-like solutions of the hydrodynamic system describing relaxing media, *Reports on Math. Phys.* **61**, pp. 381–400.

Vladimirov, V. A., Kutafina, E. V., and Pudelko, A. (2006). Constructing soliton and kink solutions of PDE models in transport and biology, *SIGMA* **2**, p. 061.

Vladimirov, V. A., Kutafina, E. V., and Zorychta, B. (2012). On the non-local hydrodynamic-type system and its soliton-like solutions, *Journal of Physics A: Mathematical and Theoretical* **45**, p. 085210.

Vladimirov, V. A. and Maczka, C. (2007). Exact solutions of the generalized burgers equation, describing traveling fronts and their interaction, *ROMP* **60**, pp. 317–328.

Vladimirov, V. A. and Skurativskyi, S. I. (2015). Solitary waves in one-dimensional pre-stressed lattice and its continual analog, in J. Awrejcewicz (ed.), *Dynamical Systems. Mechatronics and Life Sciences* (Politechnika Lodzka, Lodz), pp. 531–542.

Vladimirov, V. A. and Skurativskyi, S. I. (2020). On the spectral stability of soliton-like solutions to a non-local hydrodynamic-type model, *Communications in Nonlinear Science and Numerical Simulation* **80**, p. 104998.

Vladimirov, V. S. (1984). *Equations of Mathematical Physics* (Mir, Moscow).

Vodova, J. (2013). A complete list of conservation laws for non-integrable compacton equations of K(n, n) type, *Nonlinearity* **26**, pp. 757–762.

Yang, J., Silvestero, G., Khatri, D., De Nardo, L., and Daraio, C. (2011). Interaction of highly nonlinear solitary waves with linear elastic media, *Phys. Rev. E* **83**, p. 046606.

Zabusky, N. and Kruskal, M. (1965). Interaction of "solitons" in a collisionless plasma and the recurrence of initial states, *Physical Review Letters* **15**, 6, pp. 240–243.

Zilburg, A. and Rosenau, P. (2017). On hamiltonian formulations of the C_1(m, a, b) equations, *Phys. Lett. A* **381**, pp. 1557–1562.

Index

CPSIA information can be obtained
at www.ICGtesting.com
Printed in the USA
JSHW041437130123
36111JS00002B/4